Immunodominance

Edited by
Jeffrey A. Frelinger

Related Titles

Frosch, M., Maiden, M. C. J. (Eds.)

Handbook of Meningococcal Disease

Infection Biology, Vaccination, Clinical Management

2006

ISBN 3-527-31260-9

Lutz, M., Romani, N., Steinkasserer, A. (Eds.)

Handbook of Dendritic Cells

Biology, Diseases and Therapies

2006

ISBN 3-527-31109-2

Meager, A. (Ed.)

The Interferons

Characterization and Application

2006

ISBN 3-527-31180-7

Pollard, K. M. (Ed.)

Autoantibodies and Autoimmunity

Molecular Mechanisms in Health and Disease

2005

ISBN 3-527-31141-6

Kropshofer, H., Vogt, A.B. (Eds.)

Antigen Presenting Cells

From Mechanisms to Drug Development

2005

ISBN 3-527-31108-4

Hamann, A., Engelhardt, B. (Eds.)

Leukocyte Trafficking

Molecular Mechanisms, Therapeutic Targets, and Methods

2005

ISBN 3-527-31228-5

Kaufmann, S. H. E. (Ed.)

Novel Vaccination Strategies

2004

ISBN 3-527-30523-8

Kalden, J. R., Herrmann, M. (Eds.)

Apoptosis and Autoimmunity

From Mechanisms to Treatments

2003

ISBN 3-527-30442-8

Immunodominance

The Choice of the Immune System

Edited by Jeffrey A. Frelinger

WILEY-VCH Verlag GmbH & Co. KGaA

The Editor

Prof. Dr. Jeffrey A. Frelinger
Department of Microbiology and Immunology
University of North Carolina
CB# 7290 Jones
Chapel Hill, NC 27599-7290
USA

Cover
Shifts in repertoire and
immunodominance following primary
and secondary exposures to antigen.
For further details see Figure 6.6 on
page XXII.

Library of Congress Card No.: applied for
British Library Cataloguing-in-Publication Data
A catalogue record for this book is available from the British Library.

Bibliographic information published by Die Deutsche Bibliothek
Die Deutsche Bibliothek lists this publication in the Deutsche Nationalbibliografie; detailed bibliographic data is available in the Internet at <http://dnb.ddb.de>.

Typesetting Kühn & Weyh, Satz und Medien, Freiburg
Printing betz-druck GmbH, Darmstadt
Binding Litges und Dopf Buchbinderei GmbH, Heppenheim

Printed in the Federal Republic of Germany.
Printed on acid-free paper.

ISBN-13: 978-3-527-31274-0
ISBN-10: 3-527-31274-9

Contents

Immunodominance: The Choice of the Immune System. Edited by Jeffrey A. Frelinger

ISBN: 3-527-31274-9

Preface

The normal, intact immune system does not have equal probability of responding to every potential part of a protein. It has been known for more than 50 years that only parts of the protein that are "outside" are available for antibody binding. Yet, with the advent of Western blotting techniques, antibodies that react with the interior of the protein have been routinely produced. Although not all epitopes are equally easy to produce, or are equally protective in infection, nearly any structure can be an antibody epitope.

In contrast to antibodies, T cells must recognize fragments of proteins bound to MHC molecules. In T-cell responses against viruses, very few epitopes are easily identified. In the case of LCMV, the immune response in BALB/c mice uses only a single MHC class I protein: L. K and D are not used at all. This is not because there are no suitable peptides that can bind K and D proteins, as BALB/c mice that lack L make an excellent response to LCMV. Furthermore, only a single peptide from the LCMV genome accounts for more than 90% of the CD8 T cells responding to infection. Because LCMV has a coding size of approximately 3500 amino acids, the immune system fixates 9 of 3500 amino acids, or about 0.2% of the coding capacity.

This fixation on a small part of the potential antigenic space is not unique to LCMV in BALB/c mice. Most pathogens in inbred mice show similar immunodominance. Even in response to bacteria, where the pathogen genome size is much larger, dominance is observed. The CD8 T-cell response to Listeria monocytogenes infection is dominated by very few epitopes in both C57Bl/6 and BALB/c mice. With a genome size of almost three million base pairs, the majority of the response is restricted to two or three epitopes. The immune system is choosing only about 0.002% of the coding sequence to recognize.

Why is this so? Clearly, there are many mechanisms at work. In this volume we cover topics including (1) the mechanisms of antigen processing, i.e., how pathogen molecules are converted to molecules that are targets for cell-mediated immunity; (2) binding of processed peptides to MHC molecules, a critical step in their expression on the cell surface; and (3) the role of the pathogen itself in modifying the immune response by interfering with antigen processing and the downstream immune responses.

Immunodominance: The Choice of the Immune System. Edited by Jeffrey A. Frelinger

ISBN: 3-527-31274-9

We have not yet completed the puzzle of immunodominance, but the chapters here represent our current understanding of its pieces.

I thank Andreas Sendtko for encouragement and enthusiasm for this book.

Jeffrey A. Frelinger — Chapel Hill, October 2005

List of Contributors

Antonio Bertoletti
Institute of Hepatology
University College London
69-75 Chenies Mews
London WC1E 6HX
United Kingdom

Edward J. Collins
Department of Biochemistry and Biophysics and Department of Microbiology and Immunology
University of North Carolina at Chapel Hill
CB#7290
804 M.E. Jones Building
Chapel Hill
North Carolina 27599
USA

Sherry R. Crowe
Trudeau Institute
154 Algonquin Avenue
Saranac Lake
New York 12983
USA

James R. Drake
University of Rochester
School of Medicine and Dentistry
601 Elmwood Avenue
Box 609
Rochester
New York 14642
USA

JoAnne L. Flynn
Department of Molecular Genetics and Biochemistry
University of Pittsburgh School of Medicine
Pittsburgh
Pennsylvania 15261
USA

Michael J. Fuller
Center for Vaccines and Immunity
Columbus Children's Research Institute
700 Children's Drive
Columbus, OH 43205
USA

Immunodominance: The Choice of the Immune System. Edited by Jeffrey A. Frelinger

ISBN: 3-527-31274-9

Dalia E. Gaddis
Department of Microbiology
University of Alabama at Birmingham
845 Nineteenth Street South
446 BBRB
Birmingham
Alabama 35294
USA

John T. Harty
Interdisciplinary Program in
Immunology and
Department of Microbiology
University of Iowa
Iowa City
Iowa 52242
USA

Ann B. Hill
Oregon Heath Sciences University
Department of Molecular Microbiology
and Immunology
3181 SW Sam Jackson Park Rd
L 220
Portland OR 97281
USA

David M. Lewinsohn
Pulmonary and Critical Care Medicine
Oregon Health & Science University
Portland VA Medical Center
Portland
Oregon 97239
USA

Mala Maini
Infection and Immunity
Windeyer Institute of Medical Science
University College of London
London WC1T 4JF
United Kingdom

Peter J. Miller
Department of Biochemistry and
Biophysics
University of North Carolina at Chapel
Hill
CB#7260
405 M.E. Jones Building
Chapel Hill
North Carolina 27599
USA

Michael W. Munks
Integrated Department of Immunology
National Jewish Medical and Research
Center
1400 Jackson Street
Denver CO 80206
USA

Brandon B. Porter
Interdisciplinary Program in
Immunology
University of Iowa
Iowa City
Iowa 52242
USA

Andrea J. Sant
University of Rochester
School of Medicine and Dentistry
601 Elmwood Avenue
Box 609
Rochester
New York 14642
USA

Thomas M. Schmitt
University of Washington
Howard Hughes Medical Institute
Box 357370
Seattle, Washington 98795
USA

Alessandro Sette
La Jolla Institute for Allergy and Immunology
3030 Bunker Hill Street
Suite 326
San Diego
California 92109
USA

Roshni Sundaram
Immunosol, Inc.
10790 Roselle Street
San Diego
California 92121
USA

Masafumi Takiguchi
Division of Viral Immunology
Center for AIDS Research
Kumamoto University
2-2-1 Honjo
Kumamoto 860-0811
Japan

David L. Woodland
Trudeau Institute
154 Algonquin Avenue
Saranac Lake
New York 12983
USA

Jonathan W. Yewdell
Laboratory of Viral Diseases
National Institute of Allergy and Infectious Diseases
Building 4, Room 211
Center Drive
Bethesda
Maryland 20892
USA

Allan J. Zajac
Department of Microbiology
University of Alabama at Birmingham
845 Nineteenth Street South
446 BBRB
Birmingham
Alabama 35294
USA

Juan Carlos Zúñiga-Pflücker
University of Toronto
Sunnybrook and Women's Research Institute
2075 Bayview Avenue
Toronto
Ontario M4N 3M5
Canada

Color Plates

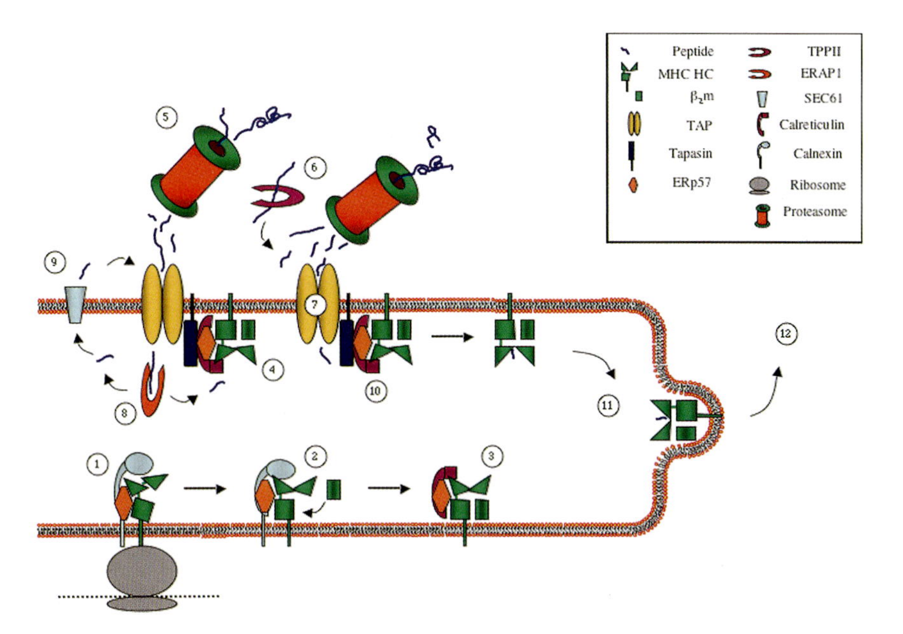

Figure 1.3 Production of peptide-loaded class I MHC. Class I MHC heavy chain and β_2m are translocated separately into the ER. Immediately when accessible, Asn86 of the heavy chain is glycosylated, allowing for recognition by the chaperone calnexin. (1) Calnexin also recruits Erp57, which mediates disulfide bond formation in $\alpha3$ and $\alpha2$ of class I heavy chain. (2) Once the heavy chain is folded and the disulfide bond in $\alpha3$ is formed, soluble β_2m binds to class I heavy chain. While the MHC class I molecule is being formed (cycling on and off calnexin), newly translated tapasin associates with calnexin and ERp57, both of which facilitate its folding and disulfide bond formation. Tapasin with calnexin–ERp57 then binds and stabilizes the TAP1/2 dimer. (3) Upon binding of β_2m, calnexin dissociates and is replaced by calreticulin. The class I heavy chain–β_2m–calreticulin–ERp57 heterocomplex then associates with the tapasin–TAP–calnexin–ERp57 heterocomplex. ERp57 remains part of the peptide-loading complex (PLC) until peptide has been loaded, but it is not known whether ERp57 comes with tapasin or the class I molecule. (4) The PLC now consists of TAP, tapasin, heavy chain, β_2m,

Immunodominance: The Choice of the Immune System. Edited by Jeffrey A. Frelinger

ISBN: 3-527-31274-9

Figure 1.3 (continued)

calreticulin, and ERp57. (5) Proteasomes in the cytosol degrade substrates, generating peptides of varying length. (6) Large proteasomal products can be further cleaved in the cytosol by aminopeptidases such as TPPII. (7) Peptides with affinity for TAP will bind, and ATP-induced conformational changes in TAP will transport it through the ER membrane. (8) ERAP1 performs amino-terminal trimming in the ER. (9) These peptides may go directly into the PLC, or they may be retrotranslocated back into the cytosol through SEC61 and have to enter the ER again via TAP. (10) If the amino acid motif of the peptide matches the class I MHC molecule, it will then bind in the peptide groove, inducing a conformational change that stabilizes peptide binding. ERp57 will mediate the disulfide formation in $\alpha 2$, "locking" the peptide in. (11) If the correct conformational change is induced, all chaperone molecules will dissociate. (12) The trimer of class I heavy chain–β_2m–peptide is shuttled from the ER to the Golgi apparatus and subsequently to the plasma membrane of the cell.
(This figure also appears on page 20.)

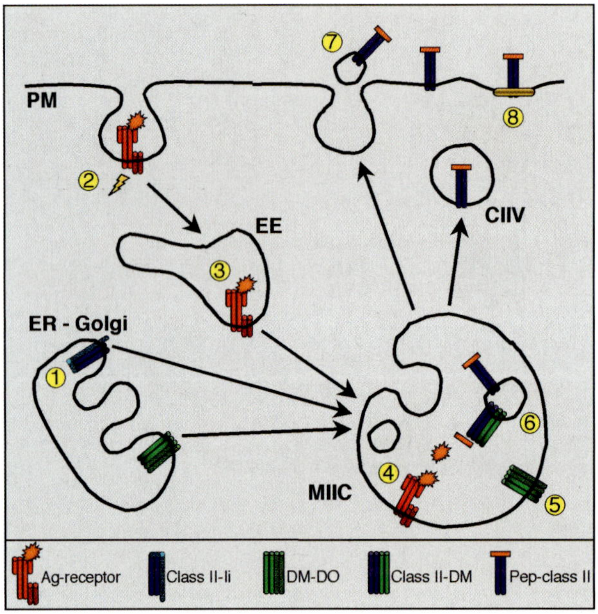

Figure 2.1 General pathway of exogenous antigen processing. Depicted is the general pathway of exogenous antigen processing and presentation. Numbered yellow circles indicate steps of the pathway at which variations can occur that will alter the hierarchy of peptide–class II complexes expressed by the APC (see text for details). (1) Relative levels of Ii isoforms (i.e., p31Ii vs. p41Ii, Section 2.2.1); (2) Effects of cell signaling (Sections 2.2.2 and 2.3); (3) Receptor-mediated antigen internalization and intracellular trafficking (Section 2.3); (4) Proteolytic processing of internalized antigen (Section 2.4.1); (5) Role of DM and DO in class II peptide loading (Section 2.4.2), (6) Intravesicular distribution of processing proteins (Section 2.4.2.4); (7) Exosomes and the cell-surface delivery of peptide–class II complexes (Section 2.4.3) (8) MHC class II signaling and partitioning of peptide–class II complexes into membrane microdomains (Section 2.4.3.2). PM: plasma membrane; EE: early endosome
(This figure also appears on page 32.).

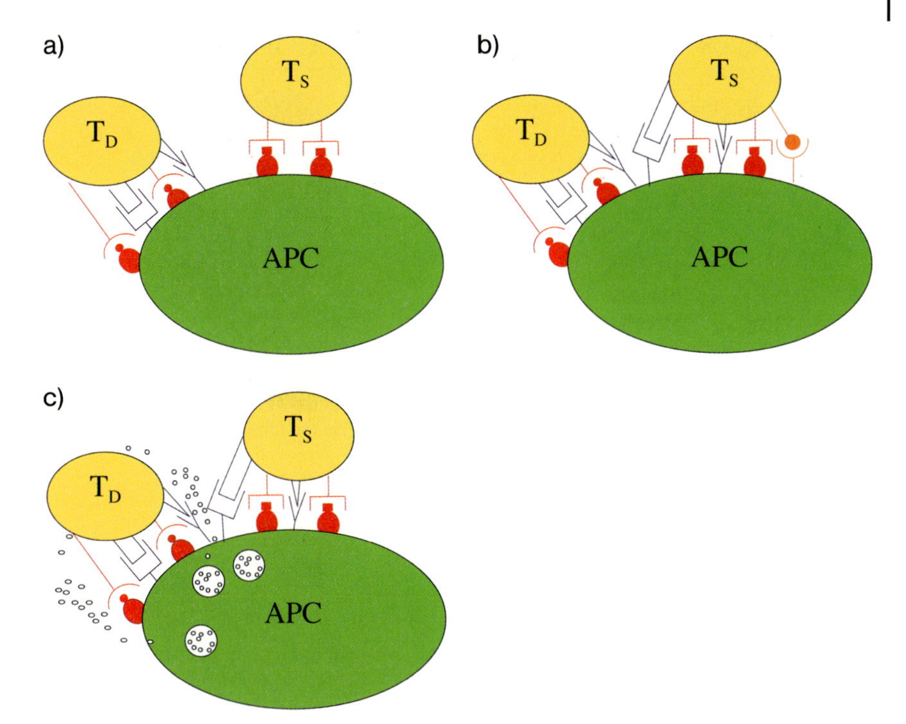

Figure 3.1 Possible mechanisms involved in immunodomination (a) Preemption of critical elements of the synapse by the dominant T cell. (b) Engagement of inhibitory receptors by the submissive T cell. (c) Lack of adequate directed secretion of APC factors. T_D = dominant T cell; T_S = submissive T cell TCR/MHC molecules are shown in red, costimulator ligands and receptors in blue, inhibitory receptors in orange, APC factors in white. (This figure also appears on page 20.)

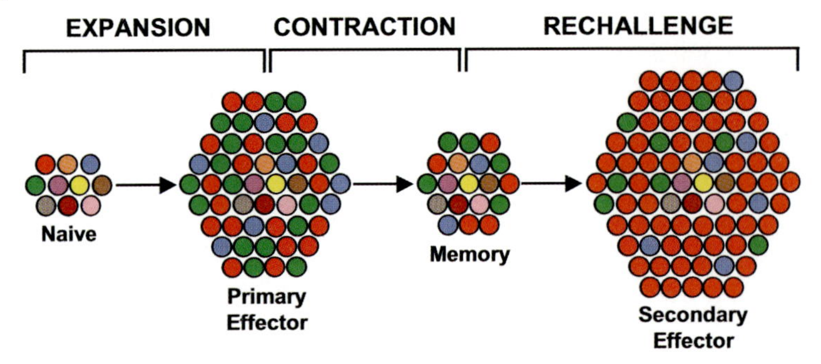

Figure 6.6 Shifts in repertoire and immunodominance following primary and secondary exposures to antigen. As primary immune responses are initiated, antigen-specific T cells become activated and expand in number. This results in a discernable shift in the T-cell repertoire as antigen-specific cells, indicated here in red, green, and blue, increase in frequency. A contraction phase ensues, following clearance of the inducing antigen; however, this downsizing is typically proportional. Consequently, the skewing of the repertoire and - the hierarchy of immunodominance that develops during the expansion phase are imprinted on the memory pool. This phenomenon is sometimes referred to as immunological scarring. Although the repertoire and hierarchies of the primary effector and memory pools are usually similar, marked differences can arise following rechallenges. In this illustration, the red responders become most dominant. (This figure also appears on page 120.)

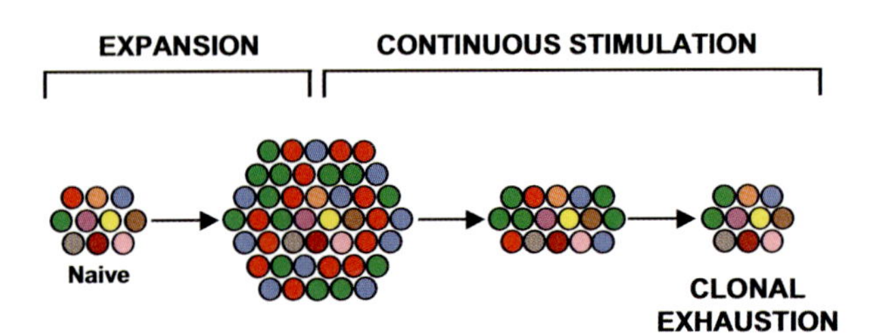

Figure 6.8 Continuous antigenic stimulation can drive responding CD8 T cells to deletion. Certain foreign antigens are not rapidly removed by the actions of the host's immune response. This is perhaps best exemplified by persistent viral infections. In these instances an initial response becomes detectable, resulting in repertoire shifts and the development of immunodominance. If the infection is not cleared by the overall immune response, then the responding cells may be subject to repetitive antigenic stimulation. Under these conditions certain clones and specificities of CD8 T cells may succumb to deletion, resulting in further changes in repertoire and epitope hierarchies. Notably, the deletion of CD8 T cells is exacerbated by the absence of CD4 T-cell help.
(This figure also appears on page 128.)

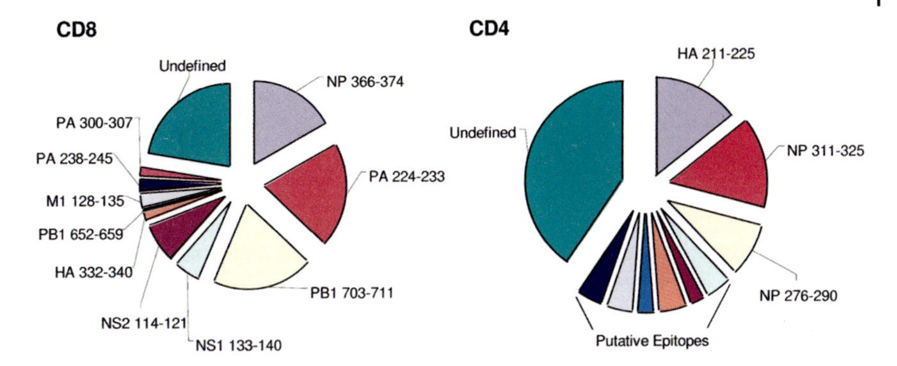

Figure 9.1 Antigen specificity in the lung airways following influenza virus infection. Three dominant epitopes and seven subdominant epitopes account for approximately 78% of the total CD8$^+$ T-cell response to influenza virus infection in the lung airways. The remaining 22% of the CD8$^+$ T-cell response is undetermined at this time. Two dominant and at least seven subdominant epitopes account for approximately 60% of the total CD4$^+$ T-cell response to influenza virus infection in the lung airways. The remaining 40% of the CD4$^+$ T-cell response is undetermined at this time. Some of the CD4$^+$ T-cell epitopes are listed as putative because detailed characterization of the epitopes has not yet been performed. (This figure also appears on page 192.)

1. Identification of motif of HLA-B*3501-binding peptides

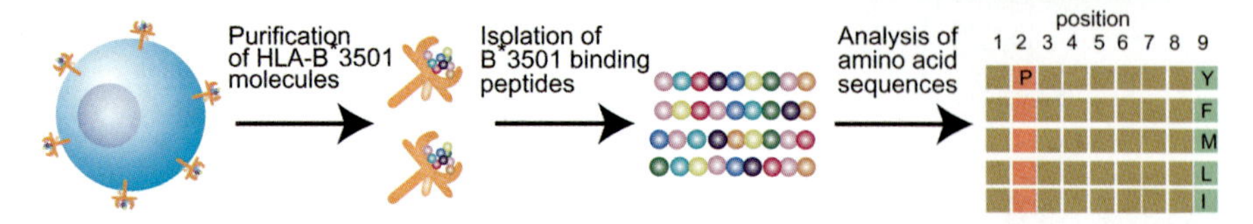

2. Synthesis of HIV-1 8-mer to 10-mer peptides carring B*3501 binding peptide motif

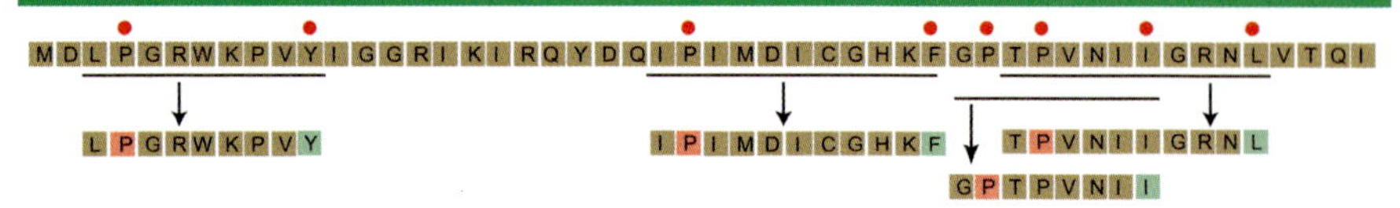

3. Identification of B*3501-binding HIV-1 peptides by a peptide binding assay

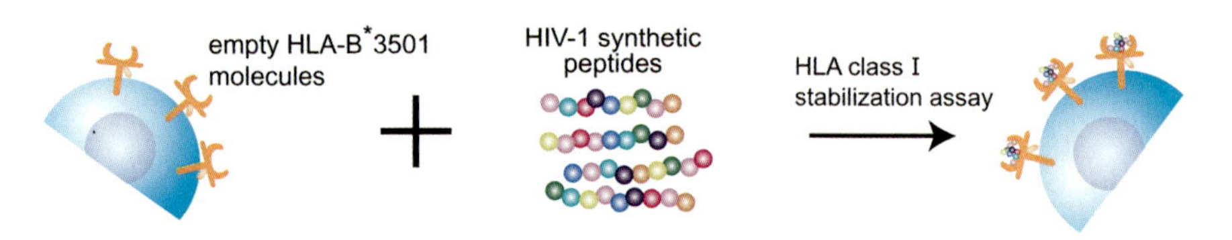

4. Induction of HIV-1 peptide-specific CTLs in PBMC from HIV-1-infected, HLA-B*3501+ Patients

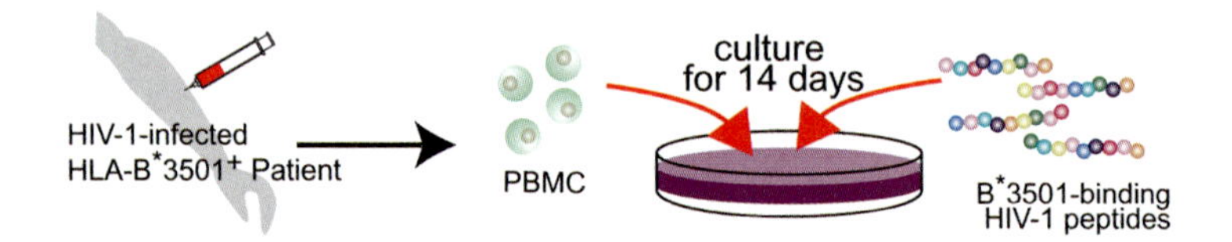

5. Killing of HIV-1-infected cells or HIV-1-recombinant vaccinia-infected cells by peptide-specific CTLs

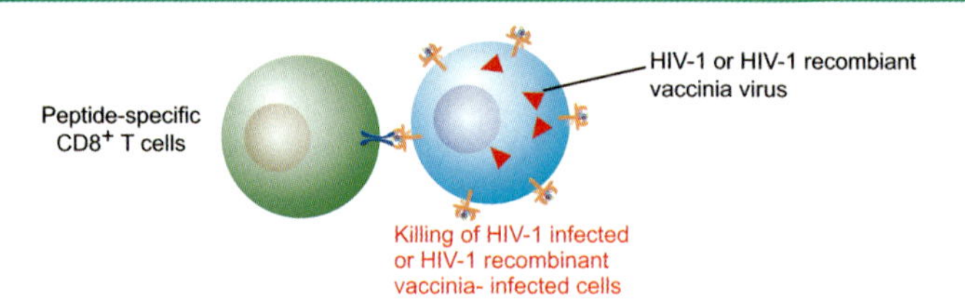

◀ **Figure 10.4** Identification of HLA-B*3501-restricted HIV-1 CTL epitopes by a method using reverse immunogenetics. (1) HLA-B*3501 molecules are isolated from HLA-negative cell lines transfected with the HLA-B*3501 gene. Peptides are eluted from isolated HLA-B*3501 molecules, and then the eluted peptides are sequenced to determine Immunodominant HIV-1 Epitopes Presented by HLA Allelesthe motif of HLA-B*3501-binding peptides. HLA-B*3501-binding peptide possessed Pro at position 2 and hydrophobic residues Tyr, Phe, Met, Leu, and Ile at the C-terminus. (2) 8-mer to 11-mer HIV-1 sequences carrying the HLA-B*3501 anchor residues at position 2 and the C-terminus are selected and synthesized. (3) Synthesized HIV-1 peptides are tested for binding to HLA-B*3501 by a peptide-binding assay such as the HLA class I stabilization assay. (4) HLA-B*3501-binding peptides are further used to induce peptide-specific CTLs from PBMCs of HLA-B*3501-positive, HIV-1-infected individuals. The PBMCs are stimulated with each HLA-B*3501-binding peptide or cocktails of the peptides. Peptide-stimulated PBMCs are cultured for approximately 14 days. Peptide-specific CTLs or $CD8^+$ T cells are identified by measuring the cytotoxic activity of cultured PBMCs toward peptide-pulsed cells or by measuring the production of IFN-γ by $CD8^+$ T cells in cultured PBMCs stimulated with peptide-pulsed cells. The peptides showing a positive response are considered epitope candidates. (5) To clarify whether the peptides are recognized as naturally occurring peptides by specific CTLs or $CD8^+$ T cells, cells infected with HIV-1 or HIV-1 recombinant vaccinia virus are used to stimulate IFN-γ production or for CTL activity. When peptide-specific T-cell clones or lines kill target cells infected with HIV-1 or HIV-1 recombinant vaccinia virus or produce IFN-γ after being stimulated with cells infected with HIV-1 or HIV-1 recombinant vaccinia virus, the peptides that these T cells recognize are concluded to be naturally occurring HIV-1 epitope peptides.
(This figure also appears on page 214.)

I
Mechanics of Antigen Processing

Immunodominance: The Choice of the Immune System. Edited by Jeffrey A. Frelinger

ISBN: 3-527-31274-9

1
Class I MHC Antigen Processing

Peter J. Miller and Edward J. Collins

1.1
Introduction

T-cell recognition of a peptide bound to class I major histocompatibility complex (MHC) molecule requires proper processing of that peptide by a cell's antigen presentation machinery. In many cases, only one peptide from a protein provides an epitope that elicits a T-cell response [1–6]. Any discussion of immunodominance must include how peptides are removed from larger precursors and loaded onto MHC. This chapter will focus on the mechanism of peptide processing and the properties of class I MHC molecules that influence those mechanisms.

1.2
Properties of MHC

1.2.1
Structure of MHC

The MHC is the most polymorphic gene family known in vertebrates. Interestingly, the crystallographic structures of class I MHC molecules are similar regardless of the allotype or species of origin [7–16]. Class I MHC molecules are heterotrimeric complexes comprised of a ~44-kDa heavy chain, a noncovalently bound 12-kDa serum protein, β_2-microglobulin (β_2m), and a small peptide (Figure 1.1a). All class I MHC molecules use their $\alpha 1$ and $\alpha 2$ domains to create a peptide-binding cleft. The cleft is flanked by two alpha helices and has a floor composed of a large beta sheet (Figure 1.1b). The $\alpha 3$ domain of class I MHC has an immunoglobulin fold and noncovalently associates with β_2m. The entire MHC protein is anchored to the membrane by a small transmembrane segment and is completed with a small intracellular tail.

The fact that the structures of class I MHC molecules are the same seems to be at odds with the observation that the MHC is the most polymorphic gene family in vertebrates. Interestingly, the distribution of polymorphic residues is not randomly scattered through the molecule; they are located in the peptide-binding cleft (Figure 1.1c).

Immunodominance: The Choice of the Immune System. Edited by Jeffrey A. Frelinger

ISBN: 3-527-31274-9

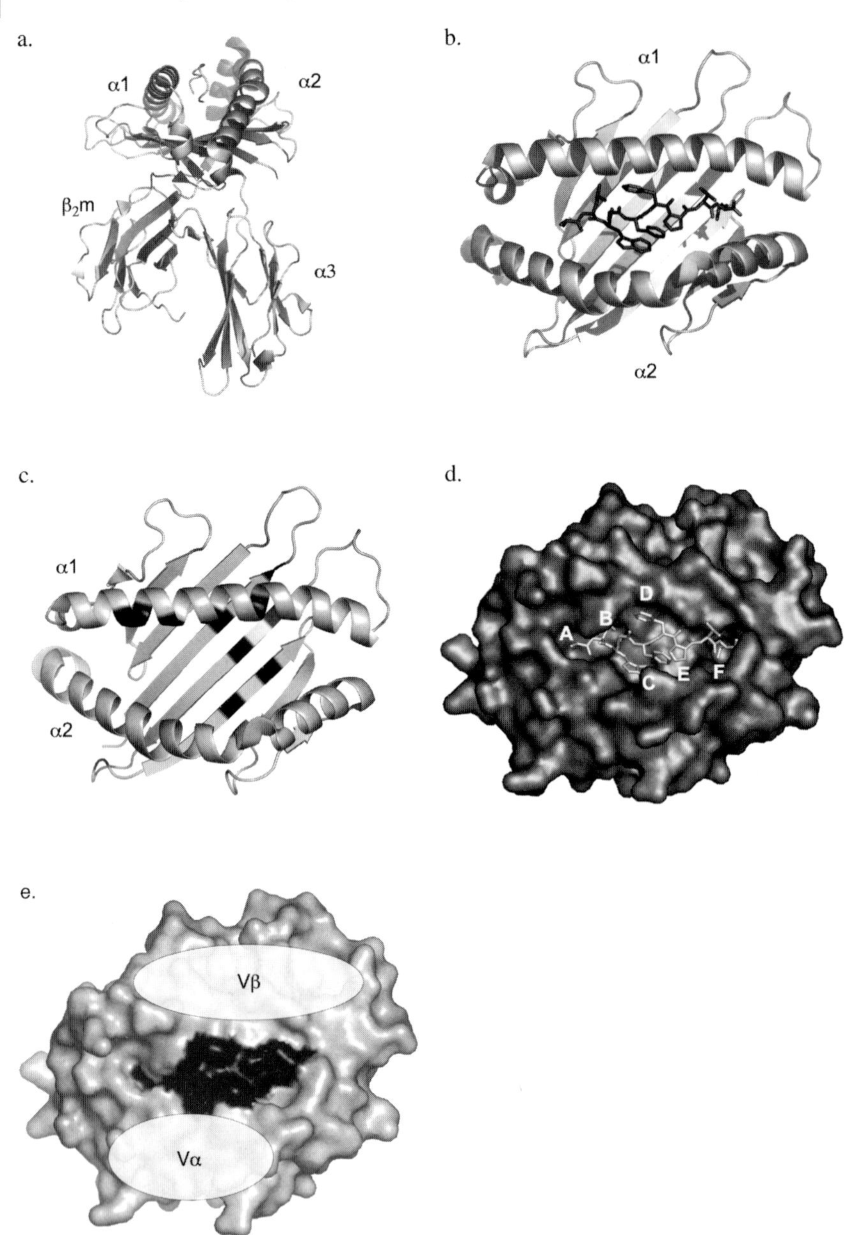

Class I MHC molecules are constitutively expressed on the plasma membrane and, in an unaltered cell, present endogenous self-peptides. In an unaltered cell, the complexity of peptides bound to class I MHC is high [17]. The presence of class I MHC molecules with those endogenous peptides on the surface of the cell is required for $CD8^+$ T-cell homeostasis [18]. The first peptides eluted from human

Figure 1.1 Class I MHC molecule. (a) Heterotrimeric complex composed of MHC class I heavy chain, β_2m, and peptide. The peptide-binding cleft created by the α1 and α2 helices and the beta-sheet floor is easily seen from the side. (b) A top view of the peptide-binding cleft of the class I MHC molecule. Bound peptide, ALWGFFPVL, is colored black. (c) The polymorphic regions of the class I MHC molecule, indicated in black, reside primarily within the α1 helix and beta-sheet floor of the peptide-binding cleft. (d) The chemical and spatial composition of the peptide-binding cleft creates six binding pockets in HLA-A2, labeled A–F. Pockets A, B, and F are especially prominent and accommodate the amino terminus, the side chain of P2, and the carboxyl-terminal leucine, respectively. (e) The molecular surface of the peptide–MHC complex is dictated by the bound peptide and the conformation of the MHC heavy chain. Peptide is highlighted in black. Ellipses represent the binding regions of TCR domains that make contact with the MHC heavy chain based on crystal structures of pMHC–TCR complexes with the Vα binding on the α2 domain and the Vβ binding more diffusely on the α1 domain. Figures were generated with PyMOL [236].

leukocyte antigen (HLA)-B27 were found to be fragments of proteins expressed at high levels in the cell, such as heat shock proteins, ribosomal proteins, histones, and HLA-B27 itself [19]. Thus, the initial hypothesis about which peptides get processed and presented was that they were linked to the expression level of the parental protein in the cell [20]. This has since proven to be simplistic, as will be described later.

1.2.2
Polymorphic Residues Generate Specificity Pockets

A close examination of the structure of HLA-A2 showed that some of the peptide side chains bound in pockets inside the peptide-binding cleft. Six binding pockets were described within the peptide-binding cleft of the class I MHC molecule (Figure 1.1d). Two pockets are comprised of conserved residues that bind the invariant positively and negatively charged peptidic termini. Pocket A is composed of conserved residues that act as hydrogen bond acceptors for the NH_3^+ of the peptide backbone. The entrance to the F pocket provides conserved hydrogen bond donors for the negatively charged carboxyl terminus of the peptide [21]. Pockets B–E and the base of pocket F are formed by a small set of conserved residues in conjunction with the polymorphic residues in the MHC. These pockets are used variably to bind peptide side chains, depending on the identity of the MHC molecule and the peptide. A comparison of HLA-A2 with HLA-B27 and HLA-Aw68 showed that the chemical nature of the pockets is complementary to the chemical nature of the side chain of the bound peptide; thus, the binding sites were described as specificity pockets [9, 13]. The primary sequence of a peptide that binds to a particular MHC molecule is a direct result of the chemical and spatial composition of these specificity pockets [22–24]. Although the majority of these polymorphic residues make contact with the bound peptide [22], the remaining polymorphic residues are solvent exposed and available for interaction with the T-cell receptor.

1.3
Properties of Peptides

1.3.1
Peptides That Bind Are Not Random Sequences

Peptides that bind to class I MHC molecules may be extracted from the MHC binding cleft by acid treatment. The peptides may be recovered by filtering the preparation through a very small pore filter to capture the peptide-sized material. Edman degradation of the eluted material showed that some of the positions in the peptide were not random distributions of amino acids [19, 24–27]. These positions were enriched for particular amino acids. Later experiments with single peptides showed that substitutions of amino acids at certain positions typically reduced peptide binding to the class I MHC molecule; therefore, these enriched positions were termed anchors [27, 28].

As technology improved, the sequences of individual peptides were determined by collision-induced dissociation mass spectrometry [17]. This allowed for a more careful determination of the distribution of bound peptides, including length. Peptides from murine H-2K^d, H-2D^b, and human HLA-A2 are all typically nonamers, whereas H-2K^b peptides are typically octamers. All peptides have a hydrophobic carboxyl-terminal end, with a predominance of leucine, isoleucine, or valine. Each allelic peptide tends to have an additional anchor residue besides the one found at the carboxyl terminus [25]. In addition to a branched hydrophobic, or aliphatic, side chain residue at the carboxyl terminus, K^d peptides tend to have a tyrosine at position 2 [25, 29]. D^b most often binds peptides with asparagine at position 5, a phenylalanine or tyrosine is present at position 5 in K^b, and A2 prefers leucine or methionine at position 2 [17, 25]. HLA-B27-bound peptides are nonameric, with a predominance of positive or hydrophobic residues at the carboxyl terminus, or position 9 (P9), and an anchor position at P2 dominated by arginine [19].

1.3.2
Peptide-binding Motifs

As a result of their specificity pockets, different MHC proteins bind different subsets of peptides. Some MHC molecules are very restrictive; B27 can bind only an arginine at position 2 [7]. Others are less restrictive and allow a larger selection, such as the P9 pocket for A2 (leucine, isoleucine, methionine). This propensity for particular amino acids at particular positions allows one to predict the types of peptides that may bind to a particular MHC. For example, a peptide that will bind to HLA-A2 should have a leucine or methionine at P2 and a valine, isoleucine, leucine, or methionine at P9. This type of description is the peptide-binding motif for HLA-A2. Table 1.1 lists a representative set of peptide-binding motifs. A complete listing up through 1994 was compiled by Rammensee et al. [28].

Table 1.1 Peptide-binding motifs of common MHC alleles.

	Position								
Allele	**1**	**2**	**3**	**4**	**5**	**6**	**7**	**8/9/10**	**Reference**
HLA-A1	–	T, S	D, E	–	–	–	–	Y	231
HLA-A2	–	L, M	–	–	–	–	–	V	17,25
HLA-A3	–	V, L, M	–	–	–	–	–	K	231
HLA-A11	–	T, V	–	–	–	–	–	K	231,232
HLA-A24	–	Y	–	–	–	–	–	F, L	231
HLA-B27	–	R	–	–	–	–	–	K	19
H-2K^d	–	Y	–	–	–	–	–	L, I	25,29
H-2D^b	–	–	–	–	N	–	–	M, I	17,25
H-2K^b	–	–	–	–	F, Y	–	–	L	17,25
H-2L^d	–	P	–	–	–	–	–	L, F, M	233
H-2D^d	–	G	P	–	–	–	–	L, I, F	234

1.3.3
Peptide Length Is Limited in Class I MHC Peptides

Peptides greater than nine amino acids long typically bind poorly to class I MHC molecules. Peptides longer than nine residues bound to H-2D^b have an approximately 100-fold greater off-rate [30]. Additionally, it was determined that peptides of 9–10 amino acids in length induce folding of free class I heavy chain *in vitro*, while peptides of greater length could not [31]. This strict requirement in length led to early predictions of class I peptides bulging in order to fit into the binding pocket of the class I MHC molecule [32]. This characteristic bulge has been confirmed by a number of crystal structures of peptide–MHC complexes [7–16]. High-affinity long peptides have been found [33–35], but the mechanism of this tight binding is not known.

1.3.4
Binding Affinity

Peptides have been reported to bind to class I MHC molecules with moderate to high affinity (K_Ds around 10^{-8} to 10^{-7} M [30, 36]) and extremely slow off-rates (tens to hundreds of hours at 37 °C [22, 30, 37]). However, there are a number of problems with describing this system as a simple equilibrium. The structure suggests that a conformational change would be required to bind or release peptide. This hypothesis was confirmed by binding experiments with an iodinated anchor residue ^{125}I-SV9. These experiments demonstrated that initial binding (on-rate) to murine H-2K^b was similar to non-iodinated wild-type peptide, but the rate of dissociation of ^{125}I-SV9 increased with temperature [38]. These data suggest that initially the heavy chain is open and can accommodate both peptides. Once binding occurs, a conformational change is induced and the MHC heavy chain closes over the peptide, trapping it in the groove. The wild-type peptide SV9 stabilizes the peptide–MHC (pMHC) complex, while the iodinated pMHC complex is unstable [38]. We can deduce then that an appropriate fit into the specificity pockets is required for the conformational change to occur. The amino and carboxyl termini of the peptide also play a crucial role in inducing a conformational change in the heavy chain and stabilizing the peptide–MHC complex [39].

A number of these earlier investigations utilized indirect measurements of peptide binding, such as competition assays [21, 30, 37, 40] and thermal denaturation [41, 42]. More recent investigations report direct measurements of peptide binding by utilizing fluorescence resonance energy transfer (FRET) [43] and fluorescence anisotropy [36]. Both studies concluded that peptide binding to and dissociation from class I MHC molecules is a biphasic, or second-order, process. The first phase is proposed as a conformational change in the MHC heavy chain from a peptide-inaccessible to peptide-accessible form, most likely induced by the association of β_2m. Rate constants (k_1) reported by these studies are 0.003 s^{-1} at 20 °C [43] and 0.008 s^{-1} at 31 °C [36]. The second phase reflects binding of peptide to the peptide-accessible heterodimer. The rate constants (k_{on}) reported for this phase are 2×10^6 M^{-1} s^{-1} at 37 °C [43] and 1×10^4 M^{-1} s^{-1} at 31 °C [36].

Peptides play an important structural role in the class I MHC molecule. They are essential in stabilizing the heavy chain during assembly of the MHC molecule in the endoplasmic reticulum (ER) [44, 45]. In their absence, class I MHC molecules can form, but they are unstable at physiological temperatures [46] and exhibit properties of molecules in a molten globule state [41]. Generally, peptides that confer the greatest stability also have the highest affinity. The fact that peptide-free MHC is extremely unstable and rapidly denatures is also a reason that it is not proper to discuss peptide binding to class I MHC as a standard equilibrium system. If the system were in simple equilibrium, Le Chatelier's principle would

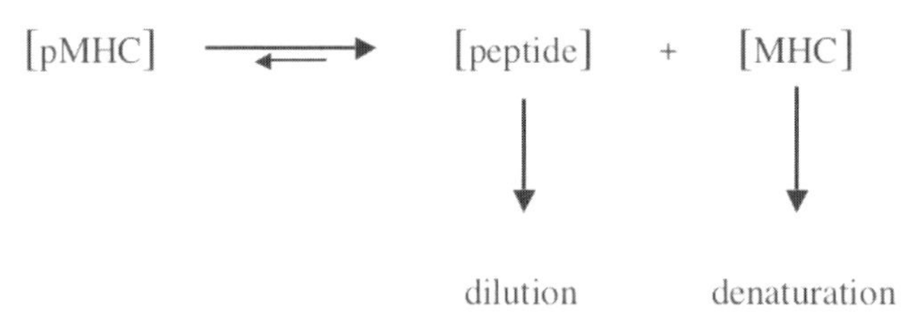

Scheme 1.1

govern. Scheme 1.1 shows that peptide-bound complex (pMHC) would be rapidly depleted because of the loss of peptide-free MHC to denaturation and free peptide to dilution.

1.3.5 Molecular Recognition

Cytotoxic T lymphocytes (CTLs) recognize the molecular surface that is formed by a combination of atoms from the peptide and the class I MHC molecule (Figure 1.1e). Mutations in both the sequence of the peptide and in the MHC molecule have significant effects on recognition. One cannot simply classify a residue as either TCR contact or MHC binding in a particular peptide. Peptide side chains bound to their complementary specificity pockets also influence the reactivity of CTLs [47–51]. Additionally, amino acid mutations in the other five pockets of both human and murine class I MHC molecules also have a significant effect on activation of CTLs [48, 52]. Presumably, these mutations cause a slight conformational change in the peptide that alters the molecular surface of the pMHC complex and, hence, recognition by CTLs. This result, combined with the fact that peptides are buried so deeply in the groove that only 100–300 $Å^2$ is exposed outside the class I MHC molecule to the CTL receptor [8], confirms that the molecular recognition of pMHC by CTL is extremely sensitive to small structural changes. A small conformational change derived from a single amino acid substitution can turn an agonist peptide into an antagonist [53].

The relationship between immunodominance and binding of peptide to MHC has been investigated extensively. Experiments using the immunodominant ovalbumin $(OVA)_{257-264}$ (SIINFEKL) determinant versus the subdominant OVA determinant OVA_{55-62} (KVVRFDKL) showed that the k_{on} for $OVA_{257-264}$ was ~10-fold faster and the k_{off} was twofold slower than the equivalent rates for the subdominant peptide. Interestingly, the anchor residues in these peptides are the same (F at P5 and L at P8) [54]. These findings and others indicate that each residue of the peptide, not just anchors, contributes to the affinity and stability of the peptide–MHC complex and aids in determining immunodominance [54–56]. However, increased immunogenicity is not guaranteed by increased affinity. Once a certain affinity threshold is reached, there is no longer a correlation between affinity and CTL reactivity [56–58].

1.3.6 Epitope Prediction

The wealth of experimental peptide-binding data has been used to create algorithms, which can predict the relative binding strengths of peptides, find peptide-binding motifs for a given MHC allele, or predict an epitope from a protein amino acid sequence. The first such algorithm was based on measuring binding of peptides using dissociation of β_2m from the complex as a marker for whether the peptide was bound (β_2m in complex) or free (β_2m free) [59]. Utilizing their experimental data for HLA-A2 (many more alleles are now available), a table of 180 coefficients (20 amino acids × 9 positions) was created. Each coefficient represents the apparent contribution that each amino acid makes toward binding at each of the nine positions. The coefficients are then multiplied together to determine a theoretical binding stability [59]. A similar approach is used by SYFPEITHI [60]. Later, computational groups created two peptide-binding prediction algorithms: the polynomial method and a neural network method [61]. A more recent Web server, ProPred1, goes one step further by utilizing the original binding database, as well as taking into account how peptides are generated in order to produce more accurate epitope predictions [62]. These prediction algorithms are powerful tools that will become increasingly reliable as more experimental data are gathered. Unfortunately, they are still unsuccessful at predicting CTL response and immunodominance of an epitope.

1.4 Cytosolic Processing

MHC-associated peptides are generated by cleavage of cellular proteins by the proteasome and other peptidases [63, 64]. These peptides are derived from proteins that are degraded at the end of their useful lifespan as part of normal protein turnover. Consistent with this thinking, the ubiquitin-dependent proteolytic pathway plays a major role in the production of peptides for MHC class I–restricted presentation [65]. The peptide pool also includes defective ribosomal products, or DRiPs, that result from errors in protein translation [20, 66, 67]. In the sections that follow, we will discuss the role that each part of the peptide-processing machinery plays in MHC peptide loading. A key factor in selection of a peptide for MHC presentation is the peptide's ability to be liberated from its precursor by proteolysis [58]. The peptide sequence must possess protease-recognition sites flanking its amino and carboxyl termini [68–70] and must lack internal cleavage sites in order to be processed successfully [71, 72].

1.4.1 The Proteasome

The proteasome is a multicatalytic proteinase complex consisting of five known proteolytic components that hydrolyze peptide bonds on the carboxyl side of basic, acidic, aromatic, branched-chain, and small amino acids. These activities are designated as trypsin-like, chymotrypsin-like, peptidyl-glutamyl peptide hydrolyzing, branched-chain amino acid preferring (BrAAP), and small neutral amino acid preferring (SNAPP) [73–77]. The proteasome contains only endoprotease activity, yielding peptides of discrete length and not single amino acids [77]. The proteasome is composed of a 20S catalytic core and a PA28 (11S) or PA700 (19S) regulator. The 20S core of the proteasome is composed of four homoheptameric rings: two composed of α-subunits and two composed of β-subunits (Figure 1.2a).

The β-subunits $\beta 1$, $\beta 2$, and $\beta 5$ are responsible for catalytic activity (reviewed in Ref. [78]). The entrance to the 20S core is closed by amino-terminal tails of the α-subunits [79, 80] (Figure 1.2b) unless opened by regulator particles [81], as is discussed below.

A role for the proteasome in generating MHC peptides was demonstrated using peptide aldehydes to inhibit the peptidase activities of the 20S and 26S (20S core with 19S regulator) particles of the proteasome. These inhibitors blocked presentation of an ovalbumin-derived peptide by class I MHC molecules [82]. Similar studies were repeated using a more efficient proteasome inhibitor, adamantine-

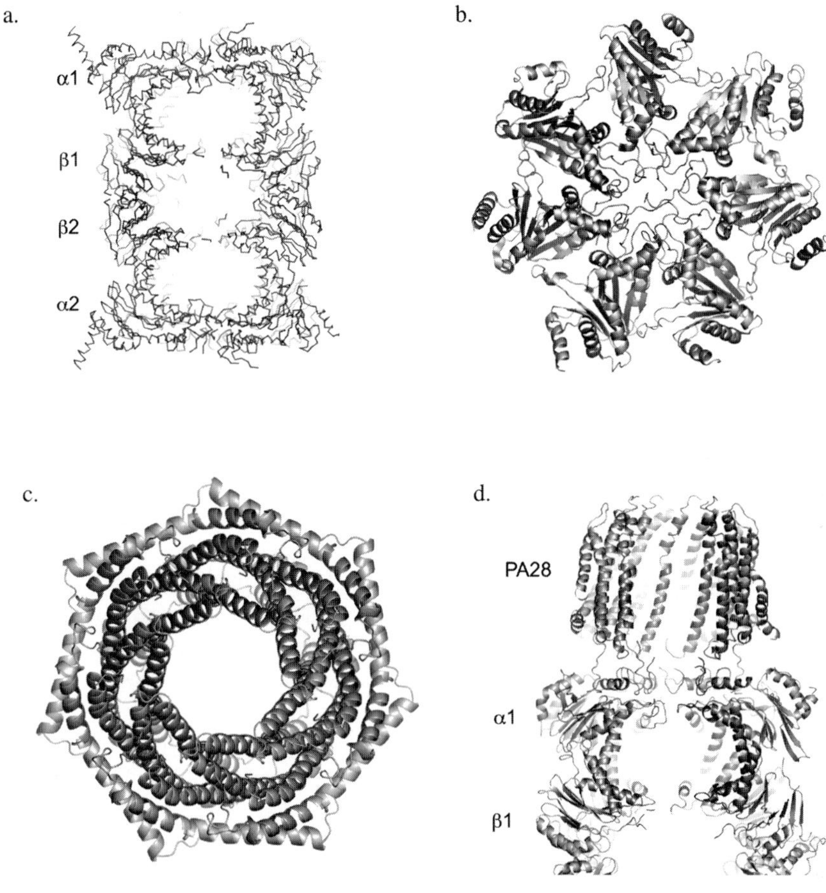

Figure 1.2 The proteasome. (a) Vertical cross-section of the 20S particle composed of four homoheptameric rings, two of α subunits and two of β subunits. The catalytic core is created by the β rings, while the α rings close the entrance to the particle. (b) Looking down on the α heptameric ring, the amino-terminal tails of the α subunits create a gate. (c) The heteroheptameric ring of the PA28 regulator is already in an open conformation. (d) PA28 associates with the α heptameric ring and binds the amino-terminal tails, opening the gate and allowing for polypeptide substrates to enter. Figures were generated with PyMOL [236].

acetyl-(6-aminohexanoyl)3-(leucinyl)3-vinyl-(methyl)-sulfone (Ada-Ahx_3-Leu_3-VS), which targets all three catalytic units of the 20S proteasome core without inhibiting cytosolic proteases. Treatment of cells with Ada-Ahx_3-Leu_3-VS did not influence cell growth characteristics, but it reduced the expression levels of class I molecules on the cell surface and inhibited presentation of endogenous tumor antigens on H-$2D^b$ [83]. This suggests that the cytosolic proteases are able to maintain the housekeeping functions required to survive but that generation of peptides for class I MHC presentation is largely handled by the proteasome.

1.4.2
The Immunoproteasome

In a pathogen-infected cell, peptide generation is increased and the peptide sequences generated are different because of the expression of an altered proteasome, the immunoproteasome. The immunoproteasome contains some of the constitutively expressed subunits of the proteasome as well as unique subunits, the low-molecular-weight proteins (LMPs). LMPs assemble into a complex that resembles the 20S proteasome particle in size and composition. LMP2 and LMP7 were mapped to the class II region of MHC in mice [84]. Similar to other MHC-encoded gene products, LMP proteins are polymorphic and conserved across species [85] and are upregulated by interferon-γ (IFN-γ) [86]. Low basal levels of LMP2 and LMP7 exist in IFN-γ-deficient mice, but IFN-γ is essential for upregulation during infection [87]. Another identified LMP, MECL-1 (LMP10), is not encoded within the *MHCI* region like LMP2 and LMP7, but MECL-1 is upregulated by IFN-γ [88, 89].

LMP2 (β1i) replaces β-type subunit Y (β1 or δ or 2) of the constitutive 20S particle. Subunit Y of the constitutive proteasome promotes cleavage on the carboxyl-terminal side of acidic residues. Incorporation of LMP2 reduces cleavage following acidic residues and increases cleavage after basic residues without affecting hydrophobic activity. LMP7 (β5i) replaces β-type subunit X (β5 or MB1 or 10 or ε). Subunit X does not seem to have a clear catalytic activity, but overexpression reduces proteolysis following hydrophobic and acidic residues. However, studies have shown that incorporation of LMP7 specifically increases cleavage after hydrophobic and basic residues without affecting acidic proteolysis [90, 91]. MECL-1 (β2i) replaces its homologous 20S subunit Z (β2) [88, 89]. RING10 and RING11 have been identified as LMP homologues in humans [92, 93].

The normal 26S proteasome is able to cleave after basic and hydrophobic residues, but not as efficiently as the proteasome with LMP2 and LMP7 [94]. Treatment of wild-type cells with IFN-γ resulted in increased proteolytic cleavage following hydrophobic and basic residues but reduced cleavage after acidic residues compared to mutants lacking LMP genes. The amount of LMP2 and LMP7 subunits in both the 20S and 26S proteasome particles increased after IFN-γ treatment, confirming that the proteasome composition changes following IFN-γ exposure [94].

The sequences of peptides recovered from the class I MHC molecules of healthy cells are different from those of peptides recovered from cells infected with Epstein-Barr virus (EBV). EBV-infected cells displayed peptides containing a glutamate residue at P2 on HLA-B40, while such peptides were missing from healthy cells. These *in vitro* findings support the idea that glutamyl activity in the proteasome is decreased during infection as a result of replacement of subunit Y by LMP2 [95].

In order to be incorporated into an immunoproteasome, the LMPs must be processed from a precursor form [96, 97]. The molecule PI31 was found to negatively regulate immunoproteasome formation by interfering with that processing. It is thought that this helps to maintain levels of constitutive proteasomes in cells when there is no infection [98]. LMP7 also appears to play a role in processing pre-LMP2 and pre-MECL1, as it is required for their incorporation into the proteasome [96].

Assembly of both the constitutive and immunoproteasome appears to be cooperative. When more X subunit is incorporated into the proteasome, then incorporation of Y is favored over LMP2, suggesting that the presence of one subunit influences the use of the other. Interestingly, IFN-γ does not suppress transcription of subunits X and Y [99]. This suggests that the use of the immuno-subunits is preferred over the constitutive subunits X and Y as the proteasome assembles [96, 99]. In addition, IFN-γ leads to reduced levels of phosphorylation of two α-subunits of the 26S proteasome. This destabilizes the constitutive proteasome complex, allowing for easier incorporation of the upregulated LMPs and the PA28 regulator [100]. Together, these data indicate that the mixing of X and Y with LMPs during proteasome assembly is disfavored [96].

1.4.3 Opening the Immunoproteasome

The catalytic core of the proteasome is closed off, and proteins or peptides may not enter the catalytic core until an 11S or 19S regulator binds to the α-heptameric ring and opens it. The 19S regulator, or PA700, is typically associated with the constitutive (housekeeping) proteasomes, while the 11S regulator, or PA28, associates with the immunoproteasome. Figure 1.2b shows that the entrance to the catalytic core is occluded by the heptameric ring of α-subunits of the 20S particle. Binding of the PA28 regulator (Figure 1.2c) or the 19S regulator induces a conformational change in the α-heptameric ring, allowing substrates to enter the catalytic core [81] (Figure 1.2d). Importantly, the 19S regulator, an ATPase, actively unfolds and guides full-length protein substrates into the proteasome in an ATP-dependent fashion, while PA28 permits shorter polypeptides to enter the catalytic core by diffusion, independent of ATP [101].

Like other members of the immunoproteasome, PA28 is upregulated by IFN-γ [102]. PA28 binds equally well to the constitutive 20S particle and the immuno-20S particle [81, 103]; however, it may aid with the incorporation of the immunosubunits, as LMP2, LMP7, and MECL-1 do not displace X, L, and Z subunits in the absence of PA28 [104]. Incorporation of the LMP subunits into the 20S particle and binding of PA28 both help to increase the variation in peptides produced for antigen presentation by class I MHC molecules and are most effective when employed coordinately [91, 105]. It was first believed that PA28 had no influence on the proteolytic activity inside the proteasome core but only enhanced uptake and release of peptides by the proteasome [91, 103]. However, PA28$^{-/-}$ cells produce 70% fewer peptides possessing a hydrophobic residue at the carboxyl terminus, and PA28 knockout (KO) mice have impaired CTL responses [104], showing that PA28 does influence the types of peptides generated. Perhaps binding of PA28 to the proteolytic core induces a conformational change that not only opens the gate but also alters the accessibility of the active sites and, consequently, the efficiency of peptide generation [106–108]. Upon infection by lymphocytic choriomeningitis virus or *Listeria monocytogenes,* liver cells in mice were found to completely replace constitutive proteasomes with immunoproteasomes and PA28 [109]. However, an immunoproteasomal pathway involving PA28 is not required for all presented peptides. PA28 KO mice launched a normal immune response against infection with influenza A virus [110].

Kinetic studies suggest that PA28 induces double cleavage of peptide substrates, allowing for the increased rate of processed antigen by the proteasome. The identified peptide products were typically 7–12 amino acids long, which fits the class I MHC ligand size, or are slightly longer, which would allow them to serve as precursors [111].

1.4.4
Peptide Trimming

The proteasome generates peptide lengths with a range of 3–22 amino acids, and less then 15% of those peptides are eight or nine residues in length [106]. In the cytoplasm, some peptides are trimmed by aminopeptidases [112] such as puromycin-sensitive aminopeptidase, bleomycin hydrolase [113], and tripeptidyl-peptidase II (TPPII) [114]. TPPII has been shown to be essential for the generation of epitopes for HLA-B51 [115], HLA-A3, and HLA-A11 [114]. However, a great deal of amino-terminal trimming takes place in the ER.

The transporter associated with antigen processing (TAP) translocates a substantial quantity of peptides into the ER that are longer than the canonical nonamer [116–120]. Delivery of a 40-amino-acid influenza peptide to the ER using a signal sequence resulted in a specific CTL response to the final nonamer in the peptide. It could not have been generated without trimming [121]. Peptides with amino acids flanking both the amino-terminal and carboxyl-terminal ends may be delivered to the ER, but only amino-terminal-flanked antigenic peptides are presented by class I molecules to CTLs [122]. Two ER aminopeptidases have recently been identified. Leukocyte-derived arginine aminopeptidase (L-RAP) has been found to cleave precursor peptides in the ER with a preference for arginine and seems to be upregulated by IFN-γ [123]. However, the primary aminopeptidase in the ER seems to be the ER aminopeptidase associated with antigen processing

(ERAAP), or ERAP1, which is also upregulated by IFN-γ [124, 125]. ERAP1, unlike other aminopeptidases, seems to limit trimming to create minimally eight-residue peptides [126]. There are a number of class I MHC allotypes that prefer peptides with proline at P2 [28, 127]; however, peptides with the X-P-X_n motif are poorly transported by TAP [117, 128]. As ERAP1 trims, it halts cleavage as it comes to X-P, resulting in an accumulation of peptides with proline at P2 [122], which can then be loaded into their respective class I MHC molecules. These trimmed peptides must then reenter the peptide-loading complex. This may occur in the ER by competing with newly translocated peptides, or these trimmed peptides may be transported out into the cytoplasm through SEC61-mediated retrotranslocation [129] and then may be transported into the ER again via TAP [130]. These studies suggest that some class I MHC molecules are able to directly bind peptides generated by the proteasome and transported via TAP, while the peptides for other class I MHC molecules require additional processing in the ER prior to loading.

1.4.5 Association of the Proteasome with the Endoplasmic Reticulum

Peptides generated in the cytosol via the constitutive proteasome, immunoproteasome, or other cytosolic proteases must be translocated into the ER for loading into a class I MHC molecule. The most prevalent pathway is via the transporter associated with antigen presentation (TAP). Early observations noted an unknown molecule (later identified as TAP) that co-immunoprecipitated with the LMP complex (later known as the immunoproteasome), suggesting a physical interaction *in vivo* [131]. The *Lmp-2* gene is physically close to *Tap1*, suggesting that both these gene products are important in antigen processing [132]. Yeast two-hybrid experiments showed that TAP interacts with both the constitutive 26S proteasome subunit X and its replacement, LMP7, of the immunoproteasome [133]. Fluorescently tagged anti-LMP2 and anti-LMP7 antibodies showed that immunoproteasomes are primarily associated with the endoplasmic reticulum in cells treated with IFN-γ, while constitutive proteasomes are evenly distributed throughout the cytosol and nucleus [134]. These results also correlate well with cellular fractionation studies [135, 136]. Taken together, these data show that upon IFN-γ stimulation, immunoproteasomes are formed and associate with TAP at the ER membrane.

1.5 Peptide Transport

The loading of peptide onto folded class I MHC molecules occurs in the ER [45]. The primary means for peptides generated in the cytosol to enter the ER is by way of the transporter associated with antigen presentation (TAP). TAP is a subunit of the peptide-loading complex (PLC). The PLC is comprised of TAP, tapasin, Erp57, calreticulin, and class I MHC molecules, each of which will be discussed individually.

1.5.1
Transport via TAP

TAP1 and TAP2, encoded within the MHC, are members of the ATP-binding cassette (ABC) family of membrane transporters [137–140]. The TAP heterodimer spans the ER membrane, allowing for movement of peptides from the cytoplasm to the lumen of the ER [141]. Mutant cells with deletions in a portion of the *MHCI* locus encompassing both *Tap* and *LMP* genes (RMA-S in mice [142], 721.174/T2 in humans [143]) do not efficiently load peptides onto class I MHC and as a result do not express stable class I MHC molecules on their surface. It was shown that presentation of peptide could be restored by transfecting these RMA-S and T2 cells with both *Tap1* and *Tap2* [144, 145] but not individually, suggesting that TAP1 and TAP2 work coordinately as a heterodimer. This association was confirmed by co-immunoprecipitation [146, 147].

Typically, ABC transporters are found as dimers, and each possesses a hydrophilic nucleotide-binding domain (NBD) and hydrophobic domain consisting of six transmembrane segments. These nucleotide-binding domains contain highly conserved sequence motifs found in many ATPases, including Walker A and B motifs [148, 149]. Like other ABC transporters, transport of peptides across the ER membrane requires ATP and/or possibly GTP [150–154]. The binding of cytosolic peptides to TAP is ATP independent [119, 150], but stabilization of the TAP dimer by association with ATP or ADP has been shown to increase affinity of TAP for peptides [153, 154]. Hydrolysis of the bound nucleotides by both TAP1 and TAP2 induces a conformational change in the transmembrane domains, which results in the transport of peptide across the membrane [153, 155–157]. The X-ray crystallographic structure of the cytoplasmic NBD of TAP1 complexed with Mg^{2+} and ADP provided support for the model in which TAP1 and TAP2 dimerize in the presence of ATP [237].

1.5.2
TAP Selectivity

TAP does not transport every peptide with the same efficiency. The selectivity of TAP is dependent on the sequence and length of peptide, not just overall charge or hydrophobicity [150–152]. The presence of proline at P1, P2, or P3 seems to be most problematic with respect to transport for TAP. Immunodominant epitopes with a proline at P3 (LCMV NP [FQPQNGQFI], Ad10 [SGPSNTPPEI], M8 [KSPWFTTL], SV9 [FAPGNYPAL]) are transported very poorly. However, if naturally occurring flanking residues are added to the amino terminus, transport of these peptides increases significantly [117].

Polymorphism present in TAP results in different transport selectivities [158]. For example, the rat expresses TAP2A and TAP2B, which display different peptide specificities [159, 160]. TAP2A is more permissive of residues at the carboxyl terminus of peptides it transports. However, TAP2B is very selective for hydrophobic residues at the carboxyl terminus [161]. This polymorphism and the subsequent properties of TAP2 affect the surface expression of the rat class I MHC molecule RT1.A^{a}, a phenomenon called class I modification (*cim*) [162]. RT1.A^{a} preferentially binds peptides with an arginine at the carboxyl terminus [163]. If a rat is homozygous for TAP2B, or *cim*^{b}, then peptide binding to RT1.A^{a} is severely impaired [164]. These biochemical data were confirmed by the crystal structures of the rat class I MHC molecules RT1.A^{a} and RT1.A1^{c}, which have substantially different F-pocket properties, leading to their preferred peptide carboxyl-terminal residues, arginine and leucine, respectively [165].

The ideal length of peptides binding to TAP is 8–11 residues, which is similar to the ideal length for binding to class I MHC molecules and seems to be the most important factor in transport. Peptides up to 25 amino acids can be transported, but TAP becomes much more selective and transport is less efficient [119]. A separate study also found that adding up to four amino acids to the amino terminus of melanoma-associated peptides improved transport into the ER [118]. These data suggest that while TAP transports peptides ready to be loaded into the class I MHC molecule, it also transports longer amino-terminally extended peptide precursors that are processed to their final length in the ER.

After length, the greatest determinant for TAP selectivity is the carboxyl-terminal residue. Peptides containing a basic or hydrophobic residues at the carboxyl terminus are translocated most effectively [119, 159, 166]. The relative importance of other residues in the peptide was determined by evaluating the affinity of TAP for over 250 peptides 9–16 residues long. Again, having hydrophobic or basic residues at the carboxyl terminus was found to be most beneficial, followed by hydrophobic residues at P3 and hydrophobic or basic residues at P2 [128]. Based on the selectivity that TAP exhibits for peptide binding and transport, it is not surprising that TAP influences peptides presented by MHC molecules to T cells [166, 167].

1.5.3 TAP-independent Peptide Transport

1.5.3.1 Endogenous Peptides

TAP is recognized as the primary transporter of peptides into the ER, but other means of peptides entering the ER in order to bind class I molecules have been identified. Some class I MHC molecules are expressed on the surface of TAP-deficient cells (T2), namely, H-2D^{p}, H-2K^{b}, and HLA-A2 [168, 169]. Peptide sequencing by mass spectrometry determined the most abundant peptide bound to HLA-A2 to be a fragment of a hydrophobic amino-terminal ER signal sequence of a constitutively expressed protein, IP30. Other peptides have been shown to be presented via class I MHC without a signal sequence. Therefore, additional TAP-independent pathways must exist for loading peptides onto class I MHC molecules [168, 170].

Another proposed alternate pathway for peptides entering the ER is via P-glycoprotein. Like TAP, P-glycoprotein is a member of the ABC transport family and was originally identified and characterized in tumor cells as a multiple drug resistance efflux pump [171]. Recent evidence suggests that it is capable of binding and transporting peptides within the size range of MHC-bound peptides with basic or hydrophobic residues at the carboxyl termini [172, 173]. P-glycoprotein has also been shown to associate with subunit X of the constitutive 26S proteasome [133].

Polymorphism between MHC alleles has been shown to affect peptide repertoire with respect to transport via TAP. HLA-B*4402 and HLA-B*4405 differ by only one amino acid in the F pocket, Asp116 and Tyr116, respectively. In addition to binding different peptides preferentially, HLA-B*4402 cell surface presentation is TAP dependent, while HLA-B*4405 presentation is TAP independent. It appears that the tyrosine residue in HLA-B*4405 causes a structural change that inhibits incorporation of the protein into the peptide-loading complex (PLC). HLA-B*4405 and other MHC alleles that are TAP independent may have an evolutionary advantage over TAP-dependent alleles during an infection that interferes with TAP activity [174].

1.5.3.2 **Exogenous Peptides**

For the most part, peptides bound to class I MHC molecules are generated endogenously or through protein synthesis occurring within the cell. This is critical because the outcome of T-cell recognition is death of the presenting cell. It would be inappropriate to lyse bystander cells that happen to pick up peptides from their neighbor. However, it is advantageous for professional antigen-presentation cells, like dendritic cells, to pick up antigen without becoming infected themselves. This process is called cross-presentation (reviewed in Ref. [175]).

Using fluorescence probes, exogenous peptides were shown to be taken up by cells in vesicles and delivered to the ER [176]. Transport into the cytosol may occur via SEC61-mediated retrotranslocation [177]. SEC61 is thought to be present in the phagosome membrane as a result of ER and cell membrane fusion during early phagocytosis [178]. If the proteins are exported from the phagosome, these exogenous molecules can then be degraded by proteasomes [179–181]. If degraded in the phagosome, these peptides may be exported to the cytosol and translocated into the ER via TAP [179]. However, because it is partly derived from ER, the phagosome may contain the same peptide-loading machinery as the ER [182–184]. Therefore, these peptides may bind to class I MHC molecules in the phagosome and shuttle to the surface through a still unknown pathway.

Endosomal compartments contain proteases that degrade proteins and are involved in class I MHC peptide presentation pathways. One such endopeptidase, cathepsin S, has been shown to play a major role in generating peptides for cross-presentation in the endosomal compartments [185]. These peptides are loaded into class I MHC molecules in post-Golgi compartments. The source of the class I MHC molecules may be the traditional endogenous pathway from the ER, or it may actually be class I MHC that has been recruited from the surface of the cell by endocytosis [186].

1.6 Class I MHC Maturation and Peptide Loading

TAP facilitates loading of peptides directly onto class I MHC molecules by its association in the ER lumen with a large complex of proteins including nascently formed class I MHC (with β_2m) [187, 188], tapasin, calreticulin, and ERp57 [20]. Peptide loading and all the molecules involved in the process are diagramed in Figure 1.3. Before peptides can be loaded onto the class I MHC molecule, they must be processed through a complicated path involving the protein-folding machinery in the cell.

1.6.1 ER Chaperones: Calnexin, Calreticulin, ERp57, and Tapasin

1.6.1.1 Calnexin

Calnexin (CNX) and its soluble relation calreticulin (CRT) are both chaperones found in the ER that promote correct folding of glycosylated proteins including class I MHC heavy chain [189–194]. Immediately following translocation of class I heavy chain into the ER membrane, the new chain acquires an N-linked oligosaccharide at asparagine 86. Calnexin, a 65-kDa membrane-bound chaperone of the ER, associates with the heavy chain almost immediately [195–198] and aids in forming the heterodimer of heavy chain– β_2m [199, 200]. As a chaperone, calnexin protects monomeric MHC heavy chains in the ER from aggregation and subsequent degradation. Calnexin also impedes egress of peptide-deficient MHC heterodimers as well as free heavy chain from the ER [201], suggesting that it contributes to the efficient expression of peptide-bound MHC molecules on the cell surface. The crystallographic structure of the lumenal portion of calnexin shows that it consists of two domains: a globular domain, which contains an oligosaccharide recognition site; and a long arm containing proline-rich repeats (P-domain), which recruits the thiol reductase ERp57 [202, 203]. This long, flexible arm may be what allows calnexin to interact so well with misfolded glycoproteins.

1.6.1.2 Tapasin

Tapasin is a 48-kDa glycoprotein encoded within the *MHCI* locus. Sequence analysis places it in the immunoglobulin family [204]. Tapasin co-immunoprecipitates with TAP and is described as the bridge between TAP and class I molecules [205]. It is a type I transmembrane glycoprotein with an ER retention signal and is essential for peptide loading of class I MHC molecules. Cells that lack expression of tapasin have no detectable class I molecules, calreticulin, or ERp57 associated with TAP [204–208]. The carboxyl-terminal 128 residues of tapasin are essential for interaction with TAP, while deletion of the amino-terminal 50 residues abrogates the ability of tapasin to stabilize the class I heavy chain–β_2m–calreticulin–ERp57 heterocomplex [209]. Mutational analysis shows that tapasin binds to the

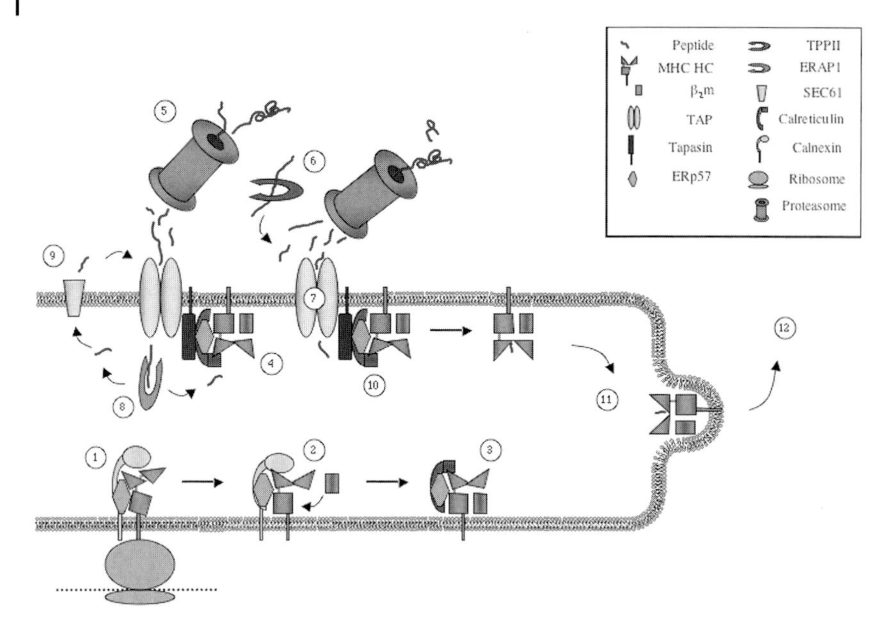

Figure 1.3 Production of peptide-loaded class I MHC. Class I MHC heavy chain and β_2m are translocated separately into the ER. Immediately when accessible, Asn86 of the heavy chain is glycosylated, allowing for recognition by the chaperone calnexin. (1) Calnexin also recruits Erp57, which mediates disulfide bond formation in $\alpha3$ and $\alpha2$ of class I heavy chain. (2) Once the heavy chain is folded and the disulfide bond in $\alpha3$ is formed, soluble β_2m binds to class I heavy chain. While the MHC class I molecule is being formed (cycling on and off calnexin), newly translated tapasin associates with calnexin and ERp57, both of which facilitate its folding and disulfide bond formation. Tapasin with calnexin–ERp57 then binds and stabilizes the TAP1/2 dimer. (3) Upon binding of β_2m, calnexin dissociates and is replaced by calreticulin. The class I heavy chain–β_2m–calreticulin–ERp57 heterocomplex then associates with the tapasin–TAP–calnexin–ERp57 heterocomplex. ERp57 remains part of the peptide-loading complex (PLC) until peptide has been loaded, but it is not known whether ERp57 comes with tapasin or the class I molecule. (4) The PLC now consists of TAP, tapasin, heavy chain, β_2m, calreticulin, and ERp57. (5) Proteasomes in the cytosol degrade substrates, generating peptides of varying length. (6) Large proteasomal products can be further cleaved in the cytosol by aminopeptidases such as TPPII. (7) Peptides with affinity for TAP will bind, and ATP-induced conformational changes in TAP will transport it through the ER membrane. (8) ERAP1 performs amino-terminal trimming in the ER. (9) These peptides may go directly into the PLC, or they may be retro-translocated back into the cytosol through SEC61 and have to enter the ER again via TAP. (10) If the amino acid motif of the peptide matches the class I MHC molecule, it will then bind in the peptide groove, inducing a conformational change that stabilizes peptide binding. ERp57 will mediate the disulfide formation in $\alpha2$, "locking" the peptide in. (11) If the correct conformational change is induced, all chaperone molecules will dissociate. (12) The trimer of class I heavy chain–β_2m–peptide is shuttled from the ER to the Golgi apparatus and subsequently to the plasma membrane of the cell. (This figure also appears with the color plates.)

$\alpha 2$ domain of class I heavy chain, an observation that is conserved between H-2D^d, H-2L^d, and HLA-A2 [210]. The stoichiometric level of tapasin to other PLC components is unknown. It could be a 1:1 ratio of TAP-1/2 and tapasin [206], or four tapasin molecules associate with TAP-1/2 [204], potentially allowing up to four class I peptide-loading complexes to associate with TAP. Tapasin-deficient mice show low cell surface expression of empty class I molecules or class I molecules bound with suboptimal, low-affinity peptides, affecting CTL response [211–214]. Tapasin-expressing cells present stably bound peptide with a long half-life, while tapasin-deficient cells present peptides with a reduced half-life [215]. Presumably, tapasin accomplishes this by selecting peptides with the greatest half-life upon conformational change of class I heavy chain [215]. The tapasin–calnexin complex also recruits the thiol reductase, ERp57, which may remain to assist with loading the class I MHC molecule [216].

1.6.1.3 **ERp57**

ERp57 is responsible for facilitating disulfide bond formation in the $\alpha 2$ and $\alpha 3$ domains of the class I MHC heavy chain [217, 218]. After being recruited by calnexin to the class I heavy chain, ERp57 remains as part of the PLC until peptide is loaded [219]. ERp57 forms an interchain disulfide bond with tapasin [220, 221]. It appears that the isomerization of the disulfide in the $\alpha 2$ domain of class I heavy chain is controlled by tapasin–ERp57 and occurs only once the proper peptide has been loaded in the groove [221]. ERp57 may have a role in quality control in the ER by reducing improperly folded class I heavy chains [222], which could be translocated back into the cytoplasm for degradation by the proteasome [223]. Following proper disulfide bond formation in the class I MHC heavy chain and association with β_2m, calnexin dissociates and calreticulin binds the MHC heterodimer [205, 218, 224].

1.6.1.4 **Calreticulin**

Calreticulin, a 46-kDa soluble protein found in the ER, displays high sequence homology to the lumenal domain of calnexin, including the P-domain [225], and functions to keep class I heavy chain–β_2m in a peptide-receptive state [226]. Calreticulin remains associated with the PLC until proper peptide loading. Deleting calreticulin resulted in reduced efficiency of peptide loading by 50–80% and impaired CTL activation [227]. Calreticulin not only binds to class I heavy chain, but its P-domain also binds ERp57 in a manner similar to that of calnexin [228], keeping ERp57 in the class I heterocomplex until peptide is loaded.

1.6.2 Peptide Loading

The timing of the events that produce a peptide-loading MHC molecule is likely to be very carefully regulated, especially during inflammatory responses. Table 1.2

shows the timing of many of the steps in antigen presentation. The rate-limiting step appears to be folding of the class I MHC molecule from the nascent chain. Once the protein is folded properly, it is not clear exactly how peptide is passed from TAP to the peptide-binding cleft to allow for the final step in maturation to occur. TAP will bind to a peptide and transport it through the ER membrane. If necessary, ERAP1 will perform any additional amino-terminal trimming if the peptide is too long. If the amino acid motif of the peptide matches the class I MHC allele, it will then bind in the peptide cleft, inducing a conformational change that stabilizes peptide binding. ERp57 will mediate the disulfide formation in $\alpha 2$, "locking" the peptide in. Tapasin then somehow evaluates the affinity and/or half-life of the peptide. If the peptide has the appropriate motif and the correct conformational change is induced, all chaperone molecules will dissociate and the trimer of class I heavy chain–β_2m–peptide will be shuttled from the ER to the Golgi apparatus and subsequently to the surface of the cell.

Table 1.2 Estimated timing of steps in the formation of peptide-loaded MHC.

Molecule	Activity	Copies per cell	Reference
Proteasome	Substrates degraded per minute per proteasome	800,000	235
Cytosolic peptidases	1 N-terminal residue per 2–3 seconds	n.d.	112
TAP	2–5 peptides per second per TAP	~10,000	151
MHC–β_2m	4 minutes post-translation of heavy chain	n.d.	200
MHC–β_2m–TAP	5–10 minutes post-translation of heavy chain	n.d.	187,200
MHC–peptide (k_{on})	2×10^6 M^{-1} s^{-1}	1000–5000 in ER	43

1.7 Immunodominance and Class I MHC Peptide Processing

Peptide processing may be the primary filter for epitope selection within the cell. Both the constitutive proteasome and the immunoproteasome selectively cleave on the carboxyl-terminal side of specific residues. Upon infection, the immunoproteasome is more adept at producing peptides preferred by TAP and class I MHC molecules: those with hydrophobic or positively charged carboxyl-terminal residues. It is likely that the proteasome, TAP, and class I MHC molecules have coevolved to generate and select peptides with similar characteristics, namely, length and carboxyl-terminal properties.

Although there have been correlations between pMHC stability or peptide binding and immunodominance, most studies have shown that the relationship between antigen processing and immunodominance is a passive one. However, if the peptide is not produced by the antigen-presenting machinery, the number of responding T cells may be significantly reduced. LMP2 knockout mice produce 50% fewer T cells to specific influenza immunodominant peptides [229]. Conversely, other studies have shown that efficient T-cell recognition of an immunodominant ovalbumin epitope remains after deliberate frameshift mutations such that the antigen-presenting cells expressed very low levels of the correct protein sequence [230]. This suggests that regardless of whatever is translated and processed, the immunodominant epitope, even when at very low levels, is still immunodominant. Ultimately, it appears that immunodominance is defined by the T-cell repertoire in the host.

References

1 Berkower, I., Matis, L. A., Buckenmeyer, G. K., Gurd, F. R., Longo, D. L., and Berzofsky, J. A. (1984) *J Immunol* **132**(3), 1370–1378
2 Finnegan, A., Smith, M. A., Smith, J. A., Berzofsky, J., Sachs, D. H., and Hodes, R. J. (1986) *J Exp Med* **164**(3), 897–910
3 Katz, M. E., Maizels, R. M., Wicker, L., Miller, A., and Sercarz, E. E. (1982) *Eur J Immunol* **12**(7), 535–540
4 Lai, M. Z., Ross, D. T., Guillet, J. G., Briner, T. J., Gefter, M. L., and Smith, J. A. (1987) *J Immunol* **139**(12), 3973–3980
5 Solinger, A. M., Ultee, M. E., Margoliash, E., and Schwartz, R. H. (1979) *J Exp Med* **150**(4), 830–848
6 Zamvil, S. S., Mitchell, D. J., Moore, A. C., Kitamura, K., Steinman, L., and Rothbard, J. B. (1986) *Nature* **324**(6094), 258–260
7 Madden, D. R., Gorga, J. C., Strominger, J. L., and Wiley, D. C. (1991) *Nature* **353**(6342), 321–325
8 Fremont, D. H., Matsumura, M., Stura, E. A., Peterson, P. A., and Wilson, I. A. (1992) *Science* **257**(5072), 919–927
9 Garrett, T. P., Saper, M. A., Bjorkman, P. J., Strominger, J. L., and Wiley, D. C. (1989) *Nature* **342**(6250), 692–696
10 Saper, M. A., Bjorkman, P. J., and Wiley, D. C. (1991) *J Mol Biol* **219**(2), 277–319
11 Meijers, R., Lai, C. C., Yang, Y., Liu, J. H., Zhong, W., Wang, J. H., and Reinherz, E. L. (2005) *J Mol Biol* **345**(5), 1099–1110
12 Kjer-Nielsen, L., Clements, C. S., Brooks, A. G., Purcell, A. W., Fontes, M. R., McCluskey, J., and Rossjohn, J. (2002) *J Immunol* **169**(9), 5153–5160
13 Guo, H. C., Jardetzky, T. S., Garrett, T. P., Lane, W. S., Strominger, J. L., and Wiley, D. C. (1992) *Nature* **360**(6402), 364–366
14 Collins, E. J., Garboczi, D. N., Karpusas, M. N., and Wiley, D. C. (1995) *Proc Natl Acad Sci U S A* **92**(4), 1218–1221
15 Zhang, W., Young, A. C., Imarai, M., Nathenson, S. G., and Sacchettini, J. C. (1992) *Proc Natl Acad Sci U S A* **89**(17), 8403–8407
16 Zhao, R., Loftus, D. J., Appella, E., and Collins, E. J. (1999) *J Exp Med* **189**(2), 359–370
17 Hunt, D. F., Henderson, R. A., Shabanowitz, J., Sakaguchi, K., Michel, H., Sevilir, N., Cox, A. L., Appella, E., and Engelhard, V. H. (1992) *Science* **255**(5049), 1261–1263
18 Kieper, W. C., and Jameson, S. C. (1999) *Proc Natl Acad Sci U S A* **96**(23), 13306–13311
19 Jardetzky, T. S., Lane, W. S., Robinson, R. A., Madden, D. R., and Wiley, D. C. (1991) *Nature* **353**(6342), 326–329

20 Reits, E. A., Vos, J. C., Gromme, M., and Neefjes, J. (2000) *Nature* **404**(6779), 774–778
21 Matsumura, M., Fremont, D. H., Peterson, P. A., and Wilson, I. A. (1992) *Science* **257**(5072), 927–934
22 Bjorkman, P. J., Saper, M. A., Samraoui, B., Bennett, W. S., Strominger, J. L., and Wiley, D. C. (1987) *Nature* **329**(6139), 512–518
23 Bjorkman, P. J., Saper, M. A., Samraoui, B., Bennett, W. S., Strominger, J. L., and Wiley, D. C. (1987) *Nature* **329**(6139), 506–512
24 Falk, K., Rotzschke, O., and Rammensee, H. G. (1990) *Nature* **348**(6298), 248–251
25 Falk, K., Rotzschke, O., Stevanovic, S., Jung, G., and Rammensee, H. G. (1991) *Nature* **351**(6324), 290–296
26 Van Bleek, G. M., and Nathenson, S. G. (1990) *Nature* **348**(6298), 213–216
27 Rammensee, H. G., Falk, K., and Rotzschke, O. (1993) *Annu Rev Immunol* **11**, 213–244
28 Rammensee, H. G., Friede, T., and Stevanoviic, S. (1995) *Immunogenetics* **41**(4), 178–228
29 Romero, P., Corradin, G., Luescher, I. F., and Maryanski, J. L. (1991) *J Exp Med* **174**(3), 603–612
30 Cerundolo, V., Elliott, T., Elvin, J., Bastin, J., Rammensee, H. G., and Townsend, A. (1991) *Eur J Immunol* **21**(9), 2069–2075
31 Elliott, T., Cerundolo, V., Elvin, J., and Townsend, A. (1991) *Nature* **351**(6325), 402–406
32 Maryanski, J. L., Verdini, A. S., Weber, P. C., Salemme, F. R., and Corradin, G. (1990) *Cell* **60**(1), 63–72
33 Probst-Kepper, M., Hecht, H. J., Herrmann, H., Janke, V., Ocklenburg, F., Klempnauer, J., van den Eynde, B. J., and Weiss, S. (2004) *J Immunol* **173**(9), 5610–5616
34 Speir, J. A., Stevens, J., Joly, E., Butcher, G. W., and Wilson, I. A. (2001) *Immunity* **14**(1), 81–92
35 Khanna, R., Burrows, S. R., and Moss, D. J. (1995) *Microbiol Rev* **59**(3), 387–405
36 Binz, A. K., Rodriguez, R. C., Biddison, W. E., and Baker, B. M. (2003) *Biochemistry* **42**(17), 4954–4961
37 Buus, S., Colon, S., Smith, C., Freed, J. H., Miles, C., and Grey, H. M. (1986) *Proc Natl Acad Sci U S A* **83**(11), 3968–3971
38 Neefjes, J. J., Dierx, J., and Ploegh, H. L. (1993) *Eur J Immunol* **23**(4), 840–845
39 Elliott, T., Elvin, J., Cerundolo, V., Allen, H., and Townsend, A. (1992) *Eur J Immunol* **22**(8), 2085–2091
40 Fahnestock, M. L., Johnson, J. L., Feldman, R. M., Tsomides, T. J., Mayer, J., Narhi, L. O., and Bjorkman, P. J. (1994) *Biochemistry* **33**(26), 8149–8158
41 Bouvier, M., and Wiley, D. C. (1998) *Nat Struct Biol* **5**(5), 377–384
42 Fahnestock, M. L., Tamir, I., Narhi, L., and Bjorkman, P. J. (1992) *Science* **258**(5088), 1658–1662
43 Gakamsky, D. M., Davis, D. M., Strominger, J. L., and Pecht, I. (2000) *Biochemistry* **39**(36), 11163–11169
44 Townsend, A., Elliott, T., Cerundolo, V., Foster, L., Barber, B., and Tse, A. (1990) *Cell* **62**(2), 285–295
45 Townsend, A., Ohlen, C., Bastin, J., Ljunggren, H. G., Foster, L., and Karre, K. (1989) *Nature* **340**(6233), 443–448
46 Ljunggren, H. G., Stam, N. J., Ohlen, C., Neefjes, J. J., Hoglund, P., Heemels, M. T., Bastin, J., Schumacher, T. N., Townsend, A., Karre, K., and et al. (1990) *Nature* **346**(6283), 476–480
47 Buxton, S. E., Benjamin, R. J., Clayberger, C., Parham, P., and Krensky, A. M. (1992) *J Exp Med* **175**(3), 809–820
48 Matsui, M., Hioe, C. E., and Frelinger, J. A. (1993) *Proc Natl Acad Sci U S A* **90**(2), 674–678
49 Santos-Aguado, J., Crimmins, M. A., Mentzer, S. J., Burakoff, S. J., and Strominger, J. L. (1989) *Proc Natl Acad Sci U S A* **86**(22), 8936–8940
50 Villadangos, J. A., Galocha, B., Lopez, D., Calvo, V., and Lopez de Castro, J. A. (1992) *J Immunol* **149**(2), 505–510
51 Winter, C. C., Carreno, B. M., Turner, R. V., Koenig, S., and Biddison, W. E. (1991) *J Immunol* **146**(10), 3508–3512
52 Rohren, E. M., Pease, L. R., Ploegh, H. L., and Schumacher, T. N. (1993) *J Exp Med* **177**(6), 1713–1721

53 Dong, T., Boyd, D., Rosenberg, W., Alp, N., Takiguchi, M., McMichael, A., and Rowland-Jones, S. (1996) *Eur J Immunol* **26**(2), 335–339
54 Chen, W., Khilko, S., Fecondo, J., Margulies, D. H., and McCluskey, J. (1994) *J Exp Med* **180**(4), 1471–1483
55 Sigal, L. J., Goebel, P., and Wylie, D. E. (1995) *Mol Immunol* **32**(9), 623–632
56 Oldstone, M. B., Lewicki, H., Borrow, P., Hudrisier, D., and Gairin, J. E. (1995) *J Virol* **69**(12), 7423–7429
57 Feltkamp, M. C., Vierboom, M. P., Kast, W. M., and Melief, C. J. (1994) *Mol Immunol* **31**(18), 1391–1401
58 Deng, Y., Yewdell, J. W., Eisenlohr, L. C., and Bennink, J. R. (1997) *J Immunol* **158**(4), 1507–1515
59 Parker, K. C., Bednarek, M. A., and Coligan, J. E. (1994) *J Immunol* **152**(1), 163–175
60 Rammensee, H., Bachmann, J., Emmerich, N. P., Bachor, O. A., and Stevanovic, S. (1999) *Immunogenetics* **50**(3–4), 213–219
61 Gulukota, K., Sidney, J., Sette, A., and DeLisi, C. (1997) *J Mol Biol* **267**(5), 1258–1267
62 Singh, H., and Raghava, G. P. (2003) *Bioinformatics* **19**(8), 1009–1014
63 Serwold, T., and Shastri, N. (1999) *J Immunol* **162**(8), 4712–4719
64 Paradela, A., Alvarez, I., Garcia-Peydro, M., Sesma, L., Ramos, M., Vazquez, J., and Lopez De Castro, J. A. (2000) *J Immunol* **164**(1), 329–337
65 Michalek, M. T., Grant, E. P., Gramm, C., Goldberg, A. L., and Rock, K. L. (1993) *Nature* **363**(6429), 552–554
66 Schubert, U., Anton, L. C., Gibbs, J., Norbury, C. C., Yewdell, J. W., and Bennink, J. R. (2000) *Nature* **404**(6779), 770–774
67 Yewdell, J. W., Reits, E., and Neefjes, J. (2003) *Nat Rev Immunol* **3**(12), 952–961
68 Del Val, M., Schlicht, H. J., Ruppert, T., Reddehase, M. J., and Koszinowski, U. H. (1991) *Cell* **66**(6), 1145–1153
69 Eisenlohr, L. C., Yewdell, J. W., and Bennink, J. R. (1992) *J Exp Med* **175**(2), 481–487
70 Eggers, M., Boes-Fabian, B., Ruppert, T., Kloetzel, P. M., and Koszinowski, U. H. (1995) *J Exp Med* **182**(6), 1865–1870
71 Niedermann, G., Butz, S., Ihlenfeldt, H. G., Grimm, R., Lucchiari, M., Hoschutzky, H., Jung, G., Maier, B., and Eichmann, K. (1995) *Immunity* **2**(3), 289–299
72 Ossendorp, F., Eggers, M., Neisig, A., Ruppert, T., Groettrup, M., Sijts, A., Mengede, E., Kloetzel, P. M., Neefjes, J., Koszinowski, U., and Melief, C. (1996) *Immunity* **5**(2), 115–124
73 Orlowski, M., Cardozo, C., and Michaud, C. (1993) *Biochemistry* **32**(6), 1563–1572
74 Orlowski, M., and Wilk, S. (1981) *Biochem Biophys Res Commun* **101**(3), 814–822
75 Wilk, S., and Orlowski, M. (1980) *J Neurochem* **35**(5), 1172–1182
76 Wilk, S., and Orlowski, M. (1983) *J Neurochem* **40**(3), 842–849
77 Dick, L. R., Moomaw, C. R., DeMartino, G. N., and Slaughter, C. A. (1991) *Biochemistry* **30**(10), 2725–2734
78 Bochtler, M., Ditzel, L., Groll, M., Hartmann, C., and Huber, R. (1999) *Annu Rev Biophys Biomol Struct* **28**, 295–317
79 Groll, M., Bajorek, M., Kohler, A., Moroder, L., Rubin, D. M., Huber, R., Glickman, M. H., and Finley, D. (2000) *Nat Struct Biol* **7**(11), 1062–1067
80 Groll, M., Ditzel, L., Lowe, J., Stock, D., Bochtler, M., Bartunik, H. D., and Huber, R. (1997) *Nature* **386**(6624), 463–471
81 Whitby, F. G., Masters, E. I., Kramer, L., Knowlton, J. R., Yao, Y., Wang, C. C., and Hill, C. P. (2000) *Nature* **408**(6808), 115–120
82 Rock, K. L., Gramm, C., Rothstein, L., Clark, K., Stein, R., Dick, L., Hwang, D., and Goldberg, A. L. (1994) *Cell* **78**(5), 761–771
83 Kessler, B., Hong, X., Petrovic, J., Borodovsky, A., Dantuma, N. P., Bogyo, M., Overkleeft, H. S., Ploegh, H., and Glas, R. (2003) *J Biol Chem* **278**(12), 10013–10021
84 Monaco, J. J., and McDevitt, H. O. (1982) *Proc Natl Acad Sci U S A* **79**(9), 3001–3005
85 Monaco, J. J., and McDevitt, H. O. (1984) *Nature* **309**(5971), 797–799

86 Yang, Y., Waters, J. B., Fruh, K., and Peterson, P. A. (1992) *Proc Natl Acad Sci U S A* **89**(11), 4928–4932
87 Barton, L. F., Cruz, M., Rangwala, R., Deepe, G. S., Jr., and Monaco, J. J. (2002) *J Immunol* **169**(6), 3046–3052
88 Hisamatsu, H., Shimbara, N., Saito, Y., Kristensen, P., Hendil, K. B., Fujiwara, T., Takahashi, E., Tanahashi, N., Tamura, T., Ichihara, A., and Tanaka, K. (1996) *J Exp Med* **183**(4), 1807–1816
89 Nandi, D., Jiang, H., and Monaco, J. J. (1996) *J Immunol* **156**(7), 2361–2364
90 Gaczynska, M., Rock, K. L., Spies, T., and Goldberg, A. L. (1994) *Proc Natl Acad Sci U S A* **91**(20), 9213–9217
91 Groettrup, M., Ruppert, T., Kuehn, L., Seeger, M., Standera, S., Koszinowski, U., and Kloetzel, P. M. (1995) *J Biol Chem* **270**(40), 23808–23815
92 Glynne, R., Powis, S. H., Beck, S., Kelly, A., Kerr, L. A., and Trowsdale, J. (1991) *Nature* **353**(6342), 357–360
93 Kelly, A., Powis, S. H., Glynne, R., Radley, E., Beck, S., and Trowsdale, J. (1991) *Nature* **353**(6345), 667–668
94 Gaczynska, M., Rock, K. L., and Goldberg, A. L. (1993) *Nature* **365**(6443), 264–267
95 Harris, P. E., Colovai, A., Liu, Z., Dalla Favera, R., and Suciu-Foca, N. (1993) *J Immunol* **151**(11), 5966–5974
96 Griffin, T. A., Nandi, D., Cruz, M., Fehling, H. J., Kaer, L. V., Monaco, J. J., and Colbert, R. A. (1998) *J Exp Med* **187**(1), 97–104
97 Nandi, D., Woodward, E., Ginsburg, D. B., and Monaco, J. J. (1997) *Embo J* **16**(17), 5363–5375
98 Zaiss, D. M., Standera, S., Kloetzel, P. M., and Sijts, A. J. (2002) *Proc Natl Acad Sci U S A* **99**(22), 14344–14349
99 Gaczynska, M., Goldberg, A. L., Tanaka, K., Hendil, K. B., and Rock, K. L. (1996) *J Biol Chem* **271**(29), 17275–17280
100 Bose, S., Stratford, F. L., Broadfoot, K. I., Mason, G. G., and Rivett, A. J. (2004) *Biochem J* **378**(Pt 1), 177–184
101 Ma, C. P., Slaughter, C. A., and DeMartino, G. N. (1992) *J Biol Chem* **267**(15), 10515–10523
102 Realini, C., Dubiel, W., Pratt, G., Ferrell, K., and Rechsteiner, M. (1994) *J Biol Chem* **269**(32), 20727–20732
103 Stohwasser, R., Salzmann, U., Giesebrecht, J., Kloetzel, P. M., and Holzhutter, H. G. (2000) *Eur J Biochem* **267**(20), 6221–6230
104 Preckel, T., Fung-Leung, W. P., Cai, Z., Vitiello, A., Salter-Cid, L., Winqvist, O., Wolfe, T. G., Von Herrath, M., Angulo, A., Ghazal, P., Lee, J. D., Fourie, A. M., Wu, Y., Pang, J., Ngo, K., Peterson, P. A., Fruh, K., and Yang, Y. (1999) *Science* **286**(5447), 2162–2165
105 Groettrup, M., Soza, A., Eggers, M., Kuehn, L., Dick, T. P., Schild, H., Rammensee, H. G., Koszinowski, U. H., and Kloetzel, P. M. (1996) *Nature* **381**(6578), 166–168
106 Kisselev, A. F., Akopian, T. N., Woo, K. M., and Goldberg, A. L. o. (1999) *J Biol Chem* **274**(6), 3363–3371
107 Li, J., Gao, X., Ortega, J., Nazif, T., Joss, L., Bogyo, M., Steven, A. C., and Rechsteiner, M. (2001) *Embo J* **20**(13), 3359–3369
108 Sun, Y., Sijts, A. J., Song, M., Janek, K., Nussbaum, A. K., Kral, S., Schirle, M., Stevanovic, S., Paschen, A., Schild, H., Kloetzel, P. M., and Schadendorf, D. (2002) *Cancer Res* **62**(10), 2875–2882
109 Khan, S., van den Broek, M., Schwarz, K., de Giuli, R., Diener, P. A., and Groettrup, M. (2001) *J Immunol* **167**(12), 6859–6868
110 Murata, S., Udono, H., Tanahashi, N., Hamada, N., Watanabe, K., Adachi, K., Yamano, T., Yui, K., Kobayashi, N., Kasahara, M., Tanaka, K., and Chiba, T. (2001) *Embo J* **20**(21), 5898–5907
111 Dick, T. P., Ruppert, T., Groettrup, M., Kloetzel, P. M., Kuehn, L., Koszinowski, U. H., Stevanovic, S., Schild, H., and Rammensee, H. G. (1996) *Cell* **86**(2), 253–262
112 Reits, E., Griekspoor, A., Neijssen, J., Groothuis, T., Jalink, K., van Veelen, P., Janssen, H., Calafat, J., Drijfhout, J. W., and Neefjes, J. (2003) *Immunity* **18**(1), 97–108
113 Stoltze, L., Schirle, M., Schwarz, G., Schroter, C., Thompson, M. W., Hersh, L. B., Kalbacher, H., Stevanovic, S., Rammensee, H. G., and Schild, H. (2000) *Nat Immunol* **1**(5), 413–418

114 Seifert, U., Maranon, C., Shmueli, A., Desoutter, J. F., Wesoloski, L., Janek, K., Henklein, P., Diescher, S., Andrieu, M., de la Salle, H., Weinschenk, T., Schild, H., Laderach, D., Galy, A., Haas, G., Kloetzel, P. M., Reiss, Y., and Hosmalin, A. (2003) *Nat Immunol* **4**(4), 375–379

115 Levy, F., Burri, L., Morel, S., Peitrequin, A. L., Levy, N., Bachi, A., Hellman, U., Van den Eynde, B. J., and Servis, C. (2002) *J Immunol* **169**(8), 4161–4171

116 Paz, P., Brouwenstijn, N., Perry, R., and Shastri, N. (1999) *Immunity* **11**(2), 241–251

117 Neisig, A., Roelse, J., Sijts, A. J., Ossendorp, F., Feltkamp, M. C., Kast, W. M., Melief, C. J., and Neefjes, J. J. (1995) *J Immunol* **154**(3), 1273–1279

118 Wang, Y., Guttoh, D. S., and Androlewicz, M. J. (1998) *Melanoma Res* **8**(4), 345–353

119 Androlewicz, M. J., and Cresswell, P. (1994) *Immunity* **1**(1), 7–14

120 Lauvau, G., Kakimi, K., Niedermann, G., Ostankovitch, M., Yotnda, P., Firat, H., Chisari, F. V., and van Endert, P. M. (1999) *J Exp Med* **190**(9), 1227–1240

121 Elliott, T., Willis, A., Cerundolo, V., and Townsend, A. (1995) *J Exp Med* **181**(4), 1481–1491

122 Serwold, T., Gaw, S., and Shastri, N. (2001) *Nat Immunol* **2**(7), 644–651

123 Tanioka, T., Hattori, A., Masuda, S., Nomura, Y., Nakayama, H., Mizutani, S., and Tsujimoto, M. (2003) *J Biol Chem* **278**(34), 32275–32283

124 Saric, T., Chang, S. C., Hattori, A., York, I. A., Markant, S., Rock, K. L., Tsujimoto, M., and Goldberg, A. L. (2002) *Nat Immunol* **3**(12), 1169–1176

125 Serwold, T., Gonzalez, F., Kim, J., Jacob, R., and Shastri, N. (2002) *Nature* **419**(6906), 480–483

126 York, I. A., Chang, S. C., Saric, T., Keys, J. A., Favreau, J. M., Goldberg, A. L., and Rock, K. L. (2002) *Nat Immunol* **3**(12), 1177–1184

127 Barber, L. D., Gillece-Castro, B., Percival, L., Li, X., Clayberger, C., and Parham, P. (1995) *Curr Biol* **5**(2), 179–190

128 van Endert, P. M., Riganelli, D., Greco, G., Fleischhauer, K., Sidney, J., Sette, A., and Bach, J. F. (1995) *J Exp Med* **182**(6), 1883–1895

129 Koopmann, J. O., Albring, J., Huter, E., Bulbuc, N., Spee, P., Neefjes, J., Hammerling, G. J., and Momburg, F. (2000) *Immunity* **13**(1), 117–127

130 Roelse, J., Gromme, M., Momburg, F., Hammerling, G., and Neefjes, J. (1994) *J Exp Med* **180**(5), 1591–1597

131 Brown, M. G., Driscoll, J., and Monaco, J. J. (1991) *Nature* **353**(6342), 355–357

132 Martinez, C. K., and Monaco, J. J. (1991) *Nature* **353**(6345), 664–667

133 Begley, G. S., Horvath, A. R., Taylor, J. C., and Higgins, C. F. (2005) *Mol Immunol* **42**(1), 137–141

134 Brooks, P., Murray, R. Z., Mason, G. G., Hendil, K. B., and Rivett, A. J. (2000) *Biochem J* **352 Pt 3**, 611–615

135 Brooks, P., Fuertes, G., Murray, R. Z., Bose, S., Knecht, E., Rechsteiner, M. C., Hendil, K. B., Tanaka, K., Dyson, J., and Rivett, J. (2000) *Biochem J* **346 Pt 1**, 155–161

136 Palmer, A., Rivett, A. J., Thomson, S., Hendil, K. B., Butcher, G. W., Fuertes, G., and Knecht, E. (1996) *Biochem J* **316 (Pt 2)**, 401–407

137 Deverson, E. V., Gow, I. R., Coadwell, W. J., Monaco, J. J., Butcher, G. W., and Howard, J. C. (1990) *Nature* **348**(6303), 738–741

138 Monaco, J. J., Cho, S., and Attaya, M. (1990) *Science* **250**(4988), 1723–1726

139 Spies, T., Bresnahan, M., Bahram, S., Arnold, D., Blanck, G., Mellins, E., Pious, D., and DeMars, R. (1990) *Nature* **348**(6303), 744–747

140 Trowsdale, J., Hanson, I., Mockridge, I., Beck, S., Townsend, A., and Kelly, A. (1990) *Nature* **348**(6303), 741–744

141 Kleijmeer, M. J., Kelly, A., Geuze, H. J., Slot, J. W., Townsend, A., and Trowsdale, J. (1992) *Nature* **357**(6376), 342–344

142 Ljunggren, H. G., and Karre, K. (1985) *J Exp Med* **162**(6), 1745–1759

143 DeMars, R., Chang, C. C., Shaw, S., Reitnauer, P. J., and Sondel, P. M. (1984) *Hum Immunol* **11**(2), 77–97

144 Momburg, F., Ortiz-Navarrete, V., Neefjes, J., Goulmy, E., van de Wal, Y.,

Spits, H., Powis, S. J., Butcher, G. W., Howard, J. C., Walden, P., and et al. (1992) *Nature* **360**(6400), 174–177

145 Powis, S. J., Townsend, A. R., Deverson, E. V., Bastin, J., Butcher, G. W., and Howard, J. C. (1991) *Nature* **354**(6354), 528–531

146 Kelly, A., Powis, S. H., Kerr, L. A., Mockridge, I., Elliott, T., Bastin, J., Uchanska-Ziegler, B., Ziegler, A., Trowsdale, J., and Townsend, A. (1992) *Nature* **355**(6361), 641–644

147 Spies, T., Cerundolo, V., Colonna, M., Cresswell, P., Townsend, A., and DeMars, R. (1992) *Nature* **355**(6361), 644–646

148 Higgins, C. F. (1992) *Annu Rev Cell Biol* **8**, 67–113

149 Schneider, E., and Hunke, S. (1998) *FEMS Microbiol Rev* **22**(1), 1–20

150 Shepherd, J. C., Schumacher, T. N., Ashton-Rickardt, P. G., Imaeda, S., Ploegh, H. L., Janeway, C. A., Jr., and Tonegawa, S. (1993) *Cell* **74**(3), 577–584

151 Neefjes, J. J., Momburg, F., and Hammerling, G. J. (1993) *Science* **261**(5122), 769–771

152 Androlewicz, M. J., Anderson, K. S., and Cresswell, P. (1993) *Proc Natl Acad Sci U S A* **90**(19), 9130–9134

153 Saveanu, L., Daniel, S., and van Endert, P. M. (2001) *J Biol Chem* **276**(25), 22107–22113

154 van Endert, P. M. (1999) *J Biol Chem* **274**(21), 14632–14638

155 Alberts, P., Daumke, O., Deverson, E. V., Howard, J. C., and Knittler, M. R. (2001) *Curr Biol* **11**(4), 242–251

156 Karttunen, J. T., Lehner, P. J., Gupta, S. S., Hewitt, E. W., and Cresswell, P. (2001) *Proc Natl Acad Sci U S A* **98**(13), 7431–7436

157 Gorbulev, S., Abele, R., and Tampe, R. (2001) *Proc Natl Acad Sci U S A* **98**(7), 3732–3737

158 Powis, S. J., Deverson, E. V., Coadwell, W. J., Ciruela, A., Huskisson, N. S., Smith, H., Butcher, G. W., and Howard, J. C. (1992) *Nature* **357**(6375), 211–215

159 Momburg, F., Roelse, J., Howard, J. C., Butcher, G. W., Hammerling, G. J., and Neefjes, J. J. (1994) *Nature* **367**(6464), 648–651

160 Heemels, M. T., Schumacher, T. N., Wonigeit, K., and Ploegh, H. L. (1993) *Science* **262**(5142), 2059–2063

161 Heemels, M. T., and Ploegh, H. L. (1994) *Immunity* **1**(9), 775–784

162 Powis, S. J., Howard, J. C., and Butcher, G. W. (1991) *J Exp Med* **173**(4), 913–921

163 Thorpe, C. J., Moss, D. S., Powis, S. J., Howard, J. C., Butcher, G. W., and Travers, P. J. (1995) *Immunogenetics* **41**(5), 329–331

164 Powis, S. J., Young, L. L., Joly, E., Barker, P. J., Richardson, L., Brandt, R. P., Melief, C. J., Howard, J. C., and Butcher, G. W. (1996) *Immunity* **4**(2), 159–165

165 Rudolph, M. G., Stevens, J., Speir, J. A., Trowsdale, J., Butcher, G. W., Joly, E., and Wilson, I. A. (2002) *J Mol Biol* **324**(5), 975–990

166 Uebel, S., Kraas, W., Kienle, S., Wiesmuller, K. H., Jung, G., and Tampe, R. (1997) *Proc Natl Acad Sci U S A* **94**(17), 8976–8981

167 Daniel, S., Brusic, V., Caillat-Zucman, S., Petrovsky, N., Harrison, L., Riganelli, D., Sinigaglia, F., Gallazzi, F., Hammer, J., and van Endert, P. M. (1998) *J Immunol* **161**(2), 617–624

168 Wei, M. L., and Cresswell, P. (1992) *Nature* **356**(6368), 443–446

169 Zweerink, H. J., Gammon, M. C., Utz, U., Sauma, S. Y., Harrer, T., Hawkins, J. C., Johnson, R. P., Sirotina, A., Hermes, J. D., Walker, B. D., and et al. (1993) *J Immunol* **150**(5), 1763–1771

170 Henderson, R. A., Michel, H., Sakaguchi, K., Shabanowitz, J., Appella, E., Hunt, D. F., and Engelhard, V. H. (1992) *Science* **255**(5049), 1264–1266

171 Raymond, M., Gros, P., Whiteway, M., and Thomas, D. Y. (1992) *Science* **256**(5054), 232–234

172 Sharom, F. J., DiDiodato, G., Yu, X., and Ashbourne, K. J. (1995) *J Biol Chem* **270**(17), 10334–10341

173 Sharom, F. J., Yu, X., DiDiodato, G., and Chu, J. W. (1996) *Biochem J* **320 (Pt 2)**, 421–428

174 Zernich, D., Purcell, A. W., Macdonald, W. A., Kjer-Nielsen, L.,

Ely, L. K., Laham, N., Crockford, T., Mifsud, N. A., Bharadwaj, M., Chang, L., Tait, B. D., Holdsworth, R., Brooks, A. G., Bottomley, S. P., Beddoe, T., Peh, C. A., Rossjohn, J., and McCluskey, J. (2004) *J Exp Med* **200**(1), 13–24

175 Heath, W. R., and Carbone, F. R. (2001) *Annu Rev Immunol* **19**, 47–64

176 Day, P. M., Yewdell, J. W., Porgador, A., Germain, R. N., and Bennink, J. R. (1997) *Proc Natl Acad Sci U S A* **94**(15), 8064–8069

177 Ackerman, A. L., and Cresswell, P. (2004) *Nat Immunol* **5**(7), 678–684

178 Gagnon, E., Duclos, S., Rondeau, C., Chevet, E., Cameron, P. H., Steele-Mortimer, O., Paiement, J., Bergeron, J. J., and Desjardins, M. (2002) *Cell* **110**(1), 119–131

179 Kovacsovics-Bankowski, M., and Rock, K. L. (1995) *Science* **267**(5195), 243–246

180 Norbury, C. C., Chambers, B. J., Prescott, A. R., Ljunggren, H. G., and Watts, C. (1997) *Eur J Immunol* **27**(1), 280–288

181 Rodriguez, A., Regnault, A., Kleijmeer, M., Ricciardi-Castagnoli, P., and Amigorena, S. (1999) *Nat Cell Biol* **1**(6), 362–368

182 Guermonprez, P., Saveanu, L., Kleijmeer, M., Davoust, J., Van Endert, P., and Amigorena, S. (2003) *Nature* **425**(6956), 397–402

183 Ackerman, A. L., Kyritsis, C., Tampe, R., and Cresswell, P. (2003) *Proc Natl Acad Sci U S A* **100**(22), 12889–12894

184 Houde, M., Bertholet, S., Gagnon, E., Brunet, S., Goyette, G., Laplante, A., Princiotta, M. F., Thibault, P., Sacks, D., and Desjardins, M. (2003) *Nature* **425**(6956), 402–406

185 Shen, L., Sigal, L. J., Boes, M., and Rock, K. L. (2004) *Immunity* **21**(2), 155–165

186 Hewitt, E. W., Duncan, L., Mufti, D., Baker, J., Stevenson, P. G., and Lehner, P. J. (2002) *Embo J* **21**(10), 2418–2429

187 Suh, W. K., Cohen-Doyle, M. F., Fruh, K., Wang, K., Peterson, P. A., and Williams, D. B. (1994) *Science* **264**(5163), 1322–1326

188 Ortmann, B., Androlewicz, M. J., and Cresswell, P. (1994) *Nature* **368**(6474), 864–867

189 Hammond, C., Braakman, I., and Helenius, A. (1994) *Proc Natl Acad Sci U S A* **91**(3), 913–917

190 Hebert, D. N., Foellmer, B., and Helenius, A. (1996) *Embo J* **15**(12), 2961–2968

191 Nauseef, W. M., McCormick, S. J., and Clark, R. A. (1995) *J Biol Chem* **270**(9), 4741–4747

192 Peterson, J. R., Ora, A., Van, P. N., and Helenius, A. (1995) *Mol Biol Cell* **6**(9), 1173–1184

193 Wada, I., Imai, S., Kai, M., Sakane, F., and Kanoh, H. (1995) *J Biol Chem* **270**(35), 20298–20304

194 Zhang, Q., Tector, M., and Salter, R. D. (1995) *J Biol Chem* **270**(8), 3944–3948

195 Degen, E., and Williams, D. B. (1991) *J Cell Biol* **112**(6), 1099–1115

196 Galvin, K., Krishna, S., Ponchel, F., Frohlich, M., Cummings, D. E., Carlson, R., Wands, J. R., Isselbacher, K. J., Pillai, S., and Ozturk, M. (1992) *Proc Natl Acad Sci U S A* **89**(18), 8452–8456

197 Hochstenbach, F., David, V., Watkins, S., and Brenner, M. B. (1992) *Proc Natl Acad Sci U S A* **89**(10), 4734–4738

198 Degen, E., Cohen-Doyle, M. F., and Williams, D. B. (1992) *J Exp Med* **175**(6), 1653–1661

199 Nossner, E., and Parham, P. (1995) *J Exp Med* **181**(1), 327–337

200 Neefjes, J. J., Hammerling, G. J., and Momburg, F. (1993) *J Exp Med* **178**(6), 1971–1980

201 Jackson, M. R., Cohen-Doyle, M. F., Peterson, P. A., and Williams, D. B. (1994) *Science* **263**(5145), 384–387

202 Schrag, J. D., Bergeron, J. J., Li, Y., Borisova, S., Hahn, M., Thomas, D. Y., and Cygler, M. (2001) *Mol Cell* **8**(3), 633–644

203 Pollock, S., Kozlov, G., Pelletier, M. F., Trempe, J. F., Jansen, G., Sitnikov, D., Bergeron, J. J., Gehring, K., Ekiel, I., and Thomas, D. Y. (2004) *Embo J* **23**(5), 1020–1029

204 Ortmann, B., Copeman, J., Lehner, P. J., Sadasivan, B., Herberg, J. A., Grandea, A. G., Riddell, S. R., Tampe, R., Spies, T., Trowsdale, J., and Cresswell, P. (1997) *Science* **277**(5330), 1306–1309

205 Sadasivan, B., Lehner, P. J., Ortmann, B., Spies, T., and Cresswell, P. (1996) *Immunity* **5**(2), 103–114

206 Li, S., Sjogren, H. O., Hellman, U., Pettersson, R. F., and Wang, P. (1997) *Proc Natl Acad Sci U S A* **94**(16), 8708–8713

207 Grandea, A. G., 3rd, Comber, P. G., Wenderfer, S. E., Schoenhals, G., Fruh, K., Monaco, J. J., and Spies, T. (1998) *Immunogenetics* **48**(4), 260–265

208 Li, S., Paulsson, K. M., Sjogren, H. O., and Wang, P. (1999) *J Biol Chem* **274**(13), 8649–8654

209 Bangia, N., Lehner, P. J., Hughes, E. A., Surman, M., and Cresswell, P. (1999) *Eur J Immunol* **29**(6), 1858–1870

210 Paquet, M. E., and Williams, D. B. (2002) *Int Immunol* **14**(4), 347–358

211 Garbi, N., Tan, P., Diehl, A. D., Chambers, B. J., Ljunggren, H. G., Momburg, F., and Hammerling, G. J. (2000) *Nat Immunol* **1**(3), 234–238

212 Grandea, A. G., 3rd, Golovina, T. N., Hamilton, S. E., Sriram, V., Spies, T., Brutkiewicz, R. R., Harty, J. T., Eisenlohr, L. C., and Van Kaer, L. (2000) *Immunity* **13**(2), 213–222

213 Purcell, A. W., Gorman, J. J., Garcia-Peydro, M., Paradela, A., Burrows, S. R., Talbo, G. H., Laham, N., Peh, C. A., Reynolds, E. C., Lopez De Castro, J. A., and McCluskey, J. (2001) *J Immunol* **166**(2), 1016–1027

214 Tan, P., Kropshofer, H., Mandelboim, O., Bulbuc, N., Hammerling, G. J., and Momburg, F. (2002) *J Immunol* **168**(4), 1950–1960

215 Howarth, M., Williams, A., Tolstrup, A. B., and Elliott, T. (2004) *Proc Natl Acad Sci U S A* **101**(32), 11737–11742

216 Hirano, N., Shibasaki, F., Sakai, R., Tanaka, T., Nishida, J., Yazaki, Y., Takenawa, T., and Hirai, H. (1995) *Eur J Biochem* **234**(1), 336–342

217 Lindquist, J. A., Jensen, O. N., Mann, M., and Hammerling, G. J. (1998) *Embo J* **17**(8), 2186–2195

218 Farmery, M. R., Allen, S., Allen, A. J., and Bulleid, N. J. (2000) *J Biol Chem* **275**(20), 14933–14938

219 Lindquist, J. A., Hammerling, G. J., and Trowsdale, J. (2001) *Faseb J* **15**(8), 1448–1450

220 Antoniou, A. N., and Powis, S. J. (2003) *Antioxid Redox Signal* **5**(4), 375–379

221 Dick, T. P., Bangia, N., Peaper, D. R., and Cresswell, P. (2002) *Immunity* **16**(1), 87–98

222 Antoniou, A. N., Ford, S., Alphey, M., Osborne, A., Elliott, T., and Powis, S. J. (2002) *Embo J* **21**(11), 2655–2663

223 Hughes, E. A., Hammond, C., and Cresswell, P. (1997) *Proc Natl Acad Sci U S A* **94**(5), 1896–1901

224 Solheim, J. C., Harris, M. R., Kindle, C. S., and Hansen, T. H. (1997) *J Immunol* **158**(5), 2236–2241

225 Michalak, M., Milner, R. E., Burns, K., and Opas, M. (1992) *Biochem J* **285 (Pt 3)**, 681–692

226 Culina, S., Lauvau, G., Gubler, B., and van Endert, P. M. (2004) *J Biol Chem* **279**(52), 54210–54215

227 Gao, B., Adhikari, R., Howarth, M., Nakamura, K., Gold, M. C., Hill, A. B., Knee, R., Michalak, M., and Elliott, T. (2002) *Immunity* **16**(1), 99–109

228 Frickel, E. M., Riek, R., Jelesarov, I., Helenius, A., Wuthrich, K., and Ellgaard, L. (2002) *Proc Natl Acad Sci U S A* **99**(4), 1954–1959

229 Chen, W., Norbury, C. C., Cho, Y., Yewdell, J. W., and Bennink, J. R. (2001) *J Exp Med* **193**(11), 1319–1326

230 Shastri, N., and Gonzalez, F. (1993) *J Immunol* **150**(7), 2724–2736

231 Kubo, R. T., Sette, A., Grey, H. M., Appella, E., Sakaguchi, K., Zhu, N. Z., Arnott, D., Sherman, N., Shabanowitz, J., Michel, H., and et al. (1994) *J Immunol* **152**(8), 3913–3924

232 Zhang, Q. J., Gavioli, R., Klein, G., and Masucci, M. G. (1993) *Proc Natl Acad Sci U S A* **90**(6), 2217–2221

233 Corr, M., Boyd, L. F., Frankel, S. R., Kozlowski, S., Padlan, E. A., and Margulies, D. H. (1992) *J Exp Med* **176**(6), 1681–1692

234 Corr, M., Boyd, L. F., Padlan, E. A., and Margulies, D. H. (1993) *J Exp Med* **178**(6), 1877–1892

235 Princiotta, M. F., Finzi, D., Qian, S. B., Gibbs, J., Schuchmann, S., Buttgereit, F., Bennink, J. R., and Yewdell, J. W. (2003) *Immunity* **18**(3), 343–354

236 DeLano, W. L. (2002) The PyMOL Molecular Graphics System. In., DeLano Scientific, San Carlos, CA, USA

237 Gaudet, R., and Wiley, D. C. (2001) *Embo J* **20**(17), 4964–4972

2
The Mechanics of Class II Processing: Establishment of a Peptide Class II Hierarchy

James R. Drake and Andrea J. Sant

2.1
General Overview

The processing and presentation of exogenous antigens by MHC class II–expressing antigen-presenting cells (APCs) starts with the internalization of extracellular antigen and ends with the presentation of cell-surface antigenic peptide–MHC class II complexes to MHC class II–restricted CD4 T cells (Figure 2.1). Antigen enters the APC via either endocytosis or phagocytosis. Within endocytic (processing) compartments, simple protein antigens are denatured and proteolytically processed to small peptides, while larger, more complex antigens (such as viruses and bacteria) undergo a more complex disassembly and subsequent processing to peptide. Concurrent with or subsequent to this proteolytic processing, antigenic peptides become associated with MHC class II molecules. This overall process is facilitated by chaperones such as the invariant chain (Ii)- and MHC-encoded DM proteins. Resulting antigenic peptide–MHC class II complexes are then delivered to the surface of the APC for presentation to $CD4^+$ T lymphocytes.

2.1.1
Immunodominance and Crypticity

It is now well accepted that during the specific immune response to infection by pathogens or immunization with a protein antigen, CD4 T lymphocytes respond to only a limited number of peptide epitopes derived from the immunogen. Historically, these peptides have been termed "immunodominant," while antigenic peptides that do not elicit an immune response after priming with the intact antigen, but that can bind to host MHC class II molecules and elicit an immune response when antigen processing is bypassed, are termed "cryptic." Considering the vast number of potential epitopes present within any antigen and the immense diversity of the T-cell repertoire, it is remarkable that CD4 T cells react to a very limited number of peptides and seemingly ignore the other potential epitopes from the same immunogen. In the last several decades, significant effort has been put forth to understand the molecular mechanisms that underlie immu-

Immunodominance: The Choice of the Immune System. Edited by Jeffrey A. Frelinger

ISBN: 3-527-31274-9

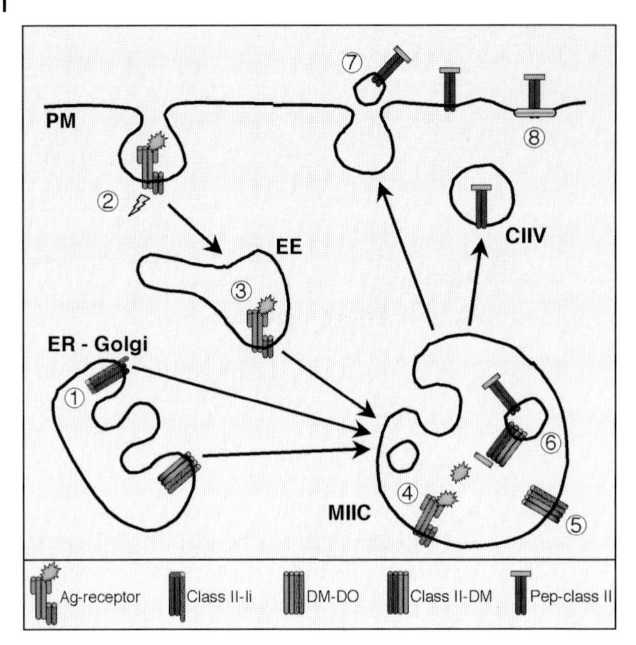

Figure 2.1 General pathway of exogenous antigen processing. Depicted is the general pathway of exogenous antigen processing and presentation. Numbered yellow circles indicate steps of the pathway at which variations can occur that will alter the hierarchy of peptide–class II complexes expressed by the APC (see text for details). (1) Relative levels of Ii isoforms (i.e., p31Ii vs. p41Ii, Section 2.2.1); (2) Effects of cell signaling (Sections 2.2.2 and 2.3); (3) Receptor-mediated antigen internalization and intracellular trafficking (Section 2.3); (4) Proteolytic processing of internalized antigen (Section 2.4.1); (5) Role of DM and DO in class II peptide loading (Section 2.4.2), (6) Intravesicular distribution of processing proteins (Section 2.4.2.4); (7) Exosomes and the cell-surface delivery of peptide–class II complexes (Section 2.4.3) (8) MHC class II signaling and partitioning of peptide–class II complexes into membrane microdomains (Section 2.4.3.2). PM: plasma membrane; EE: early endosome. (This figure also appears with the color plates.)

nodominance (reviewed in Refs. [1–5]). Many of the experiments performed to date suggest that the elements that contribute to immunodominance are highly complex and very dependent on the antigenic system under study.

The mechanisms that are thought to impact on immunodominance can be broadly divided into two categories. The first set of events that influence the hierarchy of T-cell responses relate to T-cell recognition and include such factors as precursor frequency and T-cell receptor (TCR) affinity [1, 2, 6–10]. The second major class of events that impacts immunodominance relates to antigen processing and includes such variables as antigen structure, dose, proteolytic processing, and loading of peptides onto MHC molecules. Both of these issues must be considered in any effort to reconcile the considerable literature on immunodominance and to establish a framework to understand this complex immunological event. In this chapter, we will focus primarily on the second of these two factors and examine the impact that the mechanism of antigen processing has on the generation and expression of a hierarchy of antigenic peptide–class II complexes.

2.1.2
The Impact of T-Cell Repertoire in the Experimental Analysis of Immunodominance

In order for a peptide–MHC class II complex to elicit a CD4 T cell–based immune response, there must be a threshold number of CD4 T cells within the T-cell repertoire that recognize that complex with a certain affinity. Unfortunately, both of these values are extremely difficult to measure in the unimmunized host. While the number of T cells responding to a given antigen rapidly increases during the first 3–10 days of an immune response and can approach 1 in 500 (depending on the antigen studied and assay chosen [11, 12]), clonal T-cell precursor frequencies in an unimmunized host have been estimated to be on the order of 1 in 100,000 or less [13]. TCR affinity is an even more difficult parameter to quantify. TCR affinities for antigen are estimated to be in the μM range, compared to affinities in the nM range for immunoglobulin molecules (reviewed in Ref. [14]). The low affinity and diversity of the TCRs expressed by the responding T cells also account for the paucity of knowledge in this area. To date, the affinity of the TCR for specific peptide–class II complexes has been estimated for only a few examples, using technology such as surface plasmon resonance or peptide–class II tetramers [14–16]. Therefore, our understanding of how T-cell precursor frequency and TCR affinity influence immunodominance is very limited. However, one issue is clear and important to clarify for the purpose of this discussion of immunodominance: Processing and class II–restricted presentation of self-antigens or non-self-antigens closely related to self-homologues can dramatically skew the developing T-cell repertoire that is available to respond to antigenic challenge.

It is now clear that many self-antigens gain access to the thymus and therefore can influence T-cell repertoire selection [17]. Recent data suggest that the influence of self-antigens on T-cell repertoire selection extends even to what have been historically referred to as "tissue-specific antigens" because of promiscuous expression of many peripheral antigens in a subset of thymic epithelial cells [18]. Thymic expression of self-antigens will have major consequences on T-cell repertoire development. Thymic or "central" tolerance induction during T-cell development will lead to deletion of T cells expressing TCRs of moderate to high affinity for self [19–21]. Therefore, the peripheral T-cell pool available to respond to antigenic challenge will be significantly altered. Important to our understanding of immunodominance is the realization that the efficiency of the negative selection process will be variable, depending on the degree of expression of the self-antigen in the thymus and the affinity of interaction between the TCR and the self-peptide–class II complex.

One of the best examples of the impact that self-antigen expression can have on immunodominance derives from the analysis of myelin basic protein (MBP)-specific T cells in normal mice versus mice genetically deficient in MBP production (the so-called "shiverer" mouse [21]). A comparison of the I-A^k-restricted immune response to MBP in these two strains indicates that apparent immunodominance hierarchies dramatically shift depending on the ability of the host to synthesize MBP. Similarly, the immunodominance of MBP peptides in H-2^{u} mice is biased almost exclusively towards peptides that display very low-affinity binding to the self–class II molecule [22]. In this case it appears that central tolerance induction to MBP primarily eliminates reactivity towards MBP peptides that display high affinity for the given class II molecule, but it does not eliminate T cells expressing TCR specific for MBP peptides with lower class II affinity. This type of selective tolerance induction may lead to a paradoxical focusing of the peripheral T-cell response to peptides with a low-affinity class II binding. In fact, many of the immunodominant peptides implicated in autoimmune responses display this type of low-affinity class II interaction [3, 22, 23].

For "foreign" antigens that possess close homology to host self-protein, the situation becomes even more complicated because different peptides within a foreign antigen will possess uneven sequence identity with the homologous host protein. This will result in differences in the thymic selection of clonal, potentially foreign antigen-reactive T-cell populations, thus altering the overall peptide specificity of the peripheral antigen-specific T-cell repertoire. For example, a self-peptide that has a single amino acid disparity from a foreign peptide may behave as an altered peptide ligand during thymic T-cell selection, thus paradoxically increasing, rather than decreasing, the frequency of antigen-reactive T-cell clones [24, 25].

Collectively, these considerations indicate that caution must be used in assessing the immunodominance hierarchies in response to self-antigens or antigens structurally related to host homologues, because T-cell precursor asymmetries may strongly impact the detected T-cell hierarchies. Contrastingly, in immune responses to antigens that are genetically and structurally distant from the host, one may expect that T-cell precursor frequencies to distinct antigenic peptide–class II complexes would be more similar and that other aspects of the antigen-specific immune response will dictate immunodominance. These TCR repertoire-independent elements are likely to be most important in such issues as vaccine design for protection against pathogenic organisms.

2.1.3
Different Antigen-presenting Cells Have Different Functions

Within the immune system, there are three predominant populations of MHC class II–restricted APCs: dendritic cells (DCs), B lymphocytes (B cells), and macrophages (MØs). Importantly, the immunological function of exogenous antigen processing and presentation by each type of APC is different. In their resting state, DCs (and possibly B cells) process and present self-antigens to CD4 T cells, mediating peripheral tolerance to self-antigens [26]. However, upon activation, DCs process and present foreign antigen and activate naïve T cells, allowing for the generation of effector and memory CD4 T cells [27–29]. Antigen processing and presentation by MØs serves to recruit antigen-specific effector CD4 T cells that can provide "help" to the MØ, in the way of cytokines and other factors. This activates the MØ, resulting in increased MØ effector functions such as phagocytosis and killing of intracellular organisms. Finally, antigen processing and presentation by antigen-reactive B cells serves to recruit effector CD4 T cells that are necessary to support full development of the B cell–based humoral immune response, including affinity maturation and immunoglobulin class switching.

It is also important to note that, while the general mechanism of class II–restricted antigen processing and presentation is grossly similar in all types of APCs, the precise molecular mechanisms of processing and presentation are unique for each type of APC [27, 28]. For example, B cells are the only class of APCs that express a clonally restricted, highly specific antigen receptor (that is, the B-cell receptor, BCR). Contrastingly, DCs and MØs express broadly specific pathogen recognition receptors (PRRs) such as Toll-like receptors (TLRs) and C-type lectin receptors (CLRs). Hence, while B cells can process and present noncognate antigen internalized via fluid-phase endocytosis, BCR-mediated antigen processing and presentation is the most immunologically relevant form of antigen processing by this particular class of APCs [27, 28, 30–32]. Immunologically relevant antigen processing and presentation by DCs and MØs occurs via non-clonally restricted pathogen receptors such as the CLRs. As discussed below, these differences in the mechanism of antigen binding and endocytosis will impact the spectrum of antigenic peptide–class II complexes expressed by each type of APC.

2.1.4 The Phases of Antigen Processing

Exogenous antigen processing and presentation is a complex phenomenon, and variations at many different steps along the pathway can significantly impact the hierarchy of antigenic peptide–class II complex expressed by the APC. To help organize the discussion of how these variations can impact peptide–class II expression, we will divide the pathway into three distinct, yet overlapping, phases (Table 2.1). However, it should be noted that these divisions are somewhat arbitrary and that there is a significant amount of overlap, both spatially and temporally, between each of the different phases.

Table 2.1 Phases of MHC class II–restricted antigen processing.

Phase	Subcellular localization	Variations
Phase I: Class II biosynthesis and delivery to peptide-loading compartments	ER/Golgi and endocytic pathway	Ii isoforms APC stimulation
Phase II: Antigen internalization and proteolytic processing	Plasma membrane and endocytic pathway	Receptor trafficking Differential proteolysis
Phase III: Formation and expression of peptide–class II complexes	Endocytic pathway and plasma membrane	DM/DO Membrane microdomains Exosomes

2.2
Phase I: MHC Class II Biosynthesis and Delivery to Peptide-loading Compartments

MHC-encoded α and β chain proteins are co-translationally inserted into the lipid bilayer of the endoplasmic reticulum (ER), where they subsequently assemble into α/β heterodimers with the help of a third protein, invariant chain [33, 34]. Invariant chain (Ii) is a homotrimeric type II membrane protein (Ii_3) that binds three class II heterodimers (one α/β heterodimer binding to each Ii subunit). Association of the nascent class II α/β heterodimers with Ii facilitates class II assembly and folding and blocks the binding of ER-resident peptide to the peptide-binding groove of the class II molecule. Subsequently, targeting signals within the cytoplasmic domain of Ii promotes the transport of fully assembled $(\alpha\beta)_3Ii_3$ nonameric complex to endosomal compartments.

2.2.1
Invariant Chain Isoforms

Because of differential splicing of the primary Ii RNA transcript, two different isoforms of Ii protein can be produced (see Figure 2.2). The predominant Ii isoform has a molecular mass of 31 kDa and is designated p31Ii. The second isoform has a molecular mass of 41 kDa and is designated p41Ii. Production of the p41Ii isoform is due to inclusion of exon 6b, which codes for a 64-amino-acid cysteine-rich thyroglobulin (Tg)-like domain, in the mature mRNA.

Interestingly, the Tg-like domain of p41Ii specifically interacts with cathepsin L (a protease involved in both the proteolytic processing of the Ii protein and the proteolytic processing of exogenous antigen; see below). This binding is thought to modulate cathepsin L proteolytic activity either by directly inhibiting the enzyme or by acting as a cathepsin L chaperone and potentiating the protease's activity at neutral pH [35, 36]. Interestingly, *in vitro* analysis of antigen processing by APCs capable of expressing only one of the two Ii isoforms demonstrated that the p41Ii isoform supports more efficient antigen processing and presentation [37]. Moreover, analysis of the development of CD4 T cell–dependent airway hyper-responsiveness in a murine model of

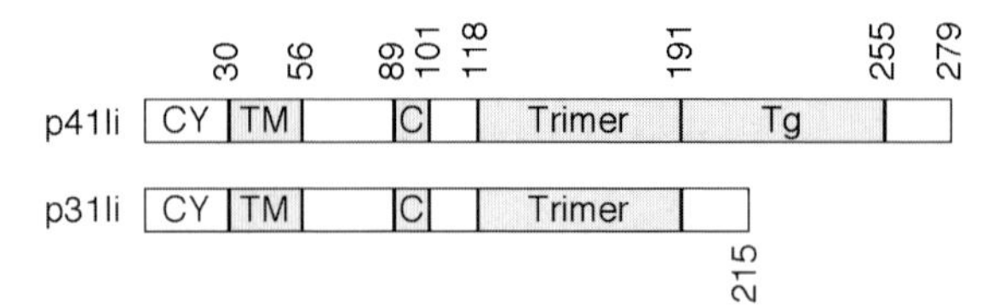

Figure 2.2 Invariant chain isoforms. The Ii protein can be produced in two isoforms. The 31-kDa p31Ii isoform possess the CLIP peptide, which associates with the class II peptide binding groove, as well as a trimerization domain. In addition to these regions, the 41-kDa p41Ii isoform contains a thyroglobulin-like domain that is encoded by exon 6b. Abbreviations: CY, N-terminal cytoplasmic domain; TM, transmembrane domain; C, CLIP region; Trimer, trimerization domain; Tg, thyroglobulin-like domain. The numbers indicate the position of the C-terminal amino acid of each domain.

asthma revealed that selective expression of the p41Ii protein, but not the p31Ii protein, supports development of all symptoms [38].

Therefore, the relative level of p31Ii versus p41Ii expressed by a particular APC will modulate the relative level of cathepsin L activity in the APC and significantly impact the spectrum of antigenic peptide–class II complexes expressed by the cell. In this context, it should be noted that p41Ii represents only 10% of the total Ii pool in B cells [39], while p41Ii can comprise upwards of 40% of the Ii pool in APCs such as DCs and MØs [40]. Hence, even if we ignore the other differences in the mechanism of antigen processing between these different APC populations (see below), the hierarchy antigenic peptide–class II complexes generated by each of these APCs will be unique.

2.2.2
Effects of Cell Signaling on MHC Class II Transport

Dendritic cells, MØs, and B cells each express one or more receptors that can mediate APC activation and elicit a change in the intracellular trafficking of MHC class II molecules that will alter the spectrum of antigenic peptides loaded onto MHC class II molecules (e.g., TLRs on DCs and the BCR on B cells). The best-studied example of this signal-induced change in class II trafficking is in DCs [27–29].

In the absence of an activating signaling, DCs are found in an immature state where nascent MHC class II molecules exist within class II–Ii complexes within intracellular compartments. Moreover, these class II–Ii complexes are not efficiently converted to antigenic peptide–class II complexes, even though the cells are capable of internalizing large amounts of extracellular antigen (see below). Immature DCs also express low levels of costimulatory molecules such as CD86. Upon stimulation by proinflammatory cytokines or microbial products such as LPS, DCs undergo a maturation process, during which the intracellular fate of class II–Ii complexes changes such that intracellular class II–Ii complexes are now efficiently converted to antigenic peptide–class II complexes and then delivered to the cell surface.

The increased surface expression of peptide–class II complexes upon DC maturation appears to be predominantly due to post-translation events, such as a change in the ability of the cells to convert protein antigens to antigenic peptides that can be use to generate peptide–class II complexes. Moreover, mature DCs also exhibit an increased ability to proteolytically convert class II–Ii complexes to class II–CLIP complexes (which can be converted to antigenic peptide–class II complexes through the action of DM; see below) as well as increased incorporation of peptide-loaded MHC class II molecules into class II vesicles (CIIV) for delivery to the cell surface. Furthermore, mature DCs express increased surface levels of costimulatory molecules such as CD86, which are delivered to the DC surface via the same transport vesicles as the peptide–class II complexes. Therefore, DC maturation not only profoundly alters the intracellular transport of MHC class II molecules and the hierarchy of peptide–class II complexes expressed on the surface of the cell but also changes the context in which these complexes are presented to CD4 T lymphocytes.

Like DCs, B cells can also exhibit a stimulation-induced change in the intracellular trafficking and expression of MHC class II molecules [41, 42]. Stimulation of B cells via the clonally restricted antigen-specific BCR results in the increased delivery of class II–Ii complexes to DM-rich late endocytic compartments and the increased cell-surface expression of antigenic peptide–class II complexes [41–43]. However, the impact of altered MHC class II trafficking on the ability of B cells to generate specific peptide–class II complexes remains unclear, as BCR signaling fails to alter the fraction of MHC class II molecules loaded with peptides derived from non-cognate antigen internalized via fluid-phase endocytosis [43]. However, an interesting unanswered question is whether the BCR stimulation-induced changes in MHC class II trafficking and MHC class II-DM colocalization result in a specific enhancement of the ability of B cells to process and present *cognate* antigen internalized via the BCR.

2.3
Phase II: Antigen Internalization and Processing

Exogenous antigen processing starts with the endocytosis or phagocytosis of extracellular antigen. While antigen processing and presentation can occur subsequent to fluid-phase antigen internalization, immunologically relevant forms of exogenous antigen processing are likely to be initiated via receptor-mediated antigen internalization. As detailed below (and summarized in Table 2.2), receptor-mediated antigen uptake by DCs and MØs occurs via proteins such as pattern recognition receptor (PRRs) and receptors for the Fc region of immunoglobulin molecules (FcRs). For B cells, immunologically relevant antigen internalization and processing occurs primarily via the clonally restricted antigen-specific BCR. Importantly, the specific receptor mediating antigen internalization and the ability of this receptor to target antigen to different intracellular compartments will significantly alter the precise molecular mechanism of antigen processing and profoundly alter the spectrum of antigenic peptide–class II complexes subsequently expressed by the APC.

Table 2.2 Receptors for antigen processing and presentation.

APC	Receptors capable of mediating antigen processing (*Family* – Examples)
Dendritic cells Macrophages	C-Type Lectin Receptors – DEC-205, DC-SIGN, Dectin-1 *Fc Receptors* – CD16, CD32 (FcγRII), CD64 *Integrins* – CR3 (CD11b/CD18)
B lymphocytes	BCR – All isotypes (e.g., IgM and IgG) *Fc Receptors* – CD32

2.3.1
BCR-mediated Antigen Internalization

BCR-mediated antigen processing is probably the most highly characterized form of receptor-mediated antigen processing and presentation. Binding of polyvalent antigen to the BCR results in BCR-mediated antigen internalization and delivery to intracellular processing compartments [27, 28]. Whether this internalization occurs via clathrin-coated pits, plasma membrane lipid rafts, or a combination of these two pathways is still a matter of debate [32, 44–46]. Nonetheless, when compared to the fluid-phase endocytosis of non-cognate antigen, internalized antigen–BCR (Ag-BCR) complexes are delivered to processing compartments with accelerated kinetics and subsequently persist within these compartments for a prolonged period of time [47, 48]. Moreover, persisting Ag-BCR complexes exhibit a higher degree of colocalization with the class II chaperone DM than with the DM inhibitor DO (see below) [43, 47]. Because of the unique trafficking of intracellular Ag-BCR complexes, antigen-specific B cells can express antigenic peptide–class II complexes for a prolonged period of time, allowing the B cell a greater window of opportunity to present antigen to CD4 T lymphocytes and thus receive the necessary signals to develop into either antibody-producing plasma cells or long-lived memory B cells and to undergo the T cell–dependent events of affinity maturation and immunoglobulin class switching.

In addition to mediating the rapid internalization of cognate antigen, the BCR can also shield portions of bound antigen from the action of proteases involved in antigen processing, or potentiate antigen processing, thus shifting the spectrum of antigenic peptides generated by the processing machinery. As demonstrated for the processing and presentation of tetanus toxoid [49], binding of antigen to the BCR can alter the spectrum of antigenic peptides generated upon antigen processing, thus altering the potential repertoire of peptides available to bind MHC class II molecules. Therefore, the precise nature of the binding between cognate antigen and the BCR may have dramatic effects on the flux of antigen through antigen-processing compartments, as well as the spectrum of antigenic peptides generated within these compartments, significantly impacting the array of resultant antigenic peptide–class II complexes expressed by the cell. This scenario suggests the possibility that individual clonal populations of B cells and T cells may interact either more or less efficiently during an immune response, depending on how BCR interaction with cognate antigen biases the array of peptide–class II complexes generated by the population of B cells. This process will in turn affect the specificity of potential CD4 T cells that can be recruited by the B cells for T-cell help.

Besides increasing the level of Ag-BCR internalization, BCR cross-linking by polyvalent antigen also results in BCR signaling events that alter multiple aspects of the antigen-processing pathway. In addition to the BCR signaling–induced alteration of MHC class II trafficking described above, BCR signaling also impacts the intracellular trafficking of internalized Ag-BCR complexes [30, 31]. Normally, internalized Ag-BCR complexes are delivered first to early endocytic compartments and then to LAMP-positive, class II–containing late endocytic compartments such as multivesicular bodies (a.k.a., MIIC), where the Ag-BCR complexes undergo processing. However, under conditions in which BCR signaling is altered or inhibited, Ag-BCR complexes are internalized [44] but ultimately fail to gain access to these class II–positive late endocytic compartments [31]. While the precise mechanism by which BCR signaling is tied to the delivery of Ag-BCR complexes to these processing compartments is not well understood, one possibility is that the association of BCR signaling molecules with the internalized Ag-BCR complexes (or post-translational modifications of the Ag-BCR complexes mediated by the associated signaling molecules) impacts the ability of the Ag-BCR complexes to gain access to MIIC [31]. Thus, various forms of antigen, capable of eliciting different levels of BCR cross-linking and BCR signaling, will be internalized, processed, and presented with various efficiencies, altering the hierarchy of antigenic peptide–class II complexes displayed by the B cell.

2.3.2
Dendritic Cells and Macrophages

In contrast to B cells, DCs and MØs do not express a clonally restricted, antigen-specific receptor. Instead, these APCs express a spectrum of PRRs, which mediate the specific recognition of pathogen-associated molecular patterns (PAMPs). Two such families of PRRs are the Toll-like receptors (TLRs) and C-type lectin receptors (CLRs). TLRs mediate the recognition of PAMPs such as lipoproteins (TLR2), lipopolysaccharides (TLR4), flagellin (TLR5), and nucleic acids (TLR3 and TLR9), whereas CLRs mediated the recognition of various mannose-type or galactose-type polysaccharides [50]. While TLRs are capable of mediating cell signaling upon ligand binding, TLRs are not generally capable of mediating efficient antigen processing and presentation [50]. Contrastingly, CLRs are capable of mediating both signaling and ligand internalization and processing and presentation [50, 51]. However, CLR signaling is not able to mediate full immune activation of DCs, such as the induction of high-level expression of costimulatory molecules. This requires TLR-based signaling. Therefore, a robust immune response to a pathogen will require engagement of both TLRs and CLRs. Moreover, because there is a significant amount of cross-talk between the TLR and CLR signaling pathways, the precise way in which a pathogen engages these two types of PRRs will ultimately determine the spectrum of antigenic peptide–class II complexes expressed by the APC and the context in which the peptide–class II complexes are displayed on the APC surface, for example, with or without costimulatory molecules. These variables have the potential to impact the spectrum of CD4 T cells recruited to the immune response by the APCs.

Interestingly, in addition to binding foreign glycoproteins, CLRs can also bind self-glycoproteins [50, 51], suggesting a mechanism by which DCs can both induce tolerance to self and initiate an immune response against non-self [26, 52]. In the absence of TLR ligation or other signaling to induce DC maturation, DCs appear to be constantly processing and presenting self-glycoprotein antigens via CLRs. However, the resultant peptide–class II complexes would be presented to naïve CD4 T cells in the absence of costimulatory molecules. Because this engagement with the APC would provide signal 1 (ligation of the TCR) but not signal 2 (costimulation), any naïve T cells capable of recognizing these self-peptide–class II complexes would likely be made tolerant. However, an encounter between a DC and a pathogen would result in the ligation of both TLRs and CLRs, which would in turn promote DC maturation (and expression of costimulatory molecules) and allow antigen internalization, processing, and presentation. Under these conditions, the DC would provide both signal 1 and signal 2 to antigen-specific naïve CD4 T cells, resulting in the activation of these cells and development of pathogen-specific effector and memory CD4 T cells.

An elegant demonstration of the *in vivo* impact of these differentiation-induced changes in DC antigen presentation comes from studies into the ability of DC-targeted antigen to elicit an immune response [29, 53]. When antigen (hen egg lysozyme, HEL, in this case) was targeted to immature DCs by conjugation to an antibody specific for the DC C-type lectin receptor DEC-205 and injected into mice, the mice mounted a transient immune response but became refractory to subsequent immunization with the same antigen, even when given with a strong adjuvant (i.e., the mice became tolerant). However, if the same DC-targeted antigen was co-injected with an anti-CD40 antibody (which, like TLR signaling, can elicit DC maturation), the immunized mice mounted a strong immune response. Therefore, the molecular form of an antigen introduced into the body, specifically, whether or not it is associated with molecules capable of eliciting DC maturation, will directly and profoundly impact the ability of DCs to present these complexes to CD4 T cells in a manner that will elicit an immune response.

2.4 Phase III: Formation and Expression of Antigenic Peptide by MHC Class II Molecules

Subsequent to the delivery of both class II–Ii complexes and internalized antigen to intracellular antigen-processing compartments, a series of proteolytic events degrades the Ii protein, leaving only a small fragment of Ii called CLIP that is associated with the class II molecule. Concurrently, proteolysis converts whole-protein antigens to antigenic peptides. After CLIP is released from the class II molecules, the formation of antigenic peptide–class II complexes can occur, a process that is facilitated by molecules such as DM and other class II chaperones (see below). The resultant peptide class II complexes are then delivered to the surface of the APC for presentation to $CD4^+$ T cells.

2.4.1 Proteolytic Antigen Processing

The ability of synthetic cryptic peptides to prime CD4 T-cell responses when offered as individual free peptides, compared to the lack of a response when the same peptide is supplied in the form of an intact antigen that requires processing, argues that the intracellular proteolytic processing events of exogenous antigen processing are critical in determining the hierarchy of peptide–class II complexes and immunodominance. The types of endosomal events that can influence this hierarchy can be divided into several subgroups. One such event is the proteolytic release of peptide from intact antigen, where variations in the efficiency of proteolysis may lead to qualitative or quantitative differences in the availability of a given peptide to complex with class II molecules and thus be presented to CD4 T cells.

The activity of endopeptidases has been demonstrated to modulate determinant display on the surface of APCs. Early studies showed that treatment of APCs with inhibitors of endosomal proteolysis could either enhance or inhibit antigen presentation, depending on the epitope under analysis [54], arguing that resident endosomal proteases can alternatively promote or diminish peptide availability. More recently, protease sensitivity has been implicated in immunodominance for the well-studied antigen hen egg lysozyme (HEL), an antigen that has been extensively studied for immunodominance patterns [2, 4, 55]. Immunodominant and cryptic peptides from HEL have been defined for many different strains of mice [9], and in H-2^d mice one of the cryptic peptides is contained within amino acids 20–35 of the protein. Schneider et al. proposed that one factor that limits the immunogenicity of this peptide is the inefficient release of the peptide from the protein, and they demonstrated that introduction of a dibasic cleavage site at the C-terminus of the peptide (within the context of the HEL protein) facilitated its presentation [56]. These data argue that inefficient release of a peptide from the intact antigen may in some cases lead to limitations in the ability to form peptide–class II complexes and ultimately result in T-cell stimulation. Conversely, there is also evidence that peptides can be destroyed by endosomal proteolysis. For example, the myelin basic protein (MBP) peptide (85–99) contains a recognition sequence for cleavage by asparagine endopeptidase and is in fact cleaved *in vitro* by this protease [57]. Moreover, inhibition of this enzyme within APCs has been shown to promote epitope expression, arguing that proteolysis normally limits the amount of this peptide available for MHC class II binding.

Collectively, these data suggest that peptides that contain a prominent protease cleavage site within the critical 10–12 amino acid segment constituting the minimal class II epitope may not effectively elicit an immune response because they are destroyed before they can bind to MHC class II molecules and be protected from more extensive proteolysis. In addition, these results suggest that antigens that have a high degree of tertiary structure (such as RNAase [58]) or proteins within large molecular complexes may sequester peptides from proteolytic release and subsequent association with class II molecules. Conversely, localization of an antigenic peptide within an extended, unstructured surface domain of a protein may promote immunodominance [59–62]. Together, these results suggest that differential release of a peptide from an intact antigen, or endosomal cleavage of a given peptide within its critical MHC-binding region, may in some cases account for the resultant hierarchy of antigenic peptide–class II complexes and thus influence immunodominance patterns in the responding population of CD4 T cells.

What remains to be established in future studies on immunodominance is how frequently the context of a peptide within an antigen accounts for immunodominance hierarchies. Stated another way, how frequently are factors such as differential unfolding of various regions of a protein and/or differences in sensitivity to proteases the factor that limits the ultimate hierarchy of responses to different peptides within a given antigen? Many early models to explain selection of immunodominant epitopes were proposed when little was known about the class II antigen-processing machinery besides that peptides were generated by endosomal proteolysis and that the class II molecules possessed a single peptide-binding pocket. Accordingly, these models in many ways reflected the view that the class II antigen processing/loading compartment might behave as a rather simple intracellular compartment that allowed colocalization of proteases, antigen, and limiting amounts of class II molecules for peptide capture. In this view, cryptic peptides were either outnumbered by or failed to compete successfully with the immunodominant peptides for class II binding. Many of the features that accounted for crypticity could then generally be grouped under the larger mechanism of "context." However, several pieces of data argue that context may not be the most critical feature of a peptide's immunodominance in the immune response. Sercarz and colleagues have shown that even though clear immunodominance patterns can be identified in the CD4 T-cell response to HEL in many different strains of mice, the peptides selected for immunodominance in one strain are different from those selected in a different strain, bearing a different set of MHC class II–presenting molecules [9] and that no region of the HEL protein is generally "cryptic." Secondly, because of advances in identification and sequences of very low-abundance peptides, it is now clear that in normal antigen-presenting cells, the vast majority of peptides presented by class II molecules are derived from internally synthesized antigens, such as the MHC class II molecule itself [63, 64]. One implication from this finding is that when peptides are released from an exogenous antigen by endosomal proteolysis, there may be an excess of endogenously produced peptides within that class II–loading compartment. Therefore, peptides from exogenous antigens are not merely competing with each other for binding to class II molecules but likely are competing against a large excess of self-peptides. Accordingly, moderate differences in the release of one peptide versus another from an intact antigen may not have a significant impact on their relative epitope density. However, pathogens or antigens that gain access to APCs by specific uptake mechanisms (see Table 2.2) may result in a higher concentration of antigenic peptides within antigen-processing compartments. In addition, it may be possible that peptides derived from receptor-internalized antigen may be loaded onto class II molecules in a directed fashion (see Section 2.4.2.4). Under these conditions, the competition among the different antigen-derived peptides and between exogenous peptides and endogenous self-peptides for binding to class II molecules may be profoundly altered, resulting in an altered hierarchy of peptide–class II complexes.

2.4.2 Class II Peptide Loading

Like the loading of antigenic peptide onto MHC class I molecules within the endoplasmic reticulum (ER), class II peptide loading within endocytic antigen-processing compartments is a protein-folding event, and like ER-based class I peptide loading, it involves chaperone molecules. In addition to the class II–associated Ii protein discussed above, one of the most highly characterized class II chaperones is the MHC-encoded intracellular protein DM [28, 65, 66]. DM is expressed in MHC class II–expressing APCs and facilitates both the release of CLIP from nascent CLIP–class II complexes and the editing of class II–associated antigenic peptides (see below). In addition, a subpopulation of class II–restricted APCs, including B cells and thymic epithelial cells, express a second MHC-encoded intracellular protein, DO, which appears to modulate the activity of the DM protein [28, 65, 66]. As we will discuss below, the relative level of each of these proteins within intracellular antigen-processing compartments can significantly alter the spectrum of antigenic peptide–class II complexes displayed on the surface of an APC.

2.4.2.1 DM

In addition to proteolytic release of peptides from antigen, a major endosomal event that has been shown to influence immunodominance is the actual formation of the peptide–class II complex. Although early experiments assessed the biology of this event by simple co-mixture of peptide with isolated class II molecules, with the identification and characterization of DM, it is now clear that peptide binding to class II is a highly regulated process [65, 67–72]. The DM protein is an MHC-encoded, noncovalently associated heterodimer that has homology with the classical class II molecules, which is now known to be a critical component of the class II presentation pathway (reviewed in Refs. [65, 66, 68]). Cell lines lacking DM accumulate class II molecules in which the Ii-derived peptide CLIP (class II–associated invariant chain peptide; see Figure 2.2) occupies the class II peptide-binding groove. Normally, CLIP must be released from class II molecules before antigenic peptide binding can occur. Biochemical studies have shown that DM binds to class II molecules and catalyzes the release of CLIP and allows the binding of antigenic peptide (reviewed in Refs. [65, 68, 73–76]). Biochemical studies involving purified DM and class II have demonstrated that DM enhances the inherent off-rate of peptides from MHC class II molecules and the subsequent loading of antigenic peptides. Thus, a currently accepted model for DM function is that it interacts directly with the class II molecule and stabilizes an "open" conformation of the peptide-binding pocket that is more likely to dissociate CLIP and to accept antigenic peptide in its place.

Soon after the discovery of the catalytic effect of DM on class II peptide loading, there was significant interest in understanding whether DM is a general catalyst of peptide binding and dissociation or whether DM displayed selectivity in its effect. Despite intensive efforts by many investigators, this question remains incompletely resolved. Most investigators have addressed this issue by probing the sensitivity of different peptide–class II complexes to DM-mediated dissociation. Early data suggested that peptides differed dramatically in regard to their sensitivity to DM-mediated dissociation, with the most kinetically stable peptide–class II complexes (such as HA-DR1) being totally resistant to DM action, while unstable complexes (such as CLIP-DR1) were highly sensitive to DM-mediated dissociation [70, 77]. Subsequently, detailed biochemical studies forced a revision of this concept and suggested that the strength of the effect of DM is relatively independent of the kinetic stability of the peptide bound to the class II molecule [72]. More recent analyses have raised the possibility that there may be some sequence specificity to DM's activity [78, 79] and demonstrate that it is difficult to predict the quantitative effect of DM editing on different class II–peptide complexes.

To put all of these experimental findings into context and consider the potential effect of DM on immunodominance, one needs to keep in mind three fairly uncontested findings. First, the time frame of peptide loading onto class II molecules within an APC is likely to be less than 6–16 hours, indicating that DM-catalyzed release of peptides from class II molecules within endosomal compartments may have negligible effects on very stable peptide–class II complexes. Second, within the time frame during which peptide loads onto class II molecules in endosomal compartments, some peptides demonstrate a net positive gain in class II loading in the presence of DM, while others exhibit a decrease in loading. Third, elution of peptides from class II molecules that are peptide loaded *in vivo* in the presence and absence of DM indicates that some peptide–class II complexes are enriched in the presence of DM while others are diminished [68, 74]. Taken together, these findings have led to the now-accepted concept of "DM editing." Functionally, this concept reflects the general observation that the hierarchy of peptide–class II complexes is affected by the catalytic activity of DM. That is to say that DM-positive cells do not simply present more antigenic and self-peptides (rather than CLIP) but rather have a different repertoire of peptides bound by their class II molecules [68]. It is likely that future experiments will allow derivation of a more precise algorithm to predict the net effect of DM editing on particular peptide–class II complexes of interest.

In light of this increasing evidence that DM effectively edits the array of peptides presented by class II molecules on APCs, the role of DM in regulating the presentation of immunodominant and cryptic peptides was examined [80]. The results of this analysis demonstrated that the class II–restricted presentation of immunodominant epitopes is dramatically enhanced by the expression of DM within APCs, while presentation of cryptic epitopes is extinguished by DM expression. Such differential editing of the repertoire of peptides presented by DM has now been extended to both self- and autoantigens [81, 82], suggesting that the intracellular regulation of MHC class II peptide loading by DM is a major event in selection of the immunodominant peptide repertoire.

2.4.2.2 **DO**

In addition to DM, B lymphocytes and thymic epithelial cells express the MHC-encoded protein DO, which is a modulator of DM activity [28, 65, 66]. However, the molecular mechanism by which DO alters the activity of DM is not completely understood. Within the ER of a DO-expressing APC, DO associates with DM and the resulting DM–DO complexes are delivered to intracellular antigen-processing compartments. If DM is produced in molar excess over DO, the APCs will contain both DM–DO complexes and free DM molecules [83]. The association of DO with DM is thought to either inhibit the ability of DM to function as a class II chaperone or decrease the optimal pH for DM action, biasing DM activity to deeper, more acidic endocytic compartments [66, 83]. Therefore, changes in the relative levels of DM and DO expressed by a B cell, like those observed during various stages of B-cell development [84, 85], will result in shifts in the spectrum of antigenic peptide–class II complexes expressed by the APC.

2.4.2.3 DO-, DM-, and BCR-Mediated Antigen Processing

Irrespective of the precise molecular mechanism of DO action, the expression of DO by B lymphocytes (the only class of APCs that expressed a clonally restricted, antigen-specific receptor, the BCR) is thought to bias antigen processing and presentation to immunologically relevant cognate antigen internalized via the BCR. This possibility is supported by the observation that the dose response of the fluid-phase processing of non-cognate antigen by B cells was not altered by ablation of DO expression, while the dose response of BCR-mediated processing of cognate antigen was shifted to higher doses [86]. However, while in most cases DO ablation decreases the efficiency of BCR-mediated antigen processing, there are exceptions to this rule, suggesting that the mechanism of DO action may be complex.

One possible mechanism by which DO could focus MHC class II peptide loading onto peptides derived from BCR-internalized cognate antigen is suggested by the observation that persisting intracellular Ag-BCR complexes within antigen-specific B cells exhibit a higher degree of colocalization with DM than with DO and that cells possessing these persisting Ag-BCR complexes can express resultant antigen peptide–class II complexes for a prolonged period of time [47]. Whether the preferential colocalization of persisting Ag-BCR complexes with free DM is due to differential colocalization of the complexes with free DM (versus DM–DO complexes) or due to the dissociation of DM–DO complexes within these late endocytic compartments (and subsequent removal of DO via export or degradation) is unclear. However, the lower degree of DO colocalization with persisting Ag-BCR complexes, when compared to DM, suggests that DO may act to focus DM-mediated class II peptide loading onto peptides derived from the processing of cognate antigen by "unleashing" free DM molecules within intracellular compartments containing persistent Ag-BCR complexes. However, while this relatively simple model for the molecular mechanism of BCR-mediated antigen processing is attractive, it ignores the potential effect that the asymmetrical distribution class II, DM, DO, and other proteins within the various membrane sub-domains of antigen-processing compartments (see below) may have on the precise mechanism of antigen processing and class II peptide loading.

2.4.2.4 The Distribution of MHC Class II and Other Proteins Within MIIC

Presently, one of the major unanswered questions in the area of MHC class II–restricted antigen processing is the precise molecular mechanism by which antigenic peptide–MHC class II complexes form *in situ*. Some interesting possibilities are suggested by the results of the fine-resolution analysis of the distribution of proteins such as class II, DM, and DO within the various membrane sub-compartments of antigen-processing vesicles [87–89].

MIIC, like multivesicular bodies (MVBs), are highly organized structures possessing two distinct membrane systems, the limiting membrane and the intralumenal vesicles (ILVs) [90]. Moreover, transport of protein molecules between these two membrane systems is a highly regulated process involving sorting platforms such as flat clathrin lattices and ESCRT proteins [90]. Furthermore, proteins involved in MHC class II processing and presentation are asymmetrically distributed between the limiting membrane and the ILVs of MVB [88, 89] and appear to interact with different efficiencies within these distinct membranes [87]. Specifically, within the limiting membrane of multivesicular MIIC, DM and DO proteins appear to interact at a molecular level, suggesting that DM within the limiting membrane is under the regulatory control of DO. Contrastingly, the DM within the ILV appears to have freed itself from DO and is now found closely associated with MHC class II molecules. This observation suggests that DM-facilitated class II peptide loading may occur most readily for MHC class II molecules within the membranes of ILVs. Because ILVs are also enriched in cholesterol, sphingolipids, and the tetraspan proteins CD63 and CD81 [27, 88]—all components of the various types of membrane microdomains that are known to be associated with the biology of peptide–class II complexes (see below)—this observation raises the interesting possibility that DM-facilitated MHC class II peptide loading within ILVs may occur within highly specialized membrane sub-domains.

Taken in total, these observations suggest that—similar to the ER-resident MHC class I peptide-loading complex that contains TAP (the transporter associated with antigen processing, which is the source of antigenic peptides for the class I loading complex), tapasin, MHC class I, and calnexin/calreticulin (see Chapter 1)—MHC class II peptide loading may take place in a peptide-loading complex consisting minimally of MHC class II and DM. In the future, it will be interesting to learn whether or not this putative class II peptide-loading complex exists and whether (analogous to the incorporation of TAP into the class I peptide-loading complex) the BCR or other antigen receptors (see Table 2.2) are able to be selectively incorporated into the complex. If so, this would establish a molecular mechanism for the enhanced processing and presentation of cognate antigen that would profoundly impact the hierarchy of antigenic peptide–class II complexes expressed by an antigen-specific APCs.

2.4.3 Cell-surface Delivery of Peptide–Class II Complexes

Once antigenic peptide–class II complexes have been formed within intracellular compartments, they are delivered to the surface of the APC for presentation to CD4 T cells. However, the mechanism by which the peptide–class II complexes reach the surface of the APC is not completely understood [33, 91]. In DCs, it appears that peptide–class II complexes reach the surface of the cell via either of two distinct mechanisms [88]. In resting DCs (which express low levels of cell-surface peptide–class II complexes), multivesicular MIIC appear to fuse directly with the plasma membrane, delivering the contents of the limiting membrane of the MIIC to the plasma membrane of the cell. This fusion of multivesicular MIIC with the plasma membrane of the APC results in the secretion of the ILVs of the MIIC, which are then termed *exosomes* (see below).

Contrastingly, in mature DCs, MIIC are remodeled by the fusion of ILVs with the limiting membrane of the compartment [88, 89]. Peptide–class II molecules within the limiting membrane of the remodeled MIIC are then delivered to the plasma membrane either by membrane tubules arising from the limiting membrane of the compartment or by the budding and translocation of transport vesicles such as class II vesicles (CIIVs). Moreover, the directionality of transport appears to be controlled such that peptide–class II complexes are delivered to the regions of the APC surface that are involved in cognate interactions with T lymphocytes [27]. The class II transport vesicles also contain additional molecules that play a key role in antigen presentation, such as the costimulatory molecules CD80 and CD86 as well as the tetraspan molecule CD81 (see below) [27, 88]. Therefore, DCs are capable of controlling the mechanism by which peptide-class II complexes are delivered to the surface of the cell, thus controlling the spatiotemporal aspects of peptide–class II expression that will profoundly impact the biology of subsequent DC–T cell interactions. Presently, the mechanism for the cell-surface delivery of peptide–class II complexes in other APCs, such as B cells and MØs, remains unclear.

2.4.3.1 Exosomes

As described above, direct fusion of multivesicular MIIC with the plasma membrane of the APC results in the release of ILVs into the extracellular environment. These secreted ILVs have been termed exosomes and have an interesting molecular composition that hints at potentially important immunoregulatory properties [88, 92].

Like ILVs, exosomes are enriched in MHC class II molecules and accessory molecules known to be involved in MHC class II–restricted antigen presentation. The tetraspan membrane proteins CD81 and CD82, as well as cholesterol and sphingomyelin (two key molecular components of lipid rafts). Because both tetraspan protein [93, 94] and lipid raft microdomains [95] have been shown to be both physically and functionally associated with MHC class II molecules on the surface of APCs, the presence of these class II interaction membrane domains within class II–bearing exosomes suggests that exosomes may be capable of functionally interacting with CD4 T cells. In addition, exosomes also contain ICAM-1 molecules, suggesting that if exosomes interact with CD4 T cells *in vivo*, these interactions may not be restricted to MHC class II–TCR interactions but may involve interactions with costimulatory and adhesion molecules on the T cell as well.

Consistent with a potential role for exosomes in regulating an immune response, exosomes released from B cells have been demonstrated to be capable of stimulating CD4 T cells *in vitro* [88, 92]. However, the evidence for an *in vivo* role for exosomes in immune modulation is much less direct [88, 92]. Therefore, while exosomes remain a potentially important mechanism of regulation for the CD4-based immune response, the *in vivo* role of exosomes in the overall immune response remains an open question.

2.4.3.2 Signaling Properties of Peptide–Class II Complexes

Once peptide–class II molecules are delivered to the surface of an APC for presentation to $CD4^+$ T cells, the class II molecules do not simply act as inert pedestals that display peptides to the T cells. Peptide–class II complexes are actually signaling-competent, cell-surface proteins that can signal APCs in many different ways [96].

Early investigations suggested that, at least in B cells, class II molecules can signal via either a cAMP-dependent pathway or a calcium/tyrosine kinase–dependent pathway, depending on the activation state of the cell. More recent molecular investigations have revealed that peptide–class II complexes can be incorporated into one or more types of plasma membrane microdomains known to be important signaling platforms [93–95]. One such type of class II–bearing membrane microdomain is formed by tetraspan proteins (for example CD81), which span the plasma membrane four times and form higher-order molecular complexes. A second type of class II–bearing membrane microdomain is the lipid raft, which is defined by a high content of cholesterol, glycosphingolipids, and signaling molecules. A subset of peptide–class II molecules are found within lipid rafts, and the lipid raft partitioning of these complexes appears to facilitate more effective T-cell stimulation at low peptide–class II levels [95]. Finally, there is a controversial report that in B cells, peptide–class II complexes can associate with the CD79 signaling heterodimer that is usually found as part of the BCR [97]. Moreover, it has been suggested that cross-linking of these CD79-associated peptide–class II complexes elicits signaling events similar to those stimulated by ligand binding to the BCR.

Therefore, while it is well established that cell-surface peptide–class II complexes are signaling-competent structures, the molecular mechanisms of signal transduction are less clear. While peptide–class II complexes are able to associate with plasma membrane lipid rafts, tetraspan protein domains, and other signaling structures, the relationship between each of these class II–bearing membrane domains and the mechanism by which class II becomes associated with the domains are unclear. Nonetheless, the ability of a particular antigenic peptide–class II complex to become associated with one or more of these membrane domains at the surface of an APC has the potential to impact the ability of these complexes to be effectively presented to antigen-specific CD4 T cells.

2.5 Conclusions

MHC class II–restricted antigen processing and the display of antigenic peptide–class II complexes on the surface of an APC is a complex biological phenomenon, and subtle variations at each step along the pathway can profoundly alter the hierarchy of peptide–class II complexes ultimately displayed by the cell (Figure 2.1). Some of the major variables that can impact this process are the activation state of the APC, the relative role of different class II chaperones, the pathway of receptor-mediated antigen endocytosis and trafficking, the pattern of antigen proteolysis, and the incorporation of peptide–class II complexes into various membrane microdomains. Therefore, any complete analysis of immunodominance in the CD4-based immune response must consider not only the effect of the repertoire and the state of CD4 T cells but also the type and physiological state of the APC participating in the response and the details of the molecular mechanism of antigen processing and presentation.

The CD4-based immune response can also be divided into two distinct phases that depend on class II–restricted antigen presentation: DC-mediated activation of naïve CD4 T cells and B-cell or MØ recruitment of effector CD4 T cells. Because these two distinct phases are dependent upon class II–restricted antigen presentation by different populations of APCs (which will express different hierarchies of peptide–class II complexes even when exposed to the same antigen), a complete understanding of immunodominance at the level of the overall immune response must account for the complex interplay between the various immunodominant and cryptic antigenic peptides of each of these phases. Critical elements contributing to the solution of this overall picture include a more precise understanding of the cell biology and molecular mechanisms of exogenous antigen processing and experimental approaches to address how subtle changes in the mechanism of antigen processing impact the hierarchy of peptide–class II complexes generated and expressed by an antigen-presenting cell.

Acknowledgments

J.R.D. is supported by NIH grants AI-46405 and AI-56320. A.J.S. is supported by NIH grants AI-51542 and AI-059898.

References

1 Blum, J.S., C. Ma, and S. Kovats, *Antigen-presenting cells and the selection of immunodominant epitopes.* Crit Rev Immunol, 1997. 17(5–6): p. 411–7.

2 Deng, H., L. Fosdick, and E. Sercarz, *The involvement of antigen processing in determinant selection by class II MHC and its relationship to immunodominance.* Apmis, 1993. 101(9): p. 655–62.

3 Fairchild, P.J., *Reversal of immunodominance among autoantigenic T-cell epitopes.* Autoimmunity, 1999. 30(4): p. 209–21.

4 Sercarz, E.E., P.V. Lehmann, A. Ametani, G. Benichou, A. Miller, and

K. Moudgil, *Dominance and crypticity of T cell antigenic determinants.* Annu Rev Immunol, 1993. 11: p. 729–66.

5 Yewdell, J.W. and J.R. Bennink, *Immunodominance in major histocompatibility complex class I-restricted T lymphocyte responses.* Annu Rev Immunol, 1999. 17: p. 51–88.

6 Gapin, L., Y. Bravo de Alba, A. Casrouge, J.P. Cabaniols, P. Kourilsky, and J. Kanellopoulos, *Antigen presentation by dendritic cells focuses T cell responses against immunodominant peptides: studies in the hen egg-white lysozyme (HEL) model.* J Immunol, 1998. 160(4): p. 1555–64.

7 Lo-Man, R., J.P. Langeveld, P. Martineau, M. Hofnung, R.H. Meloen, and C. Leclerc, *Immunodominance does not result from peptide competition for MHC class II presentation.* J Immunol, 1998. 160(4): p. 1759–66.

8 Ma, C., P.E. Whiteley, P.M. Cameron, D.C. Freed, A. Pressey, S.L. Chen, B. Garni-Wagner, C. Fang, D.M. Zaller, L.S. Wicker, and J.S. Blum, *Role of APC in the selection of immunodominant T cell epitopes.* J Immunol, 1999. 163(12): p. 6413–23.

9 Moudgil, K.D., D. Sekiguchi, S.Y. Kim, and E.E. Sercarz, *Immunodominance is independent of structural constraints: each region within hen eggwhite lysozyme is potentially available upon processing of native antigen.* J Immunol, 1997. 159(6): p. 2574–9.

10 Safley, S.A., P.E. Jensen, P.A. Reay, and H.K. Ziegler, *Mechanisms of T cell epitope immunodominance analyzed in murine listeriosis.* J Immunol, 1995. 155(9): p. 4355–66.

11 McHeyzer-Williams, M.G., J.D. Altman, and M.M. Davis, *Tracking antigen-specific helper T cell responses.* Curr Opin Immunol, 1996. 8(2): p. 278–84.

12 McHeyzer-Williams, M.G., J.D. Altman, and M.M. Davis, *Enumeration and characterization of memory cells in the TH compartment.* Immunol Rev, 1996. 150: p. 5–21.

13 Stetson, D.B., M. Mohrs, V. Mallet-Designe, L. Teyton, and R.M. Locksley, *Rapid expansion and IL-4 expression by Leishmania-specific naive helper T cells in vivo.* Immunity, 2002. 17(2): p. 191–200.

14 Wilson, D.B., D.H. Wilson, K. Schroder, C. Pinilla, S. Blondelle, R.A. Houghten, and K.C. Garcia, *Specificity and degeneracy of T cells.* Molecular Immunology, 2004. 40(14–15): p. 1047–55.

15 Savage, P.A. and M.M. Davis, *A kinetic window constricts the T cell receptor repertoire in the thymus.* Immunity, 2001. 14(3): p. 243–52.

16 Margulies, D.H., D. Plaksin, S.N. Khilko, and M.T. Jelonek, *Studying interactions involving the T-cell antigen receptor by surface plasmon resonance.* Curr Opin Immunol, 1996. 8(2): p. 262–70.

17 Kyewski, B. and J. Derbinski, *Self-representation in the thymus: an extended view.* Nat Rev Immunol, 2004. 4(9): p. 688–98.

18 Anderson, M.S., E.S. Venanzi, L. Klein, Z. Chen, S.P. Berzins, S.J. Turley, H. von Boehmer, R. Bronson, A. Dierich, C. Benoist, and D. Mathis, *Projection of an immunological self shadow within the thymus by the aire protein.* Science, 2002. 298(5597): p. 1395–401.

19 Gebe, J.A., B.A. Falk, K.A. Rock, S.A. Kochik, A.K. Heninger, H. Reijonen, W.W. Kwok, and G.T. Nepom, *Low-avidity recognition by CD4+ T cells directed to self-antigens.* European Journal of Immunology, 2003. 33(5): p. 1409–17.

20 Liu, G.Y., P.J. Fairchild, R.M. Smith, J.R. Prowle, D. Kioussis, and D.C. Wraith, *Low avidity recognition of self-antigen by T cells permits escape from central tolerance.* Immunity, 1995. 3(4): p. 407–15.

21 Targoni, O.S. and P.V. Lehmann, *Endogenous myelin basic protein inactivates the high avidity T cell repertoire.* J Exp Med, 1998. 187(12): p. 2055–63.

22 Harrington, C.J., A. Paez, T. Hunkapiller, V. Mannikko, T. Brabb, M. Ahearn, C. Beeson, and J. Goverman, *Differential tolerance is induced in T cells recognizing distinct epitopes of myelin basic protein.* Immunity, 1998. 8(5): p. 571–80.

23 Wucherpfennig, K.W., *Autoimmunity in the central nervous system: mechanisms of antigen presentation and recognition.* Clin Immunol Immunopathol, 1994. 72(3): p. 293–306.

24 Williams, C.B., D.L. Engle, G.J. Kersh, J. Michael White, and P.M. Allen, *A*

kinetic threshold between negative and positive selection based on the longevity of the T cell receptor-ligand complex. Journal of Experimental Medicine, 1999. 189(10): p. 1531–44.

25 Hogquist, K.A., S.C. Jameson, W.R. Heath, J.L. Howard, M.J. Bevan, and F.R. Carbone, *T cell receptor antagonist peptides induce positive selection.* Cell, 1994. 76(1): p. 17–27.

26 Steinman, R.M., D. Hawiger, and M.C. Nussenzweig, *Tolerogenic dendritic cells.* Annu Rev Immunol, 2003. 21: p. 685–711.

27 Boes, M., A. Cuvillier, and H. Ploegh, *Membrane specializations and endosome maturation in dendritic cells and B cells.* Trends Cell Biol, 2004. 14(4): p. 175–83.

28 Bryant, P. and H. Ploegh, *Class II MHC peptide loading by the professionals.* Curr Opin Immunol, 2004. 16(1): p. 96–102.

29 Mellman, I. and R.M. Steinman, *Dendritic cells: specialized and regulated antigen processing machines.* Cell, 2001. 106(3): p. 255–8.

30 Pierce, S.K., *Lipid rafts and B-cell activation.* Nat Rev Immunol, 2002. 2(2): p. 96–105.

31 Clark, M.R., D. Massenburg, K. Siemasko, P. Hou, and M. Zhang, *B-cell antigen receptor signaling requirements for targeting antigen to the MHC class II presentation pathway.* Curr Opin Immunol, 2004. 16(3): p. 382–7.

32 Cheng, P.C., A. Cherukuri, M. Dykstra, S. Malapati, T. Sproul, M.R. Chen, and S.K. Pierce, *Floating the raft hypothesis: the roles of lipid rafts in B cell antigen receptor function.* Semin Immunol, 2001. 13(2): p. 107–14.

33 Hiltbold, E.M. and P.A. Roche, *Trafficking of MHC class II molecules in the late secretory pathway.* Curr Opin Immunol, 2002. 14(1): p. 30–5.

34 Ceman, S. and A.J. Sant, *The function of invariant chain in class II-restricted antigen presentation.* Semin Immunol, 1995. 7(6): p. 373–87.

35 Guncar, G., G. Pungercic, I. Klemencic, V. Turk, and D. Turk, *Crystal structure of MHC class II-associated p41 Ii fragment bound to cathepsin L reveals the structural basis for differentiation between cathepsins L and S.* Embo J, 1999. 18(4): p. 793–803.

36 Fineschi, B., K. Sakaguchi, E. Appella, and J. Miller, *The proteolytic environment involved in MHC class II-restricted antigen presentation can be modulated by the p41 form of invariant chain.* J Immunol, 1996. 157(8): p. 3211–5.

37 Peterson, M. and J. Miller, *Antigen presentation enhanced by the alternatively spliced invariant chain gene product p41.* Nature, 1992. 357(6379): p. 596–8.

38 Ye, Q., P.W. Finn, R. Sweeney, E.K. Bikoff, and R.J. Riese, *MHC class II-associated invariant chain isoforms regulate pulmonary immune responses.* J Immunol, 2003. 170(3): p. 1473–80.

39 Strubin, M., C. Berte, and B. Mach, *Alternative splicing and alternative initiation of translation explain the four forms of the Ia antigen-associated invariant chain.* Embo J, 1986. 5(13): p. 3483–8.

40 Kampgen, E., N. Koch, F. Koch, P. Stoger, C. Heufler, G. Schuler, and N. Romani, *Class II major histocompatibility complex molecules of murine dendritic cells: synthesis, sialylation of invariant chain, and antigen processing capacity are down-regulated upon culture.* Proc Natl Acad Sci U S A, 1991. 88(8): p. 3014–8.

41 Zimmermann, V.S., P. Rovere, J. Trucy, K. Serre, P. Machy, F. Forquet, L. Leserman, and J. Davoust, *Engagement of B cell receptor regulates the invariant chain-dependent MHC class II presentation pathway.* J Immunol, 1999. 162(5): p. 2495–502.

42 Lankar, D., H. Vincent-Schneider, V. Briken, T. Yokozeki, G. Raposo, and C. Bonnerot, *Dynamics of major histocompatibility complex class II compartments during B cell receptor-mediated cell activation.* J Exp Med, 2002. 195(4): p. 461–72.

43 McGovern, E.M., A.E. Moquin, A. Caballero, and J.R. Drake, *The effect of B cell receptor signaling on antigen endocytosis and processing.* Immunol Invest, 2004. 33(2): p. 143–56.

44 Putnam, M.A., A.E. Moquin, M. Merrihew, C. Outcalt, E. Sorge, A. Caballero, T.A. Gondre-Lewis, and J.R. Drake, *Lipid raft-independent B cell receptor-mediated antigen internalization and intracellular trafficking.* J Immunol, 2003. 170(2): p. 905–12.

45 Stoddart, A., A.P. Jackson, and F.M. Brodsky, *Plasticity of B Cell Receptor*

Internalization upon Conditional Depletion of Clathrin. Mol Biol Cell, 2005.

46 Stoddart, A., M.L. Dykstra, B.K. Brown, W. Song, S.K. Pierce, and F.M. Brodsky, *Lipid rafts unite signaling cascades with clathrin to regulate BCR internalization.* Immunity, 2002. 17(4): p. 451–62.

47 Gondre-Lewis, T.A., A.E. Moquin, and J.R. Drake, *Prolonged antigen persistence within nonterminal late endocytic compartments of antigen-specific B lymphocytes.* J Immunol, 2001. 166(11): p. 6657–64.

48 Aluvihare, V.R., A.A. Khamlichi, G.T. Williams, L. Adorini, and M.S. Neuberger, *Acceleration of intracellular targeting of antigen by the B-cell antigen receptor: importance depends on the nature of the antigen-antibody interaction.* Embo J, 1997. 16(12): p. 3553–62.

49 Watts, C., A. Antoniou, B. Manoury, E.W. Hewitt, L.M. McKay, L. Grayson, N.F. Fairweather, P. Emsley, N. Isaacs, and P.D. Simitsek, *Modulation by epitope-specific antibodies of class II MHC-restricted presentation of the tetanus toxin antigen.* Immunol Rev, 1998. 164: p. 11–6.

50 McGreal, E.P., L. Martinez-Pomares, and S. Gordon, *Divergent roles for C-type lectins expressed by cells of the innate immune system.* Mol Immunol, 2004. 41(11): p. 1109–21.

51 Koppel, E.A., K.P. van Gisbergen, T.B. Geijtenbeek, and Y. van Kooyk, *Distinct functions of DC-SIGN and its homologues L-SIGN (DC-SIGNR) and mSIGNR1 in pathogen recognition and immune regulation.* Cell Microbiol, 2005. 7(2): p. 157–65.

52 Steinman, R.M., D. Hawiger, K. Liu, L. Bonifaz, D. Bonnyay, K. Mahnke, T. Iyoda, J. Ravetch, M. Dhodapkar, K. Inaba, and M. Nussenzweig, *Dendritic cell function in vivo during the steady state: a role in peripheral tolerance.* Ann N Y Acad Sci, 2003. 987: p. 15–25.

53 Hawiger, D., K. Inaba, Y. Dorsett, M. Guo, K. Mahnke, M. Rivera, J.V. Ravetch, R.M. Steinman, and M.C. Nussenzweig, *Dendritic cells induce peripheral T cell unresponsiveness under steady state conditions in vivo.* J Exp Med, 2001. 194(6): p. 769–79.

54 Vidard, L., K.L. Rock, and B. Benacerraf, *The generation of immunogenic peptides can be selectively increased or decreased by proteolytic enzyme inhibitors.* J Immunol, 1991. 147(6): p. 1786–91.

55 Nelson, C.A., N.J. Viner, and E.R. Unanue, *Appreciating the complexity of MHC class II peptide binding: lysozyme peptide and I-Ak.* Immunological Reviews, 1996. 151: p. 81–105.

56 Schneider, S.C., J. Ohmen, L. Fosdick, B. Gladstone, J. Guo, A. Ametani, E.E. Sercarz, and H. Deng, *Cutting edge: introduction of an endopeptidase cleavage motif into a determinant flanking region of hen egg lysozyme results in enhanced T cell determinant display.* Journal of Immunology, 2000. 165(1): p. 20–3.

57 Watts, C., C.X. Moss, D. Mazzeo, M.A. West, S.P. Matthews, D.N. Li, and B. Manoury, *Creation versus destruction of T cell epitopes in the class II MHC pathway.* Ann N Y Acad Sci, 2003. 987: p. 9–14.

58 Wlodawer, A., R. Bott, and L. Sjolin, *The refined crystal structure of ribonuclease A at 2.0 A resolution.* J Biol Chem, 1982. 257(3): p. 1325–32.

59 Thai, R., G. Moine, M. Desmadril, D. Servent, J.L. Tarride, A. Menez, and M. Leonetti, *Antigen stability controls antigen presentation.* Journal of Biological Chemistry, 2004. 279(48): p. 50257–66.

60 Dai, G., S. Carmicle, N.K. Steede, and S.J. Landry, *Structural basis for helper T-cell and antibody epitope immunodominance in bacteriophage T4 Hsp10. Role of disordered loops.* Journal of Biological Chemistry, 2002. 277(1): p. 161–8.

61 Dai, G., N.K. Steede, and S.J. Landry, *Allocation of helper T-cell epitope immunodominance according to three-dimensional structure in the human immunodeficiency virus type I envelope glycoprotein gp120.* Journal of Biological Chemistry, 2001. 276(45): p. 41913–20.

62 Brown, S.A., J. Stambas, X. Zhan, K.S. Slobod, C. Coleclough, A. Zirkel, S. Surman, S.W. White, P.C. Doherty, and J.L. Hurwitz, *Clustering of Th cell epitopes on exposed regions of HIV envelope despite defects in antibody activity.* Journal of Immunology, 2003. 171(8): p. 4140–8.

63 Chicz, R.M., R.G. Urban, J.C. Gorga, D.A. Vignali, W.S. Lane, and

J.L. Strominger, *Specificity and promiscuity among naturally processed peptides bound to HLA-DR alleles.* J Exp Med, 1993. 178(1): p. 27–47.

64 Hunt, D.F., H. Michel, T.A. Dickinson, J. Shabanowitz, A.L. Cox, K. Sakaguchi, E. Appella, H.M. Grey, and A. Sette, *Peptides presented to the immune system by the murine class II major histocompatibility complex molecule I-Ad.* Science, 1992. 256(5065): p. 1817–20.

65 Busch, R., R.C. Doebele, N.S. Patil, A. Pashine, and E.D. Mellins, *Accessory molecules for MHC class II peptide loading.* Curr Opin Immunol, 2000. 12(1): p. 99–106.

66 Brocke, P., N. Garbi, F. Momburg, and G.J. Hammerling, *HLA-DM, HLA-DO and tapasin: functional similarities and differences.* Curr Opin Immunol, 2002. 14(1): p. 22–9.

67 Denzin, L.K. and P. Cresswell, *HLA-DM induces CLIP dissociation from MHC class II alpha beta dimers and facilitates peptide loading.* Cell, 1995. 82(1): p. 155–65.

68 Kropshofer, H., A.B. Vogt, G. Moldenhauer, J. Hammer, J.S. Blum, and G.J. Hammerling, *Editing of the HLA-DR-peptide repertoire by HLA-DM.* Embo J, 1996. 15(22): p. 6144–54.

69 Morris, P., J. Shaman, M. Attaya, M. Amaya, S. Goodman, C. Bergman, J.J. Monaco, and E. Mellins, *An essential role for HLA-DM in antigen presentation by class II major histocompatibility molecules.* Nature, 1994. 368(6471): p. 551–4.

70 Sloan, V.S., P. Cameron, G. Porter, M. Gammon, M. Amaya, E. Mellins, and D.M. Zaller, *Mediation by HLA-DM of dissociation of peptides from HLA-DR.* Nature, 1995. 375(6534): p. 802–6.

71 Stebbins, C.C., G.E. Loss, Jr., C.G. Elias, A. Chervonsky, and A.J. Sant, *The requirement for DM in class II-restricted antigen presentation and SDS-stable dimer formation is allele and species dependent.* J Exp Med, 1995. 181(1): p. 223–34.

72 Weber, D.A., B.D. Evavold, and P.E. Jensen, *Enhanced dissociation of HLA-DR-bound peptides in the presence of HLA-DM.* Science, 1996. 274(5287): p. 618–20.

73 Alfonso, C. and L. Karlsson, *Nonclassical MHC class II molecules.* Annual Review of Immunology, 2000. 18: p. 113–42.

74 Kropshofer, H., G.J. Hammerling, and A.B. Vogt, *How HLA-DM edits the MHC class II peptide repertoire: survival of the fittest?* Immunology Today, 1997. 18(2): p. 77–82.

75 Jensen, P.E., D.A. Weber, W.P. Thayer, X. Chen, and C.T. Dao, *HLA-DM and the MHC class II antigen presentation pathway.* Immunol Res, 1999. 20(3): p. 195–205.

76 Jensen, P.E., D.A. Weber, W.P. Thayer, L.E. Westerman, and C.T. Dao, *Peptide exchange in MHC molecules.* Immunol Rev, 1999. 172: p. 229–38.

77 van Ham, S.M., U. Gruneberg, G. Malcherek, I. Broker, A. Melms, and J. Trowsdale, *Human histocompatibility leukocyte antigen (HLA)-DM edits peptides presented by HLA-DR according to their ligand binding motifs.* J Exp Med, 1996. 184(5): p. 2019–24.

78 Chou, C.L. and S. Sadegh-Nasseri, *HLA-DM recognizes the flexible conformation of major histocompatibility complex class II.* J Exp Med, 2000. 192(12): p. 1697–706.

79 Belmares, M.P., R. Busch, E.D. Mellins, and H.M. McConnell, *Formation of two peptide/MHC II isomers is catalyzed differentially by HLA-DM.* Biochemistry, 2003. 42(3): p. 838–47.

80 Nanda, N.K. and A.J. Sant, *DM determines the cryptic and immunodominant fate of T cell epitopes.* J Exp Med, 2000. 192(6): p. 781–8.

81 Lich, J.D., J.A. Jayne, D. Zhou, J.F. Elliott, and J.S. Blum, *Editing of an immunodominant epitope of glutamate decarboxylase by HLA-DM.* J Immunol, 2003. 171(2): p. 853–9.

82 Lightstone, L., R. Hargreaves, G. Bobek, M. Peterson, G. Aichinger, G. Lombardi, and R. Lechler, *In the absence of the invariant chain, HLA-DR molecules display a distinct array of peptides which is influenced by the presence or absence of HLA-DM.* Proc Natl Acad Sci U S A, 1997. 94(11): p. 5772–7.

83 Denzin, L.K., D.B. Sant'Angelo, C. Hammond, M.J. Surman, and P. Cresswell, *Negative regulation by HLA-DO of MHC class II-restricted antigen processing.* Science, 1997. 278(5335): p. 106–9.

84 Chen, X., O. Laur, T. Kambayashi, S. Li, R.A. Bray, D.A. Weber, L. Karlsson, and

P.E. Jensen, *Regulated expression of human histocompatibility leukocyte antigen (HLA)-DO during antigen-dependent and antigen-independent phases of B cell development.* J Exp Med, 2002. 195(8): p. 1053–62.

85 Glazier, K.S., S.B. Hake, H.M. Tobin, A. Chadburn, E.J. Schattner, and L.K. Denzin, *Germinal center B cells regulate their capability to present antigen by modulation of HLA-DO.* J Exp Med, 2002. 195(8): p. 1063–9.

86 Alfonso, C., G.S. Williams, J.O. Han, J.A. Westberg, O. Winqvist, and L. Karlsson, *Analysis of H2-O influence on antigen presentation by B cells.* J Immunol, 2003. 171(5): p. 2331–7.

87 Zwart, W., A. Griekspoor, C. Kuijl, M. Marsman, J. van Rheenen, H. Janssen, J. Calafat, M. van Ham, L. Janssen, M. van Lith, K. Jalink, and J. Neefjes, *Spatial separation of HLA-DM/HLA-DR interactions within MIIC and phagosome-induced immune escape.* Immunity, 2005. 22(2): p. 221–33.

88 Murk, J.L., W. Stoorvogel, M.J. Kleijmeer, and H.J. Geuze, *The plasticity of multivesicular bodies and the regulation of antigen presentation.* Semin Cell Dev Biol, 2002. 13(4): p. 303–11.

89 Kleijmeer, M., G. Ramm, D. Schuurhuis, J. Griffith, M. Rescigno, P. Ricciardi-Castagnoli, A.Y. Rudensky, F. Ossendorp, C.J. Melief, W. Stoorvogel, and H.J. Geuze, *Reorganization of multivesicular bodies regulates MHC class II antigen presentation by dendritic cells.* J Cell Biol, 2001. 155(1): p. 53–63.

90 Raiborg, C., T.E. Rusten, and H. Stenmark, *Protein sorting into multivesicular endosomes.* Curr Opin Cell Biol, 2003. 15(4): p. 446–55.

91 Wubbolts, R. and J. Neefjes, *Intracellular transport and peptide loading of MHC class II molecules: regulation by chaperones and motors.* Immunol Rev, 1999. 172: p. 189–208.

92 Stoorvogel, W., M.J. Kleijmeer, H.J. Geuze, and G. Raposo, *The biogenesis and functions of exosomes.* Traffic, 2002. 3(5): p. 321–30.

93 Kropshofer, H., S. Spindeldreher, T.A. Rohn, N. Platania, C. Grygar, N. Daniel, A. Wolpl, H. Langen, V. Horejsi, and A.B. Vogt, *Tetraspan microdomains distinct from lipid rafts enrich select peptide-MHC class II complexes.* Nat Immunol, 2002. 3(1): p. 61–8.

94 Hammond, C., L.K. Denzin, M. Pan, J.M. Griffith, H.J. Geuze, and P. Cresswell, *The tetraspan protein CD82 is a resident of MHC class II compartments where it associates with HLA-DR, -DM, and -DO molecules.* J Immunol, 1998. 161(7): p. 3282–91.

95 Poloso, N.J. and P.A. Roche, *Association of MHC class II-peptide complexes with plasma membrane lipid microdomains.* Curr Opin Immunol, 2004. 16(1): P. 103–7.

96 Al-Daccak, R., N. Mooney, and D. Charron, *MHC class II signaling in antigen-presenting cells.* Curr Opin Immunol, 2004. 16(1): p. 108–13.

97 Lang, P., J.C. Stolpa, B.A. Freiberg, F. Crawford, J. Kappler, A. Kupfer, and J.C. Cambier, *TCR-induced transmembrane signaling by peptide/MHC class II via associated Ig-alpha/beta dimers.* Science, 2001. 291(5508): p. 1537–40.

3
The Phenomenon of Immunodomination: Speculations on the Nature of Immunodominance

Alessandro Sette and Roshni Sundaram

3.1
Introduction

Our group has had a long-standing interest in studying immunodominance, in terms of understanding the molecular basis of this phenomenon, of exploring practical strategies to modulate immunodominance, and, more recently, of understanding the molecular basis of immunodomination, defined as the process by which expression and/or recognition of one epitope influences the recognition of a second, distinct epitope. Several different independent investigations suggest that MHC binding, cellular processing, and the repertoire of TCR specificities all play major roles in determining immunodominance. The influence of these variables alone, however, appears to be insufficient to fully account for either the paucity of distinct epitopes recognized in naturally occurring adaptive immune responses in general or the phenomenon of immunodomination in particular. Herein we review a number of different studies that suggest, in our opinion, that immunodomination (1) may involve the active participation of CD8 T cells; (2) may be related to lack of expansion, not lack of priming of the subdominant T cells; (3) is not operative if different APCs present antigen; and (4) does not require killing of APCs. Furthermore, there have been reports that suggest that immunodomination is the result of the dominating T cell affecting the capacity of APCs to present antigen effectively to the subdominate or submissive T cell. We hypothesize that APC engagement by a dominating T cell results in sequestration in the immune synapse of certain critical molecules involved in T-cell signaling, such as antigen–MHC complexes, adhesion, and costimulatory molecules. We further hypothesize that upon APC engagement by the submissive T cell, a defective synapse may be formed that is deficient in one or more of these molecules, and thus incomplete signals are delivered to the submissive T cell. It is also possible that immunodomination is caused by the induction in the dominating cell of regulatory and inhibitory signals. Our laboratory is currently testing these hypotheses by utilizing dominant and subdominant epitope pairs.

Immunodominance: The Choice of the Immune System. Edited by Jeffrey A. Frelinger

ISBN: 3-527-31274-9

3.2
MHC Binding, Cellular Processing, and T-Cell Repertoire are Major Determinants of Immunodominance

Naturally occurring immune responses do not recognize all possible epitopes but instead are commonly focused on relatively few of all potential epitopes. This phenomenon is defined as immunodominance. Immunodominance can be very profound, resulting in only one or a few specificities dominating the response to a given complex pathogen, with a large fraction of activated T cells being specific for a single epitope–MHC combination [1–4]. It is well accepted that three main factors have a profound influence in determining immunodominance: (1) the efficiency with which an epitope is generated by cellular processing, (2) its capacity to bind the MHC molecules present in the host, and (3) the existence of a repertoire of specific TCR available to recognize the epitope–MHC complex once formed.

Several studies have documented a correlation between the abundance of various potential epitopes that are generated by cellular processing, and, in the case of class I–restricted epitopes, the efficiency of their transport across the ER membrane and the phenomenon of immunodominance [5]. In recent studies, Yewdell and Bennink's group performed studies to quantitate protein synthesis, degradation, and endogenous antigen presentation as it relates to the immune response following immunization with influenza virus [6]. In recent years, the enzymatic and processing events linked to class I–restricted antigen presentation has been greatly clarified by several excellent studies [7–9]. Whether these findings on immunodominance can be generalized to the MHC class II system is still the object of some debate. In fact, Unanue and coworkers have clearly shown that in the lysozyme system, the hierarchy of class II–restricted immunodominant epitopes is not directly related to chemical abundance [10].

Several studies have documented the crucial role of MHC binding in restricting the number of possible potential epitopes to which immune responses are directed [5]. Peptides with the highest binding affinity for MHC also tend to be the most immunogenic, possibly because they can be efficiently captured from MHC molecules even if produced in relatively low copy numbers as a result of cellular processing.

Finally, several independent studies point out the importance of T-cell repertoire composition and previous antigenic encounters in shaping immunodominance [11, 12]. Significant amounts of data suggest an interplay between T cells that can result in the preferential recognition of one (dominant) epitope over another (subdominant) epitope. One mechanism by which this can occur is the presence of a greater number of T cells that recognize the dominant epitope relative to T cells capable of recognizing the subdominant epitope. This can occur as a result of selection events that take place in the thymus or as a result of prior antigenic exposure to the dominant antigen or a cross-relative antigen that results in expansion of the number and possible avidity of T cells capable of recognizing the dominant epitope ("original antigenic sin"). For example, Brehm and coworkers pointed out that immunodominance can be the result of exposure to cross-reactive antigens present in different viruses to which the host had been previously exposed [13]. Likewise, it has been shown that increasing the magnitude of T-cell responses directed to a subdominant epitope through, for example, pre-priming might lead to the diminution of the response directed against an otherwise dominant epitope [14]. It is widely appreciated that in relative terms, deletion, tolerization, or silencing of T-cell responses against a dominant T-cell epitope can lead to the appearance of a previously undetectable response directed against subdominant epitopes. Furthermore, Slifka showed that differential selection in the thymus could alter immunodominance [15]. A recent excellent study by Nielsen also brings into focus the structural constraints operating at the level of TCR molecules [16].

In conclusion, the immune system "evaluates" the immunogenic potential of the various potential epitopes, and the ultimate result is a relative ranking of epitope potency. Taken together, the studies referred to above demonstrate the importance of antigen processing, affinity for MHC molecules, and the nature and size of the TCR repertoire in establishing the potential of MHC–peptide complexes for immunogenicity in absolute terms. The influence of these variables alone, however, appears to be insufficient to fully account for the paucity of distinct epitopes recognized in naturally occurring adaptive immune responses. Following the nomenclature of Yewdell and coworkers, we define immunodomination as the process by which expression and/or recognition of an epitope influences the recognition of a second, distinct epitope. After a brief review of some studies from our group and others related to the study of immunodominance, we present a series of reflections, hypotheses, and speculations relating to the possible molecular mechanisms involved in immunodomination.

3.3
Previous Systematic Analysis of Immunodominance by Our Group

Our group has in the past performed a systematic analysis of immunodominance in two different experimental murine models of viral infection, namely, lymphocytic choriomeningitis virus (LCMV) and influenza. In this section we briefly summarize the results of these analyses.

It was previously known that the cytotoxic T-cell response against LCMV in BALB/c (H-2d) mice is predominantly directed against a single immunodominant Ld-restricted epitope from the viral nucleoprotein (NP.118–126). In our study [2] we reported that the immunodominance of this peptide could be at least in part attributed to its very high affinity for Ld class I molecules. By employing motif searches and binding assays with purified MHC molecules, we also identified five Kd-binding peptides in the viral nucleoprotein and glycoprotein among a list of 16 peptides that contained Kd-binding motifs. Of the 18 Db motif–positive peptides from nucleoprotein and glycoprotein sequences, two were found to also bind Dd, but with low affinity. Two of the Kd-binding peptides, spanning residues 99–108 and 283–291 of the LCMV glycoprotein, were identified as subdominant epitopes, as demonstrated by the fact that although these peptides did not sensitize target cells for direct *ex vivo* killing by primary antiviral cytotoxic T lymphocytes (CTLs), secondary *in vitro* responses against these peptides were readily detected in BALB/c mice after acute LCMV infection. BALB/c mice that had cleared a chronic LCMV infection showed sustained CTL responses against these epitopes, suggesting that subdominant responses might play a role in clearance of chronic infection. One of the subdominant epitopes, GP.283–291, conferred partial protection against persistent viral infection after peptide vaccination, following either peptide or lipopeptide immunization. In a parallel study [17], we undertook a systematic analysis of the influenza (Flu) PR8 determinants recognized by H-2b mice. Of a total of 23 Flu PR8–derived peptides that bound either Kb or Db molecules *in vitro*, 16 were immunogenic following peptide immunization of C57BL/6 mice. However, CTLs induced by peptide immunization recognized PR8-infected target cells only in the case of the NP.366 and NS2.114 epitopes, indicating the importance of antigen processing in the immune response. Confirming this was the observation that CTL responses following whole PR8 virus immunization were detected only for these determinants. CTLs recognizing dominant epitopes had high avidity for peptide-pulsed target cells, with 5–200 pM of peptide being required for 30% specific lysis. In contrast, most (80%) of the remaining epitopes were recognized with lower avidity, in the range of 0.4–50 nM. Repeated *in vitro* stimulation of primary CTL cultures revealed one additional Kb-restricted epitope, M1.128, that bound Kb with high affinity and induced CTLs that effectively recognized infected cells. Thus, while this epitope is produced by natural processing in relatively high amounts, low precursor frequency appears to be related to its subdominant status. Taken together, these studies demonstrated the crucial contributions of MHC-binding capacity, antigen processing, and T-cell repertoire availability to the shaping of the repertoire of CTL specificities for Flu and LCM viruses.

As described above, our group has performed a systematic analysis of the impact of MHC binding on immunodominance in the Flu and LCMV experimental systems. These studies should be put in the context of several other excellent studies analyzing immunodominance in these two experimental model systems. Herein, we briefly summarize the key features of these antigenic systems such as natural processing efficiency, size and avidity of specific repertoires, and other relevant information associated with the specific epitopes derived from these two viruses. The LCMV system is of considerable interest and is a paradigmatic example of immunodominance in CTL responses. Several epitopes have been defined that are restricted by the H-2^b and H-2^d alleles. TCR-transgenic mice directed against the dominant Gp33 epitope exist, and various mutant viral strains have been described and characterized in detail [18, 19]. In fact, the LCMV system is arguably the best-characterized system of experimental infection in mice utilizing a natural murine pathogen. In general, vigor of responses correlates with the abundance of each epitope that is produced in the course of natural processing [20]. Protective efficacy against viral challenge seems to correlate best with the avidity of the responding T-cell populations [20]. Most importantly, the LCMV system provides good evidence and multiple examples of immunodomination. In collaboration with Ahmed's group, we have shown that silencing a dominant epitope results in enhanced responses against subdominant epitopes. By using dm2 mutants of Balb/c mice, which do not express the Ld molecule, we have documented that in the absence of the NP118 response, the subdominant GP283 epitope becomes immunodominant [16]. According to an exemplary study of Rodriguez and coworkers [21], in NP pre-primed mice other codominant, GP-derived epitopes are rendered subdominant. However, in IFN-γ KO mice this effect is abrogated, leading to equivalent responses against GP and NP in the NP pre-primed mice, suggesting an active suppression of the NP response that is IFN-γ mediated. Slifka's data [14] also show that deletion of T cells specific for the immunodominant NP118 epitope by negative selection leads to increased response to the otherwise subdominant epitopes, GP283 and NP313. LCMV responses in the H-2^b background have also been analyzed by taking advantage of Kb and Db knockout mice [22]. Van der Most [23] reports that the absence of restricted Db responses (including the response to the dominant NP396 and GP33 epitopes) renders GP34, NP205, and GP118 codominant. In Kb knockout mice the response of the dominant Db-restricted NP396 epitope was further increased. Taken together, these results suggest that in the majority of cases immunodominance is readily demonstrable in the LCMV system. However, two exceptions to this general rule have been reported. In a study by Bevan's group [24], the effect of transferring TCR-transgenic cells specific for the dominant GP33 epitope on the response to the NP396 and GP276 subdominant epitopes was assessed. In this particular case no inhibition of subdominant responses was observed. In another report [21], deletion of a dominant LCMV epitope only slightly increased responses to the subdominant Ld-restricted NP313 epitope, generated with poor efficiency by natural processing. The marginal increase observed in this case may be related to the intrinsic poor immunogenicity of this epitope.

As mentioned above, the murine model of influenza (Flu) infection has also been intensely studied and well characterized. The dominant H2K^d-restricted NP366 epitope was in fact the first CTL epitope to be rigorously defined, almost 20 years ago by Townsend and coworkers [25]. Approximately 15 epitopes have now been identified in the H-2^b and H-2^d haplotypes. Vitiello and coworkers identified two subdominant epitopes restricted by the Kb molecule. Additional epitopes were identified by several other groups [26, 27].

Seminal studies by Yewdell and Bennick's group defined the molecular characteristics of several Flu epitopes recognized in the context of H-2^d molecules. The HA462 epitope is the epitope most abundantly generated in infected cells, and also binds well to H-2K^d, but is subdominant because of a limited TCR repertoire. NP218 is also a subdominant epitope, in this case presumably because of its poor binding affinity to H-2K^d. The NP39 and HA518 epitopes are produced with low efficiency by natural processing, but if their yield is increased by the use of vaccinia-delivered minigenes, sizeable increases in responses are observed. In this system, clear evidence of immunodomination was also obtained [1].

Of particular relevance for the topic of this review, pre-priming with specific epitopes and adoptive transfer of T-cell lines specific for either dominant or subdominant epitopes inhibited specific responses for each of the other determinants tested. The word "immunodomination" was coined to describe this phenomenon. Interestingly, selectivity in sensitivity to immunodomination was noted. In general, T cells specific for the most dominant epitopes were most effective in immunodomination and least susceptible to immunodomination from T cells specific for other less dominant epitopes. These results are of importance because they suggest a possible explanation for the variable reports on sensitivity to immunodomination noted in the LCMV system, as described above. Furthermore, they suggest that the Flu system may also be well suited to the definition of the molecular basis for immunodomination. Taken together, these studies raise the question of what might be the molecular basis for immunodomination involving T cells specific for different determinants. It is possible that the intrinsic requirement for costimulatory activity, CD8, adhesion molecules, synapse geometry, and composition might explain the differential susceptibility to immunodomination of T cells specific for different epitopes, as discussed in more detail in the following sections.

Recently, Woodland and coworkers [28, 29] showed that the two immunodominant MHC class I–restricted influenza epitopes NP366–374 and PA224–233 are differentially expressed on infected cells, with the NP epitope strongly expressed on all infected cells and PA strongly expressed on only non-dendritic cells. This difference in epitope expression correlated with preferential expansion of NP366–374/Db-specific $CD8^+$ memory T cells in secondary infection. Most interestingly, vaccination with the PA224–233 peptide actually had a detrimental effect on the clearance of a subsequent influenza virus infection. Yewdell and coworkers [30] investigated these issues further and suggested that the inability of non-DCs to generate the PA(224–232) epitope is relative and that the preferential use of cross-priming in secondary responses could also account for the observed revised hierarchy. In their study, active immunodomination of PA(224–233)-specific $CD8^+$ T cells by nucleoprotein 366–374-specific $CD8^+$ T cells was shown to play an unimportant role in the phenomenon, and immunodomination was unlikely to be mediated by lysis of APCs or other cells.

3.4 Cellular and Molecular Events in Immunodomination

Klaus Karre and his group have provided pivotal evidence relating to the mechanisms of immunodomination (or interference, in that author's nomenclature) operating at the level of APCs. In one study [31], it was shown that the CD8 responses against a mixture of five synthetic peptides corresponding to well-defined immunogenic epitopes in B6 mice were directed against two dominant epitopes. Obviously, lack of immunogenicity of the epitopes by themselves or lack of processing could not explain the results, and it was also shown that insufficient presentation could not account for the results either. Although the exact mechanisms involved were not elucidated, an important observation was made that this interference was dependent on simultaneous presentation of dominant and subdominant epitopes by the same antigen-presenting cell. When separate APCs presented the dominant and subdominant epitopes, no domination occurred. Additional studies by the same group [32] showed that this phenomenon did not depend on CD4 cells, that there was no long-lasting unresponsiveness to the subdominant peptides, and that this interference or immunodominance could be broken *in vivo* and *in vitro* by the addition of excess dendritic cells as APCs.

Immunodominance has also been recently demonstrated in the context of the immune response to hepatitis B surface antigen (HBsAg). Studies by Reimann and coworkers have shown that the Ld-restricted response against the S28–30 epitope suppresses the CTL responses against all other epitopes irrespective of the processing pathway that generates the epitope or the MHC restriction of the epitopes. Because the binding affinities for all of these epitopes were determined to be similar, immunodominance of the S28–30 epitope was attributed to the difference in the affinity of the respective TCR for MHC-peptide complex, although this was not directly addressed in their studies [33].

Further seminal studies implicating the role of APCs in immunodomination were published by Kedl and collaborators [34–36]. By using the SIINFEKL dominant and KVVRFDKL subdominant epitopes of ovalbumin, it was shown that transfer of TCR transgenic–derived T cells specific for SIINFEKL directly interfered with the endogenous responses against either of the two peptides, although the interference with the endogenous SIINFEKL response was more dramatic. The phenomenon was not due to accelerated APC clearance or killing. Rather, as in the Karre experiment, it appeared to in some way be related to APC competition for APCs, because it could be overcome by the injection of a large number of additional antigen-presenting cells into the animals at the time of immunization. These studies are of fundamental importance in that they demonstrate a critical involvement of the antigen-presenting cell in determining immunodomination. The molecular mechanism responsible for the effects detected remains to be elucidated. The authors speculated, as Karre did, that somehow competition for the APC surface, or APC-derived factors or costimulation, might explain the observed phenomena [37].

3.5 Speculations on the Mechanism of Immunodomination

3.5.1 Involvement of APCs

In our view, competition for APC surface is unlikely. When the same number of APCs pulsed separately with a dominant and subdominant epitope are used, epitope dominance is not observed. Thus, APC "space" is clearly sufficient to support the priming and expansion of both T-cell populations. Furthermore, seminal papers by Cerundolo and coworkers [38] and Whitton and coworkers [21] demonstrated that competition from CTL narrows the responses in prime-boost protocols. Interestingly, the competition appears to takes place at the level of expansion of the various CTL specificities and not at the level of priming, as different CTL specificities can be expanded by the use of constructs separately encoding the various epitopes. Recent studies from Auphan-Anezin [39] also suggested that priming of CD8 T cells in the absence of cell division might be induced by suboptimal stimulation. These data suggest that competition for APC space is an unlikely explanation for immunodomination. Our hypothesis is that an APC that presents antigen simultaneously to both a dominating and a submissive T cell imports distinct activation signals that result in differential proliferation or differentiation programs and, ultimately, in immunodomination.

Finally, Chen and Yewdell reported that mice lacking CD4 or perforin maintain immunodominance [26]. These results seem to point at intrinsic features of the CD8 T-cell level as crucial in the establishment of immunodominance and seem to eliminate APC killing or CD4-mediated help as major determinants of immunodominance. Consistent with these studies are also the results from Whitton and coauthors [21], who show that in their system establishment of immunodominance is IFN-γ mediated, and also indicate that co-expression of both dominant and subdominant determinants by the same APC is important in establishing immunodominance. In conclusion, the available data presented in this and previous sections suggest that APCs are involved in immunodomination. They also suggest that the phenomenon does not involve CD4 T cells, nor does it involve killing of APCs, but may involve IFN-γ and the active participation of CD8 T cells. Finally, immunodomination appears to be related to lack of expansion, not lack of priming of the submissive T cells.

What could be a possible mechanism for immunodomination at the cellular (APC) level? It is possible to hypothesize a “reverse-licensing” mechanism, in which a productive T-cell encounter would alter or diminish APC function for a subsequent encounter. In this scenario, the dominant T cells may get activated before subdominant T cells, perhaps because of their greater sensitivity to antigen or preferential expression of the epitope. While physically killing the APC seems to be ruled out based on the results described above, an encounter with T cells may render the APC temporarily “stunned.” Based on the observations quoted above, it is postulated that the stunned APC would still be capable of priming subdominant T cells but would not be capable of giving an expansion signal. However, it remains to be determined whether the observed priming reflects no expansion or expansion but no effector function. The molecular mechanism associated with the phenomenon could be passive, such as downregulation of MHC or costimulatory molecules, or active, such as the expression of molecules associated with negative regulation of responses (e.g., CTLA4 or PD1 family members). However, little data are currently available to support this possibility. Most data suggest T cell–induced activation of dendritic cells and other APCs, with a subsequent positive feedback loop that leads to further T-cell activation, a phenomenon usually referred to as APC “licensing” of T cells [40]. In this context, it should be noted that any licensing in which dominant T cells engage APC in a positive feedback loop would impede immunodomination, because the APC would simultaneously present both dominant and subdominant epitopes. By contrast, a decreased ability of APC function is required to explain immunodomination. Furthermore, reverse licensing (or worse, killing of the APC) would be wasteful in that it would only allow an APC to activate one or a few T cells, and it is also arguable that this type of mechanism would not allow for continued engagement of the APC with specific T cells, a factor important in determining fitness of T cells in terms of survival and expansion. Based on these considerations, we favor a model in which the APCs simultaneously maintain maximum activating capacity for the dominating T cell, while being capable of only suboptimal presentation of subdominant epitopes.

3.5.2
Possible Involvement of the Immune Synapse in Immunodomination

One possible mechanism of APC-mediated immunodomination is that APC engagement by a dominating T cell results in sequestration and concentration in the resulting immunologic synapse of molecules. This is critical for immune responses such as antigen–MHC complexes, adhesion molecules, and costimulatory molecules and/or intracellular signaling molecules. If such an APC also engages a T cell specific for a subdominant epitope, a defective synapse will form between the submissive T cell and the APC, which results in less effective or incomplete activation signals being delivered to the submissive T cell. This postulation assumes that normally the number of APCs capable of presenting antigen following immunization is limiting and a single APC presents multiple epitopes to T cells of differing specificities. The phenomenon of immunodomination takes place when one T cell recognizing a dominant epitope is preferentially able to form a mature functioning synapse, while another T cell recognizing a subdominant epitope presented by the same APC is prevented from forming a normal functionally mature synapse because of the preemption of critical elements of the synapse by the dominant T cell (Figure 3.1a).

On the T-cell side, the synapse is composed of T-cell receptors, CD4 and CD8 molecules, adhesion molecules, and costimulatory molecules. Several molecules associated with T-cell activation also actively colocalize at the immune synapse (IS). Other molecules, such as CD43 (and CD45), are actively excluded, which is mediated at least in part by molecules from the ERM family of cytoskeletal-associated proteins [41]. It is possible that alterations such as removal or exclusion of these molecules from the IS of submissive T cells are in part responsible for immunodomination. At the APC side, the IS encompasses MHC loaded with both relevant and irrelevant complexes; adhesion and costimulatory ligands such as CD80 and ICAM-1; lipid rafts and their associated tetraspan proteins such as CD9, CD63, CD81, or CD82; and molecules such as LAT, Thg1, and other GP proteins. Once again, it is possible that alterations in the amount or distribution of these molecules in the IS of the APC contacting submissive T cells are related to immunodomination, leading to partial activation (Figure 3.1b). A set of receptor

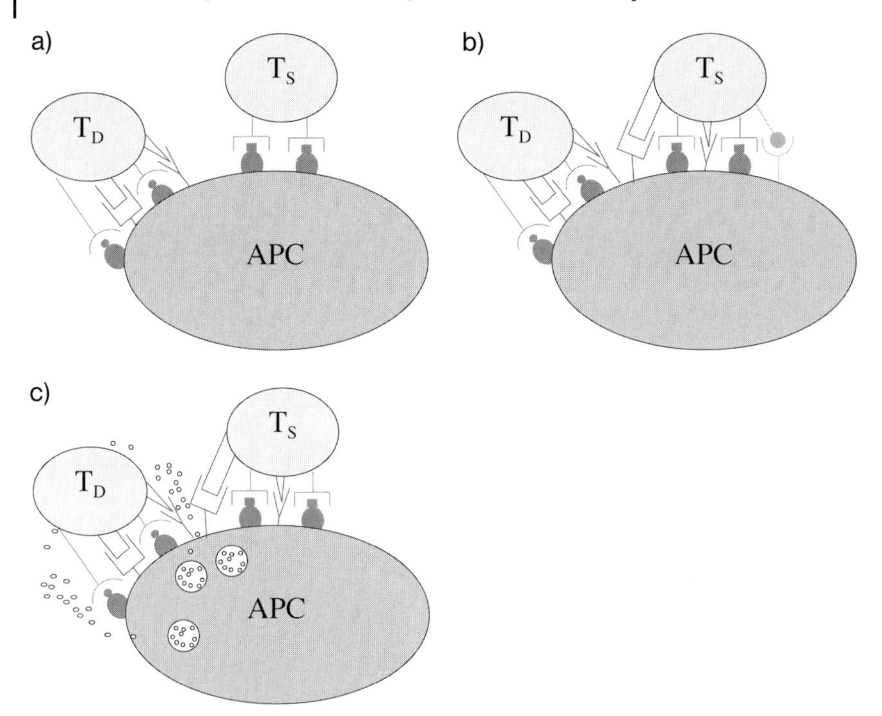

Figure 3.1 Possible mechanisms involved in immunodomination (a) Preemption of critical elements of the synapse by the dominant T cell. (b) Engagement of inhibitory receptors by the submissive T cell. (c) Lack of adequate directed secretion of APC factors. T_D = dominant T cell; T_S = submissive T cell TCR/MHC molecules are shown in red, costimulator ligands and receptors in blue, inhibitory receptors in orange, APC factors in white. (This figure also appears with the color plates.)

ligands of possible relevance in this regard are B7:CTLA4 and the recently described ICOS:B7H and PD1:PD-L1 and PD-L2 [42]. Since the PD-L1 and -L2 molecules, which are expressed on APCs, appear to be involved in downregulation of T-cell responses, their selective exclusion from the dominant IS (or selective inclusion in the submissive IS) may contribute to immunodomination [43]. Although evidence from planar membrane studies suggests that the APC side of IS is formed passively (at least for MHC and ICAM), some studies indicate that in DCs, an active cytoskeletal rearrangement is critical for clustering and activation of resting T cells [5].

An altered synapse may lead to a less optimal (or simply phenotypically different) activation of T cells. An example of an altered synapse may be provided by the stimulation of T cells by low-affinity altered ligands. In this case a synapse is formed, but it is devoid of the p-SMAC/c-SMAC features and is associated with incomplete signaling [44]. It has also been described that DCs can form antigen-independent synapses that are associated with weak biochemical activation (Ca^{++} and tyrosine phosphorylation) and weak proliferation and survival [45].

Interference with CD80/CD28 or ICAM-1/LFA-1 interactions impedes synapse morphology and leads to a reduction in the area and density of accumulated complexes, thus resulting in reduced proliferation [46]. Interestingly, other aspects of T-cell activation such as CD69 and CD25 expression and TCR down-modulation remained unaffected. In this respect, formation of the mature synapse is correlated with cytokine production and proliferation, while activation without it is not.

Finally, formation of a mature IS stops migration and thus helps amplify the signal by allowing sustained signaling through the T-cell receptor and costimulatory molecules [47]. This stop signal may be actually reversed by IFN-γ. Thus, a mature IS plus interferon (or other inflammatory milieus) reinstates the lymphocyte capacity to detach from APC and migrate [48]. It is also possible that immunodomination is related to the induction from the dominating T cell of regulatory and inhibitory signals. In this light it is particularly striking that several studies report that a lack of IFN-γ abolishes or reduces the hierarchy of immunodominance. This effect could be mediated by the anti-proliferative properties of interferon (especially in a suboptimally stimulated T cell). This mechanism might provide an explanation for why submissive T cells are apparently primed, but do not proliferate, and why the lack of IFN-γ abolishes the phenomenon of domination.

3.5.3 The Potential Role of MTOC in Immunodomination

A recent report describes the induction of MHC class II–containing endosomal tubular structures within the dendritic cells that are oriented towards the T-cell contact point. This induction was dependent on the formation of an immunological synapse and the clustering of membrane proteins as well as the ligation of adhesion molecules such as LFA-1 and CD2. It was also shown that activated T cells were more effective in inducing these tubular structures [50]. Although there is no direct evidence regarding the role of the MTOC in the directed secretion of cytokines such as IL-12 towards the dominating T cell, it is possible to speculate that IS formation leads to MTOC polarization within the APC, that this event is crucial for directed secretion of factors from the APC towards the T-cell interface, and that these factors are important in determining full T-cell activation. APC engagement by a dominant T cell may render it impossible for the MTOC to polarize towards a second subdominant T cell also engaging the same APC, and the consequent lack of adequate directed secretion of APC factors may have a decisive role in determining immunodomination (Figure 3.1c). The finding that exogenous administration of IL-12 leads to abolishment of immunodominance would be consistent with this hypothesis [49].

3.6 Significance of Studying Immunodominance for Vaccine Development

In the context of epitope-based vaccines, a study by Livingston and coworkers demonstrated that responses against multiple class I–restricted determinants can be obtained as a result of balancing the processing yield of the epitopes themselves [51]. Le et al. report on the use of polytopes to achieve the induction of multiple protective responses [52]. Peter and Corradin reported the induction of multiple responses by the use of equimolar pools of peptides in IFA, although the responses were weaker in response to the peptide pool, as compared to the isolated peptides [53]. Finally, multiple responses were obtained in a clinical setting utilizing a mixture of four immunogenic peptides loaded onto dendritic cells.

There are several advantages to the development of multi-epitope vaccines for infectious diseases as well as cancer. Some of these include broad population coverage and, more importantly, a multi-pronged attack against the infectious agent or antigen that would also prevent the emergence of escape variants. In this scenario, it is important to ensure that strong, robust responses are generated against all epitopes included in the vaccine construct and that the phenomenon of immunodomination does not exclude any of the selected epitopes. Understanding the molecular mechanisms of immunodominance and immunodomination may allow the development of rational strategies to achieve these goals, by interfering with key molecular mediators of immunodomination. One such strategy might be the inhibition of interferon production during the priming phase [21]. However, this particular strategy may be associated with drawbacks such as diminished overall immunogenicity.

3.7 Conclusions

Taken together, the results presented herein underline how, in our opinion, a systematic evaluation of the relationship between subdominant and dominant T cells on the one hand and synapse formation, activation signals, and the resulting differentiation patterns and programs on the other may be of significant importance. Such an evaluation might help to clarify the mechanisms involved in immunodomination and, in a more general sense, contribute to our understanding of T-cell activation and differentiation. Firstly, it is likely that "altered synapses" form physiologically as a result of the differential nature and state of activation of different APCs. In this light, a systematic study of the functional outcomes associated with formation by T cells and synapses of different or altered composition (in terms of accessory, costimulatory, and adhesion molecules, as well as other factors) may have relevance for our understanding of T-cell activation and regulation in general, just as altered peptide ligands have been utilized to study the molecular events associated with antigen recognition, T-cell activation, positive and negative selection, and the like. Furthermore, if the concept of altered synapses can be verified experimentally, it may provide the foundation for the development of practical ways to modulate the breadth, quality, and quantity of immune responses following natural infection, vaccination, or immunotherapeutic interventions.

References

1 W. Chen, L. C. Anton, J. R. Bennink, J. W. Yewdell, Dissecting the multifactorial causes of immunodominance in class I-restricted T cell responses to viruses. Immunity **2000**, *12*, 83–93.

2 R. G. van der Most, A. Sette, C. Oseroff, J. Alexander, K. Murali-Krishna, L. L. Lau, S. Southwood, J. Sidney, R. W. Chesnut, M. Matloubian, R. Ahmed, Analysis of cytotoxic T cell responses to dominant and subdominant epitopes during acute and chronic lymphocytic choriomeningitis virus infection. J. Immunol. **1996**, *157*, 5543–5554.

3 P. A. Wentworth, A. Vitiello, J. Sidney, E. Keogh, R. W. Chesnut, H. Grey, A. Sette, Differences and similarities in the A2.1-restricted cytotoxic T cell repertoire in humans and human leukocyte antigen-transgenic mice. Eur. J. Immunol. **1996**, *26*, 97–101.

4 J. W. Yewdell, J. R. Bennink, Immunodominance in major histocompatibility complex class I-restricted T lymphocyte responses. Annu. Rev. Immunol. **1999**, *17*, 51–88.

5 M. M. Al-Alwan, G. Rowden, T. D. Lee, K. A. West, The dendritic cell cytoskeleton is critical for the formation of the immunological synapse. J. Immunol. **2001**, *166*, 1452–1456.

6 M. F. Princiotta, D. Finzi, S. B. Qian, J. Gibbs, S. Schuchmann, F. Buttgereit, J. R. Bennink, J. W. Yewdell, Quantitating protein synthesis, degradation, and endogenous antigen processing. Immunity **2003**, *18*, 343–354.

7 E. Reits, A. Griekspoor, J. Neijssen, T. Groothuis, K. Jalink, P. van Veelen, H. Janssen, J. Calafat, J. W. Drijfhout, J. Neefjes, Peptide diffusion, protection, and degradation in nuclear and cytoplasmic compartments before antigen presentation by MHC class I. Immunity **2003**, *18*, 97–108.

8 T. Serwold, F. Gonzalez, J. Kim, R. Jacob, N. Shastri, ERAAP customizes peptides for MHC class I molecules in the endoplasmic reticulum. Nature **2002**, *419*, 480–483.

9 I. A. York, A. X. Mo, K. Lemerise, W. Zeng, Y. Shen, C. R. Abraham, T. Saric, A. L. Goldberg, K. L. Rock, The cytosolic endopeptidase, thimet oligopeptidase, destroys antigenic peptides and limits the extent of MHC class I antigen presentation. Immunity **2003**, *18*, 429–440.

10 C. Velazquez, I. Vidavsky, K. van der Drift, M. L. Gross, E. R. Unanue, Chemical identification of a low abundance lysozyme peptide family bound to I-Ak histocompatibility molecules. J. Biol. Chem. **2002**, *277*, 42514–43522.

11 R. A. Singh, J. R. Rodgers, M. A. Barry, The role of T cell antagonism and original antigenic sin in genetic immunization. J. Immunol. **2002**, *169*, 6779–6786.

12 R. M. Welsh, L.K. Selin, No one is naive: the significance of heterologous T-cell immunity. Nat. Rev. Immunol. **2002**, *2*, 417–426.

13 M. A. Brehm, A. K. Pinto, K. A. Daniels, J. P. Schneck, R. M. Welsh, L. K. Selin, T cell immunodominance and maintenance of memory regulated by unexpectedly cross-reactive pathogens. Nat. Immunol. **2002**, *3*, 627–634.

14 R. G. van der Most, R. J. Concepcion, C. Oseroff, J. Alexander, S. Southwood, J. Sidney, R. W. Chesnut, R. Ahmed, A. Sette, Uncovering subdominant cytotoxic T-lymphocyte responses in lymphocytic choriomeningitis virus-infected BALB/c mice. J. Virol. **1997**, *71*, 5110–5104.

15 M. K. Slifka, J. N. Blattman, D. J. Sourdive, F. Liu, D. L. Huffman, T. Wolfe, A. Hughes, M. B. Oldstone, R. Ahmed, M. G. Von Herrath, Preferential escape of subdominant $CD8^+$ T cells during negative selection results in an altered antiviral T cell hierarchy. J. Immunol. **2003**, *170*, 1231–1239.

16 L. Kjer-Nielsen, C. S. Clements, A. W. Purcell, A. G. Brooks, J. C. Whisstock, S. R. Burrows, J. McCluskey, J. Rossjohn, A structural basis for the selection of dominant alphabeta T cell receptors in antiviral immunity. Immunity **2003**, *18*, 53–64.

17 A. Sette, A. Vitiello, P. Farness, J. Furze, J. Sidney, J. M. Claverie, H. M. Grey, R. Chesnut, Random association between the peptide repertoire of A2.1 class I and several different DR class II molecules. J. Immunol. **1991**, *147*, 3893–3900.

18 M. T. Puglielli, A. J. Zajac, R. G. van der Most, J. L. Dzuris, A. Sette, J. D. Altman, R. Ahmed, *In vivo* selection of a lymphocytic choriomeningitis virus variant that affects recognition of the GP33–43 epitope by H-2Db but not H-2Kb. J. Virol. **2001**, *75*, 5099–5107.

19 H. A. Lewicki, M. G. Von Herrath, C. F. Evans, J. L. Whitton, M. B. Oldstone, CTL escape viral variants. II. Biologic activity *in vivo*. Virology **1995**, *211*, 443–450.

20 A. Gallimore, H. Hengartner, R. Zinkernagel, Hierarchies of antigen-specific cytotoxic T-cell responses. Immunol. Rev. **1998**, *164*, 29–36.

21 F. Rodriguez, S. Harkins, M. K. Slifka, J. L. Whitton, Immunodominance in virus-induced CD8(+) T-cell responses is dramatically modified by DNA immunization and is regulated by gamma interferon. J. Virol. **2002**, *76*, 4251–4259.

22 B. Perarnau, M. F. Saron, B. R. San Martin, N. Bervas, H. Ong, M. J. Soloski, A. G. Smith, J. M. Ure, J. E. Gairin, F. A. Lemonnier, Single H2Kb, H2Db and double H2KbDb knockout mice: peripheral $CD8^+$ T cell repertoire and anti-lymphocytic choriomeningitis virus cytolytic responses. Eur. J. Immunol. **1999**, *29*, 1243–1252.

23 R. G. van der Most, K. Murali-Krishna, J. G. Lanier, E. J. Wherry, M. T. Puglielli, J. N. Blattman, A. Sette, R. Ahmed, Changing immunodominance patterns in antiviral CD8 T-cell responses after loss of epitope presentation or chronic antigenic stimulation. Virology **2003**, *315*, 93–102.

24 E. A. Butz, M. J. Bevan, Massive expansion of antigen-specific $CD8^+$ T cells during an acute virus infection. Immunity **1998**, *8*, 167–175.

25 A. R. Townsend, J. Rothbard, F. M. Gotch, G. Bahadur, D. Wraith, A. J. McMichael, The epitopes of influenza nucleoprotein recognized by cytotoxic T lymphocytes can be defined with short synthetic peptides. Cell **1986**, *44*, 959–968.

26 W. Chen, J. R. Bennink, P. A. Morton, J. W. Yewdell, Mice deficient in perforin, $CD4^+$ T cells, or CD28-mediated signaling maintain the typical immunodominance hierarchies of $CD8^+$ T-cell responses to influenza virus. J. Virol. **2002**, *76*, 10332–10337.

27 G. T. Belz, W. Xie, J. D. Altman, P. C. Doherty, A previously unrecognized H-2D(b)-restricted peptide prominent in the primary influenza A virus-specific CD8(+) T-cell response is much less apparent following secondary challenge. J. Virol. **2000**, *74*, 3486–3493.

28 S. R. Crowe, S. C. Miller, R. M. Shenyo, D. L. Woodland, Vaccination with an acidic polymerase epitope of influenza virus elicits a potent antiviral T cell response but delayed clearance of an influenza virus challenge. J. Immunol. **2005**, *174*, 696–701.

29 S. J. Turner, S. C. Miller, A. D. Roberts, R. A. Rappolo, P. C. Doherty, K. H. Ely, D. F. Woodland, Differential antigen presentation regulates the changing patterns of $CD8^+$ T cell immunodominance in primary and secondary influenza virus infections. Crowe SR, J Exp. Med. **2003**, *198*, 399–410.

30 W. Chen, K. Pang, K. A. Masterman, G. Kennedy, S. Basta, N. Dimopoulos, F. Hornung, M. Smyth, J. R. Bennink, J. W. Yewdell, Reversal in the immunodominance hierarchy in secondary $CD8^+$ T cell responses to influenza A virus: roles for cross-presentation and lysis-independent immunodomination. J. Immunol. **2004**, *173*, 5021–5027.

31 P. Grufman, J. K. Sandberg, E. Z. Wolpert, K. Karre, Immunization with dendritic cells breaks immunodominance in CTL responses against minor histocompatibility and synthetic peptide antigens. J. Leukoc. Biol. **1999**, *66*, 268–271.

32 P. Grufman, E. Z. Wolpert, J. K. Sandberg, K. Karre, T cell competition for the antigen-presenting cell as a model for immunodominance in the cytotoxic T lymphocyte response against minor histocompatibility antigens. Eur. J. Immunol. **1999**, *29*, 2197–2204.

33 R. Schirmbeck, D. Stober, S. El-Kholy, P. Riedl, J. Reimann, The immunodominant, Ld-restricted T cell response to hepatitis B surface antigen (HBsAg) efficiently suppresses T cell priming to multiple Dd-, Kd-, and Kb-restricted HBsAg epitopes. J. Immunol. **2002**, *168*, 6253–6262.

34 R. M. Kedl, J. W. Kappler, P. Marrack, Epitope dominance, competition and T cell affinity maturation. Curr. Opin. Immunol. **2003**, *15*, 120–127.

35 R. M. Kedl, B. C. Schaefer, J. W. Kappler, P. Marrack, T cells down-modulate peptide-MHC complexes on APCs *in vivo*. Nat. Immunol. **2002**, *3*, 27–32.

36 R. M. Kedl, W. A. Rees, D. A. Hildeman, B. Schaefer, T. Mitchell, J. Kappler, P. Marrack, T cells compete for access to antigen-bearing antigen-presenting cells. J. Exp. Med. **2000**, *192*, 1105–1113.

37 A. Lanzavecchia, Lack of fair play in the T cell response. Nat. Immunol. **2002**, *3*, 9–10.

38 M. J. Palmowski, E. M. Choi, I. F. Hermans, S. C. Gilbert, J. L. Chen, U. Gileadi, M. Salio, A. Van Pel, S. Man, E. Bonin, P. Liljestrom, P. R. Dunbar, V. Cerundolo, Competition between CTL narrows the immune response induced by prime-boost vaccination protocols. J. Immunol. **2002**, *168*, 4391–4398.

39 N. Auphan-Anezin, G. Verdeil, A. M. Schmitt-Verhulst, Distinct thresholds for CD8 T cell activation lead to functional heterogeneity: CD8 T cell priming can occur independently of cell division. J. Immunol. **2003**, *170*, 2442–2448.

40 G. J. van Mierlo, A. T. den Boer, J. P. Medema, E. I. van der Voort, M. F. Fransen, R. Offringa, C. J. Melief, R. E. Toes, CD40 stimulation leads to effective therapy of CD40(–) tumors through induction of strong systemic cytotoxic T lymphocyte immunity. Proc. Natl. Acad. Sci. USA **2002**, *99*, 5561–5566.

41 A. S. Shaw, FERMing up the synapse. Immunity **2001**, *15*, 683–686.

42 L. Liang, W. C. Sha, The right place at the right time: novel B7 family members regulate effector T cell responses. Curr. Opin. Immunol. **2002**, *14*, 384–390.

43. P. Loke, J. P. Allison, PD-L1 and PD-L2 are differentially regulated by Th1 and Th2 cells. Proc Natl Acad Sci USA **2003**, *100*, 5336–5341.

44 C. Wulfing, C. Sumen, M. D. Sjaastad, L. C. Wu, M. L. Dustin, M. M. Davis, Costimulation and endogenous MHC ligands contribute to T cell recognition. Nat. Immunol. **2002**, *3*, 42–47.

45 P. Revy, M. Sospedra, B. Barbour, A. Trautmann, Functional antigen-independent synapses formed between T cells and dendritic cells. Nat. Immunol. **2001**, *2*, 925–931.

46 S. A. Wetzel, T. W. McKeithan, D. C. Parker, Live-cell dynamics and the role of costimulation in immunological synapse formation. J. Immunol. **2002**, *169*, 6092–6101.

47 M. L. Dustin, Coordination of T cell activation and migration through formation of the immunological synapse. Ann. N.Y. Acad. Sci. **2003**, *987*, 51–59.

48 S. S. Tay, A. McCormack, C. Lawson, M. L. Rose, IFN-gamma reverses the stop signal allowing migration of antigen-specific T cells into inflammatory sites. J. Immunol. **2003**, *170*, 3315–3322.

49 G. Eberl, B. Kessler, L. P. Eberl, M. J. Brunda, D. Valmori, G. Corradin, Immunodominance of cytotoxic T lymphocyte epitopes co-injected *in vivo* and modulation by interleukin-12. Eur. J. Immunol. **1996**, *26*, 2709–2716.

50 N. Bertho, J. Cerny, Y. M. Kim, E. Fiebiger, H. Ploegh, M. Boes, Requirements for T cell-polarized tubulation of class II+ compartments in dendritic cells. J. Immunol. **2003**, *171*, 5689–5696.

51 B. D. Livingston, M. Newman, C. Crimi, D. McKinney, R. Chesnut, A. Sette, Optimization of epitope processing enhances immunogenicity of multiepitope DNA vaccines. Vaccine **2001**, *19*, 4652–4660.

52 T. T. Le, D. Drane, J. Malliaros, J. C. Cox, L. Rothel, M. Pearse, T. Woodberry, J. Gardner, A. Suhrbier, Cytotoxic T cell polyepitope vaccines delivered by ISCOMs. Vaccine **2001**, *19*, 4669–4675.

53 K. Peter, Y. Men, G. Pantaleo, B. Gander, G. Corradin, D. Drane, J. Malliaros, J. C. Cox, L. Rothel, M. Pearse, T. Woodberry, J. Gardner, A. Suhrbier, T-cell response to HIV-1 proteins with short synthetic peptides and human compatible adjuvants. Vaccine **2001**, *19*, 4121–4129.

II
Proteosome Specificity and Immuno-Proteosomes

Immunodominance: The Choice of the Immune System. Edited by Jeffrey A. Frelinger

ISBN: 3-527-31274-9

4
Endogenous Antigen Processing

Jonathan W. Yewdell

4.1
Unbottling the Genie

We should never forget just how amazing CD8 T cells (T_{CD8+}) are. They are endowed with receptors that are capable of recognizing in a highly specific manner any virus that ever existed, exists now, or will exist. They are capable of recognizing virus-infected cells with incredible sensitivity and of detecting just a few (maybe even one) MHC class I molecules bearing the complementary viral peptide in a sea of hundreds of thousands of class I molecules bearing irrelevant peptides. They exist in a virtual state of suspended animation until activated and then divide madly at a rate that well exceeds any other cell type in the body. Once activated, they home to sites of viral infections and kill infected cells before the virus can fully complete its replication cycle.

As with genies, the magical powers of T_{CD8+} need to be carefully controlled. Armed with the power to kill, T_{CD8+} have the potential to wreak havoc in the event that self-reactive T_{CD8+} become activated. Control is principally achieved by thymic selection. Like every biological process, however, this is not perfect, and additional mechanisms are required to control autoimmunity. A critical checkpoint is that naïve T_{CD8+} cannot be activated by just any virus-infected cell. Indeed, it appears that under most circumstances, naïve T_{CD8+} are activated exclusively by dendritic cells (DCs), which are present in the right place (lymphoid tissue) and express the right molecules (CD80 and other “costimulatory” factors) to activate T_{CD8+}.

This creates a problem. Viruses are highly mutable and prove to be insightful students of the vertebrate immune system. They could easily remain under the T_{CD8+} radar simply by not expressing their proteins in DCs, either by not entering DCs or by blocking their gene expression in DCs. But the immune system is up to the challenge. DCs are adept at acquiring viral antigens from infected cells and using this material to activate T_{CD8+}. This pathway, termed cross-priming, probably plays an important role in activation of anti-viral T_{CD8+}. The key word here is *probably*, because the role of cross-priming vs. direct priming, i.e., priming by virus-infected cells, is not clearly established and in fact is a highly contentious issue [1,

Immunodominance: The Choice of the Immune System. Edited by Jeffrey A. Frelinger

ISBN: 3-527-31274-9

2]. It is likely that both pathways are important in priming anti-viral T_{CD8+}, with the relative contributions varying widely depending on the following factors:

- identity of the host,
- identity of the virus,
- route of viral infection,
- dose of viral infection,
- identity of viral protein that provides the determinant, and
- identity of the determinant itself.

Sorting out the relative contributions of these factors is still in its infancy, but there is reasonably solid evidence for the contribution of both direct priming and cross-priming depending on the system employed [3].

At the request of the editor, I have been tasked in this chapter to describe cell biology that underlies direct presentation of determinants from endogenously synthesized viral proteins to T_{CD8+}.

4.2 DRiPs to the Rescue

As the ultimate obligate intracellular parasite, viruses are obliged to produce their proteins on cellular ribosomes. Because functional ribosomes are located in cytosol (at least according to dogma: there is controversial evidence for nuclear translation [4]), viral proteins begin their existence in the cytosol (with the interesting exception of proteins co-translationally targeted to the endoplasmic reticulum [ER]). There are two general possibilities for generating a peptide from a viral mRNA: either the information is initially decoded into a peptide-sized product, or the peptide is generated from the intended protein (or at least a longer polypeptide) by proteolysis. The former was originally proposed by Van Pel and Boon as the "pepton" hypothesis [5], which, while brilliant in concept, has found little experimental validation. Rather, there is considerable evidence that peptides are generated from degradation of standard translation products.

But how are these translation products converted to MHC class I peptide ligands (hereafter referred to as peptide ligands)? Most viral proteins are extremely stable in cells, frequently with multi-day half-lives of degradation. Yet T_{CD8+} specific for peptides from these proteins can recognize cells within an hour of *adding* virus to cells. This conundrum led to the DRiP hypothesis (TDH) of antigen presentation: peptides derive from defective ribosomal products, defective forms of proteins that are recognized as aberrant by cellular quality-control (QC) devices and degraded [6].

Numerous types of errors can occur in the translation of information from viral (or cell) nucleic acids to the finished product, a functional protein. Bases can be misincorporated into mRNA, or, in the case of viruses that produce mRNA in the nucleus, mRNA can be inappropriately spliced. Amino acids can be misincorporated into proteins by misloading of tRNA or by ribosomes selecting the wrong tRNA. Proteins can be terminated prematurely or extended through intended stop codons. Proteins can even be spliced post-translationally [7], adding another potential error. For proteins that are members of multi-subunit complexes, there must be imbalances in synthesis resulting in excess amounts of subunits that are unstable in the absence of their normal partners. Protein targeting can go awry, resulting in protein misdelivery to an inappropriate organelle. Even more basic, nascent proteins emerge from the ribosome into a cytosol packed with other proteins and macromolecules. Folding in this environment is challenging, particularly for long polypeptides [8].

All these mistakes can be lumped together in TDH [6]. Presumably, cells would destroy DRiPs relatively rapidly, as they are worse than useless because they can potentially interfere with functional cellular components. For T_{CD8+}, the use of DRiPs as a source of peptide ligands would also enable class I molecules to monitor protein synthesis rates in cells and not protein concentrations, offering the possibility of rapid detection of virus-infected cells. Thus, viral proteins would be detected on a more than equal footing with host proteins, because viruses often hijack a considerable fraction of ribosomes for their own use. Another significant advantage is that peptides could be generated from all types of proteins despite their ultimate organellar destination in the cell or their apparent metabolic stability.

4.3 The Ubiquitin–Proteasome Pathway

Before contemplating the experimental evidence for TDH, I need to briefly outline the ubiquitin–proteasome pathway (UPP). The importance of the UPP to biology and human health was recognized in 2004 by the awarding of the Nobel Prize for Chemistry to Hershko, Ciechanover, and Rose for their work in demonstrating the importance of ubiquitin in protein degradation. Ubiquitin is the principal signal that targets proteins to proteasomes, which degrades the vast majority of intracellular proteins under normal circumstances [9]. Proteasomes are evolutionarily ancient and are present in all eukaryotes and in many prokaryotic species as well. Because proteasomes are essential for viability in yeast, it caused quite a stir when it was reported that mouse cells adapted to grow in the presence of a proteasome inhibitor were able to propagate while lacking functional proteasomes [10]. Subsequently, however, evidence was presented that these cells possess enzymatically active proteasomes that are essential for degrading ubiquitylated proteins and for continued cell viability [11]. Thus, it appears that other cellular proteases cannot completely substitute for proteasomes.

A common misconception is that proteasomes are located exclusively in the cytosol. In fact, proteasomes are present in the nucleus at similar or even higher concentrations [12]. Among cytosolic proteasomes, a substantial number are in close proximity to the ER. The long sought-after association of proteasomes with TAP has yet to be reported, though tantalizingly, immunoproteasomes are enriched at the ER [13]. Clearly, proteasomes on the cytosolic face of the ER have more to do than provide antigenic peptides to TAP, because yeast also have ER-associated proteasomes, in fact, at a higher density than in human cells [14, 15]. It is likely that these proteasomes, and at least a fraction of mammalian ER-associated proteasomes, are involved in ER-associated degradation (ERAD), the process by which ER proteins are translocated from the ER to the cytosol for proteasome degradation.

Proteasomes comprise two major structural assemblies, a catalytic core known as the 20S proteasome and regulatory subunits that attach to each end of the 20S proteasome. Together, a 20S regulator and the 19S regulator at its each end form the 26S proteasome. 20S proteasomes are barrel-shaped structures that comprise 14 distinct subunits arrayed in a four-ring structure of the type $\alpha_7\beta_7\beta_7\alpha_7$. Three of the β subunits in each inner ring are known to be catalytically active, with their active sites facing a central proteolytic chamber. 20S proteasomes are closed at both ends and must bind to a regulatory particle for protein substrates to gain access to the central chamber. The major type of regulatory particle is the 19S regulator, which functions to recognize potential substrates bearing polyUb chains, to deubiquitylate the substrate, and to unfold and feed the substrate, spaghetti-like, into a barrel of the 20S proteasome via a narrow portal that is opened as part of the process [16].

Ub is a 76-residue protein that is remarkably conserved among eukaryotes (three amino acid substitutions, all conservative, between vertebrates and yeast). It is abundant in cells (~10^8 copies per HeLa cell) [17] and has an ever-expanding repertoire of functions that revolve around its covalent conjugation to the ε-NH_2 groups of Lys residues via its COOH terminus. Proteins can be monoubiquitylated (attachment of a single Ub at a single site), polyubiquitylated (attachment of a polyUb chain), or multiubiquitylated (mono- or polyubiquitylation at multiple sties) [18]. Monoubiquitylation serves as a signal for other processes and is not involved in protein degradation. Indeed, polyUb chains must reach four Ub subunits for substrates to be efficiently recognized by 19S caps and fed to the 20S protease. For no obvious reason, protein ubiquitylation is highly heterogeneous, and trees of different sizes can be added to multiple substrate sites in what appears to be a highly irregular manner, such that polyubiquitylated proteins usually migrate as a ladder (or smear) in SDS-PAGE. Ub is removed from proteins by the numerous ubiquitin hydrolases present in cells, which are extremely active. Thus, ubiquitylation is completely and rapidly reversible, though the reasons for this are obscure, particularly with substrates that will eventually be degraded [19]. Ubiquitylation is not an absolute prerequisite for protein degradation by proteasomes, and there is an increasing list of proteins that are degraded in a Ub-independent manner by 26S proteasomes and apparently even by 20S proteasomes acting without regulators [20–22].

The contributions of proteasomes to mammalian cell physiology have been defined largely through the use of membrane-permeant, low-molecular-weight pharmacological inhibitors of the proteolytic activities of 26S proteasomes [23, 24]. These include natural products of microorganisms and synthetic oligopeptidebased compounds. While the value of proteasome inhibitors is indisputable, the secondary effects of blocking proteasomes on cell physiology can seriously confound experimental analysis. These effects include rapid depletion of the pool of free Ub available for conjugation [23, 25], induction of molecular chaperones [26], and a general inhibition of protein synthesis. That some or all of these effects are likely induced by an accumulation of ubiquitylated substrates awaiting destruction does not simplify matters.

The introduction of short interfering RNAs (siRNAs) to specifically reduce levels of complementary mRNAs has provided another approach to modulate proteasome function. While powerful, this approach suffers from the same limitations of chemical inhibitors in inducing secondary effects as a result of blocking proteasome function.

Proteasomes exist in multiple forms in cells. Indeed, the initial evidence implicating proteasomes in antigen processing was the discovery that the MHC encodes two proteasome subunits and that their (and a third subunit) expression is controlled by cytokines released by activated T cells [27]. When induced, these subunits replace constitutively expressed subunits in newly assembled proteasomes to create 20S immunoproteasomes, which are somewhat more adept at producing peptides favored by class I molecules [28]. At the same time, some peptides are produced more efficiently by standard proteasomes than by immunoproteasomes. Complicating matters further for *in vivo* studies, knocking out immunoproteasome subunits in mice can alter the repertoire of responding T_{CD8+} [29]. Such mice also exhibit defects in T_{CD8+} and other immune-cell responsiveness unrelated to antigen presentation. Although immunoproteasomes clearly influence antigen presentation, their major function in immunity may be independent of antigen presentation.

Other modifications of proteasomes occur. The functional consequences of these alterations are uncertain. Subunits can be phosphorylated or produced from alternatively spliced transcripts. Cytokine-exposed cells also produce 11S regulators, which take the place of 19S regulators at one end of the 20S proteasome and seem to favor antigen presentation through an undefined mechanism [30–32].

Proteasomes are capable of generating the precise peptide ligands presented by class I molecules, but this is probably infrequent. A recent report suggests that most proteasome-generated peptides have to be further cleaved by tripeptidyl protease II, another large multi-subunit protease present in the cytosol with endo- and exopeptidase activities [33]. Amino-terminal trimming is further mediated by cytosolic aminopeptidases, and particularly by ER-associated aminopeptidase (ERAP). ERAP is required for trimming most peptides and appears to be largely, if not exclusively, dedicated to antigen processing [34, 35]. ER trimming is possible because TAP, the MHC-encoded heterodimer that functions to transport peptides across the ER membrane to nascent class I molecules, transport peptides between 8 and 17 residues long with similar efficiency [36–38].

4.4
Pressing TDH Questions

TDH raises two immediate questions:

1. What is the efficiency of protein synthesis, i.e., what fraction of ribosomal products do not reach a functional form and are targeted for more rapid degradation (the "DRiP rate")?
2. What fraction of peptide ligands derive from DRiPs vs. once-functional proteins?

It is important to recognize that the DRiP rate and the fraction of peptide ligands that arise from DRiPs might be largely independent of each other. The relationship depends on how proteasome-generated ligands are handled by cells. It is possible that heterogeneity exists among proteasomes such that their products are delivered to class I molecules with significantly different efficiencies.

4.4.1
Answer to Question 1

The overall efficiency of protein synthesis in mammalian cells was first addressed more than 20 years ago. Extending the prior findings of Wibo and Poole [39], Wheatley and colleagues found that following brief pulse labeling (30 sec to 5 min) of HeLa cells with [^{3}H]-Leu, ~30% of proteins were degraded within an hour of their synthesis. [40] Though there was contemporary supporting evidence for a high DRiP rate for individual proteins (40% of collagen degraded within minutes of synthesis [41]), Wheatley's work was not widely appreciated. ,

Unaware of Wheatley's work, Schubert et al. basically repeated the pulse-labeling experiments but with some important modifications, including the use of proteasome inhibitors (at the time of Wheatley's experiments, the proteasome had yet to be discovered), the measurement of protein degradation by SDS-PAGE to resolve the M_r of the degraded proteins, and the fractionation of labeled proteins into soluble and insoluble fractions by high-speed centrifugation of cells disrupted by freeze-thawing [42]. These methods revealed that for HeLa cells, bone marrow–derived dendritic cells, and activated lymph node cells tested *ex vivo*, 30–40% of newly synthesized proteins are degraded with a $t_{1/2}$ on the order of a few minutes. Most of the rapidly degraded proteins migrated with a M_r of greater than 50 kDa and therefore could not simply represent NH_2-terminal targeting sequences that are cleaved and degraded rapidly. Such a high fraction of rapidly degraded nascent proteins implies that newly synthesized proteins are the principal substrate for proteasomes; indeed, immunoblotting with a polyUb-specific mAb revealed that polyUb proteins are depleted within 1 min of adding cycloheximide to cells to block protein synthesis.

Notably, the shortest-lived proteins known (ornithine decarboxylase is probably the champion) are degraded with a $t_{1/2}$ of 10–20 min, and most truly "short-lived" proteins (SLiPs) are degraded with half-lives on the order of a few hours. It seems likely that the bulk of the extremely rapidly degraded proteins are DRiPs. Nevertheless, to avoid drawing premature conclusions, it is more accurate to refer to the rapidly degraded cohort as rapidly degraded polypeptides (RDPs), which include DRiPs, SLiPs, and, when measured by TCA precipitation, probably also NH_2-terminal leader sequences, many of which are acid insoluble despite their small size.

4.4.2 The Real World

It is important to think about cells in terms of actual numbers for its constituents and rates of various processes, rather than fractional or relative values [43]. Cells are governed by the law of mass action, and it is necessary to know the concentrations of various reactants to determine the physiological relevance of interactions and pathways established in cell-free systems. Taking an economics approach to protein synthesis and degradation, Princiotta et al. could account for the steady-state level of total L-K^b cell proteins given the measured rates of protein synthesis,

Table 4.1 The protein economy of L-K^b cells.

Production	
Proteins per cell	2.6×10^9
Ribosomes per cell	6×10^6
Translation rate	4×10^6 min^{-1}
Percentage required for cell division (24 h)	43%
Percentage secreted or released	12%
Percentage of ribosomes engaged in translation	>95%
Percentage of cellular metabolism devoted to protein synthesis	45%
Destruction	
Proteasomes per cell	8×10^5
Overall degradation rate	1.8×10^6 min^{-1}
Newly synthesized proteins	1.3×10^6 min^{-1}
Long-lived proteins	5×10^5 min^{-1}
Proteasome activity	2.5 substrates degraded min^{-1}
Efficiency of SIINFEKL-K^b complex production	1/440 to 1/3000 substrates degraded per complex created

secretion, degradation, and cell division (Table 4.1) [44]. This revealed that RDPs amount to more than a million substrates degraded per minute, accounting for ~55% of proteasome substrates, with the remaining substrates derived from slowly degraded polypeptides (SDPs), which exhibited an average $t_{1/2}$ of 1.6 d under the conditions employed. This explained the rapid drop in polyubiquitylated proteins observed by Schubert et al. [42] upon inhibiting protein synthesis.

Remarkably, protein synthesis consumed approximately half of the ATP generated by L-K^b cells, meaning that RDPs consume more than 10% of the cellular energy. In more tangible terms, if this number is generally applicable to mammalian cells, if you happen to be eating a pack of M&M's while reading this chapter, two of the 20 M's in the package will go towards feeding the DRiPs generated by your cells. This may not seem like much, but if you were just barely scraping by on your M&M diet (because of funding difficulties, perhaps), should you lose those two M's, this 10% expenditure would mean the difference between life and death.

So, what gives? As discussed previously [45], the true DRiP rate *in vivo* may be considerably lower than what we measure *in vitro*. Cells did not evolve to grow on plastic in synthetic media with 10% serum from newborn cows. Oxygen levels *in vivo* are closer to 3% partial pressure than to the 20% levels present in room air and CO_2 incubators. Still, we entertain the possibility that high DRiP rates are a general feature of cells. Protein biosynthesis may simply be a difficult task, and investments in increasing efficiency may not be worth the cost. It may be wiser for cells to err on the side of caution in protein triage decisions and to destroy potentially misfolded proteins rather than risk the havoc they could wreak by inappropriately associating with functional cellular components.

4.4.3
Answer to Question 2

The relative contribution of DRiPs to the generation of peptide ligands was first addressed experimentally well before TDH was conceived; these studies date back to the dawn of class I antigen processing as a discipline. Once it became possible to study the recognition of individual viral proteins through recombinant gene technology, it became clear that viral membrane proteins are at least as well represented as T_{CD8+} antigens as are other viral proteins [46]. With few exceptions, peptide ligands are generated from membrane proteins in a TAP- and proteasome-dependent manner, demonstrating that processing occurs in the cytosol. There are two defined pathways for ER-targeted proteins to reach the cytosol: either they don't reach ER because of a failure in targeting, or they are transported from the ER via retrotranslocation. The available evidence suggests that the latter is the predominant route of antigen processing [47, 48], but in both scenarios, peptides derive from DRiPs and not from functional viral glycoproteins, which by definition are exported from the ER (note that peptide ligands present in cleavable NH_2-terminal signal sequences would be classified as quasi-DRiPs because the signal is naturally short-lived in most cases).

Schubert et al. more directly addressed the overall contribution of DRiPs to peptide ligand generation by showing that protein synthesis inhibitors retarded class I export from the ER. Because class I export requires peptide association, this finding implied that DRiPs are an important source of peptide ligands [42]. The requirement for protein synthesis in peptide ligand generation was more directly examined by Reits et al. [49]. This study utilized a fiendishly clever assay for peptide ligand generation based on the mobility of a TAP1–GFP fusion protein in the ER membrane as determined by fluorescent recovery after photobleaching. Establishing that TAP mobility is inversely proportional to peptide availability, Reits et al. found that 15–30 min of blocking protein synthesis is sufficient to deplete cells of TAP-transported peptides, in either influenza A virus–infected or uninfected cells.

This remarkable paper implies that nearly all TAP-transported peptides derive from newly synthesized proteins. Kinetic studies of cell surface class I complex expression using model viral antigens encoded by recombinant vaccinia virus are consistent with the intimate relationship between protein synthesis and peptide ligand generation [44].

At this point, the numbers in Table 4.1 provide crucial insight into potential mechanisms of peptide generation. Although RDPs provide the bulk of proteasome substrates, ~25% of substrates derive from SDPs whose degradation is not affected by protein synthesis inhibitors. Yet this pool does not seem to represent a significant source of peptide ligands. This implies that there is great heterogeneity in the ability of proteasomes to generate peptide ligands, which is not so surprising. There is evidence that immunoproteasomes are better at producing class I binding ligands than are standard proteasomes, although this effect appears to be relatively modest (indeed, for some determinants, standard proteasomes are the preferred generators). As mentioned above, there is increasing evidence for proteasome heterogeneity resulting from translational modification of their subunits and use of alternatively spliced genes. An increasing number of substrates are reported to be degraded in a Ub-independent manner by 20S proteasomes (i.e., the core proteasome acting without regulators) [50].

So, what if a subset of proteasomes were devoted to generating antigenic peptides from nascent proteins? Taking this a step further, what if a subset of ribosomes ("immunoribosomes") were devoted to making substrates for said subset of proteasomes? This would make good sense for the immune system, because it would provide a direct mechanism for rapid detection of viral proteins synthesized into a sea of self-proteins.

Further, this could explain the abysmal overall efficiency of antigen processing, which is on the order of one peptide–class I complex expressed on the cell surface per 10,000 peptides generated by proteasomes [51]. If the efficiency of peptide generation from immunoribosomes were on the order of 1% (i.e., one complex generated for 100 proteins degraded) then only one in a hundred gene products would need to be translated on immunoribosomes to account for antigen presentation. An obvious mechanism for increasing the efficiency of antigen presentation is to tether immunoribosomes to TAP-bound proteasomes. Another possibility is the targeting of immunoribosome products to TAP-bound proteasomes or to proteasomes whose products are delivered to TAP via some specific targeting mechanism.

4.5
What Does This Have to Do With Immunodominance?

In all honesty, maybe everything and maybe nothing. Maybe everything, in the sense that inasmuch generation of sufficient numbers of class I peptide complexes is limiting for ~50% of viral determinants [52], and naïve T_{CD8+} are activated by direct priming, it is essential to understand the underlying mechanisms responsible for peptide ligand generation. Maybe nothing, inasmuch as the relative contributions of cross-priming and direct priming to anti-viral T_{CD8+} are uncertain.

One of the strangest aspects of immunodominance is that the immunogenicity of individual determinants in cross-priming parallels the ability of the endogenous processing pathway to generate the determinant. Recent studies point to the dependence of cross-presentation on the workings of a specialized organelle, the ERgosome, generated by the fusion of endosomes or phagosomes with the ER [53]. Could the ERgosome processing somehow recapitulate the function of the immunoribosome?

As we learn more and more about antigen processing, the questions become more and more intriguing. The cellular immune system is a perfect stage for nature to demonstrate its amazing creativity.

References

1 Zinkernagel, R. M. 2002. On cross-priming of MHC class I-specific CTL: rule or exception? *Eur. J Immunol* 32:2385–2392.

2 Chen, W., K. A. Masterman, S. Basta, S. M. Haeryfar, N. Dimopoulos, B. Knowles, J. R. Bennink, and J. W. Yewdell. 2004. Cross-priming of CD8+ T cells by viral and tumor antigens is a robust phenomenon. *Eur J Immunol* 34:194–199.

3 Yewdell, J. W. and S. M. Haeryfar. 2005. Understanding presentation of viral antigens to CD8(+) T cells in vivo: The key to rational vaccine design. *Annu. Rev. Immunol.* 23:651–682.

4 Iborra, F. J., D. A. Jackson, and P. R. Cook. 2004. The case for nuclear translation. *J. Cell Sci.* 117:5713–5720.

5 Van Pel, A. and T. Boon. 1989. T cell-recognized antigenic peptides derived from the cellular genome are not protein degradation products but can be generated directly by transcription and translation of short subgenic regions. A hypothesis. *Immunogen* 29:75–79.

6 Yewdell, J. W., L. C. Antón, and J. R. Bennink. 1996. Defective ribosomal products (DRiPs). A major source of antigenic peptides for MHC class I molecules? *J. Immunol.* 157:1823–1826.

7 Hanada, K., J. W. Yewdell, and J. C. Yang. 2004. Immune recognition of a human renal cancer antigen through post-translational protein splicing. *Nature* 427:252–256.

8 Frydman, J. 2001. Folding of newly translated proteins in vivo: the role of

molecular chaperones. *Annu. Rev. Biochem.* 70:603–647.
9 Rock, K. L. and A. L. Goldberg. 1999. Degradation of cell proteins and the generation of MHC class I-presented peptides. *Annu. Rev. Immunol.* 17:739–779.
10 Glas, R., M. Bogyo, J. S. McMaster, M. Gaczynska, and H. L. Ploegh. 1998. A proteolytic sysytem that compensates for loss of proteasome function. *Nature* 392:618–622.
11 Princiotta, M. F., U. Schubert, W. Chen, J. R. Bennink, J. Myung, C. M. Crews, and J. W. Yewdell. 2001. Cells adapted to the proteasome inhibitor 4-hydroxy-5-iodo-3- nitrophenylacetyl-Leu-Leu-leucinal-vinyl sulfone require enzymatically active proteasomes for continued survival. *Proc. Natl. Acad. Sci. U. S. A* 98:513–518.
12 Reits, E. A. J., A. M. Benham, B. Plougastel, J. Neefjes, and J. Trowsdale. 1997. Dynamics of proteasome distribution in living cells. *EMBO J.* 16:6087–6094.
13 Rivett, A. J., S. Bose, P. Brooks, and K. I. Broadfoot. 2001. Regulation of proteasome complexes by gamma-interferon and phosphorylation. *Biochimie* 83:363–366.
14 Enenkel, C., A. Lehmann, and P. M. Kloetzel. 1998. Subcellular distribution of proteasomes implicates a major location of protein degradation in the nuclear envelope-ER network in yeast. *EMBO J.* 17:6144–6154.
15 Wilkinson, C. R., M. Wallace, M. Morphew, P. Perry, R. Allshire, J. P. Javerzat, J. R. McIntosh, and C. Gordon. 1998. Localization of the 26S proteasome during mitosis and meiosis in fission yeast. *EMBO J.* 17:6465–6476.
16 Glickman, M. H., D. M. Rubin, H. Fu, C. N. Larsen, O. Coux, I. Wefes, G. Pfeifer, Z. Cjeka, R. Vierstra, W. Baumeister, V. Fried, and D. Finley. 1999. Functional analysis of the proteasome regulatory particle. *Mol. Biol. Rep.* 26:21–28.
17 Haas, A. L. and P. M. Bright. 1985. The immunochemical detection and quantitation of intracellular ubiquitin-protein conjugates. *J. Biol. Chem.* 260:12464–12473.
18 Thrower, J. S., L. Hoffman, M. Rechsteiner, and C. M. Pickart. 2000. Recognition of the polyubiquitin proteolytic signal. *EMBO J.* 19:94–102.
19 Wilkinson, K. D. 2000. Ubiquitination and deubiquitination: targeting of proteins for degradation by the proteasome. *Semin. Cell Dev. Biol.* 11:141–148.
20 Coffino, P. 2001. Regulation of cellular polyamines by antizyme. *Nat. Rev. Mol. Cell Biol.* 2:188–194.
21 Benaroudj, N., E. Tarcsa, P. Cascio, and A. L. Goldberg. 2001. The unfolding of substrates and ubiquitin-independent-protein degradation by proteasomes. *Biochimie* 83:311–318.
22 Gruendler, C., Y. Lin, J. Farley, and T. Wang. 2001. Proteasomal Degradation of Smad1 Induced by Bone Morphogenetic Proteins. *J. Biol. Chem.* 276:46533–46543.
23 Lee, D. H. and A. L. Goldberg. 1998. Proteasome inhibitors: valuable new tools for cell biologists. *Trends Cell Biol.* 8:397–403.
24 Bogyo, M., M. Gaczynska, and H. L. Ploegh. 1997. Proteasome inhibitors and antigen presentation. *Biopolymers* 43:269–280.
25 Mimnaugh, E. G., C. Chavany, and L. Neckers. 1996. Polyubiquitination and proteasomal degradation of the $p185^{c\text{-}erbB\text{-}2}$ receptor protein-tryrosine kinase induced by geldanamycin. *J. Biol. Chem.* 271:22796–22801.
26 Bush, K. T., A. L. Goldberg, and S. K. Nigam. 1997. Proteasome inhibition leads to a heat-shock response, induction of endoplasmic reticulum chaperones, and thermotolerance. *J. Biol. Chem.* 272:9086–9092.
27 Monaco, J. J. and H. O. McDevitt. 1982. Identification of a fourth class of proteins linked to the murine major histocompatibility complex. *Proc. Natl. Acad. Sci. USA* 79:3001–3005.
28 Tanaka, K. and M. Kasahara. 1998. The MHC class I ligand-generating system: roles of immunoproteasomes and the interferon-gamma-inducible proteasome activator PA28. *Immunol. Rev.* 163:161–176.
29 Chen, W., C. C. Norbury, Y. Cho, J. W. Yewdell, and J. R. Bennink. 2001.

Immunoproteasomes shape immunodominance hierarchies of antiviral CD8(+) t cells at the levels of T cell repertoire and presentation of viral antigens. *J. Exp. Med.* 193:1319–1326.

30 Preckel, T., W. P. Fung-Leung, Z. Cai, A. Vitiello, L. Salter-Cid, O. Winqvist, T. G. Wolfe, M. Von Herrath, A. Angulo, P. Ghazal, J. D. Lee, A. M. Fourie, Y. Wu, J. Pang, K. Ngo, P. A. Peterson, K. Fruh, and Y. Yang. 1999. Impaired immunoproteasome assembly and immune responses in PA28-/- mice. *Science* 286:2162–2165.

31 Rechsteiner, M., C. Realini, and V. Ustrell. 2000. The proteasome activator 11 S REG (PA28) and class I antigen presentation. *Biochem. J.* 345 Pt 1:1–15.

32 Murata, S., H. Udono, N. Tanahashi, N. Hamada, K. Watanabe, K. Adachi, T. Yamano, K. Yui, N. Kobayashi, M. Kasahara, K. Tanaka, and T. Chiba. 2001. Immunoproteasome assembly and antigen presentation in mice lacking both PA28alpha and PA28beta. *EMBO J.* 20:5898–5907.

33 Reits, E., J. Neijssen, C. Herberts, W. Benckhuijsen, L. Janssen, J. W. Drijfhout, and J. Neefjes. 2004. A major role for TPPII in trimming proteasomal degradation products for MHC class I antigen presentation. *Immunity.* 20:495–506.

34 Serwold, T., F. Gonzalez, J. Kim, R. Jacob, and N. Shastri. 2002. ERAAP customizes peptides for MHC class I molecules in the endoplasmic reticulum. *Nature* 419:480–483.

35 York, I. A., S. C. Chang, T. Saric, J. A. Keys, J. M. Favreau, A. L. Goldberg, and K. L. Rock. 2002. The ER aminopeptidase ERAP1 enhances or limits antigen presentation by trimming epitopes to 8–9 residues. *Nat. Immunol* 3:1177–1184.

36 Elliott, T. 1997. Transporter associated with antigen processing. *Advanced in Immunology* 65:47–109.

37 Momburg, F. and G. J. Hammerling. 1998. Generation and TAP-mediated transport of peptides for major histocompatibility complex class I molecules. *Adv. Immunol.* 68:191–256:191–256.

38 Lankat-Buttgereit, B. and R. Tampe. 2002. The transporter associated with antigen processing: function and implications in human diseases. *Physiol Rev.* 82:187–204.

39 Poole, B. and M. Wibo. 1973. Protein degradation in cultured cells. The effect of fresh medium, fluoride, and iodoacetate on the digestion of cellular protein of rat fibroblasts0. *J. Biol. Chem.* 248:6221–6226.

40 Wheatley, D. N., S. Grisolia, and J. Hernandez-Yago. 1982. Significance of the rapid degradation of newly synthesized proteins in mammalian cells: a working hypothesis. *J. Theor. Biol.* 98:283–300.

41 Bienkowski, R. S., B. J. Baum, and R. G. Crystal. 1978. Fibroblasts degrade newly synthesised collagen within the cell before secretion. *Nature* 276:413–416.

42 Schubert, U., L. C. Anton, J. Gibbs, C. C. Norbury, J. W. Yewdell, and J. R. Bennink. 2000. Rapid degradation of a large fraction of newly synthesized proteins by proteasomes. *Nature* 404:770–774.

43 Yewdell, J. W. 2001. Not such a dismal science: the economics of protein synthesis, folding, degradation and antigen processing. *Trends Cell Biol.* 11:294–297.

44 Princiotta, M. F., D. Finzi, S. B. Qian, J. Gibbs, S. Schuchmann, F. Buttgereit, J. R. Bennink, and J. W. Yewdell. 2003. Quantitating protein synthesis, degradation, and endogenous antigen processing. *i* 18:343–354.

45 Yewdell, J. W., U. Schubert, and J. R. Bennink. 2001. At the crossroads of cell biology and immunology: DRiPs and other sources of peptide ligands for MHC class I molecules. *J. Cell Sci.* 114:845–851.

46 Bennink, J. R. and J. W. Yewdell. 1990. Recombinant vaccinia viruses as vectors for studying T lymphocyte specificity and function. *Curr. Top. Microbiol. Immunol.* 163:153–184.

47 Bacik, I., H. L. Snyder, L. C. Anton, G. Russ, W. Chen, J. R. Bennink, L. Urge, L. Otvos, B. Dudkowska, L. Eisenlohr, and J. W. Yewdell. 1997. Introduction of a glycosylation site into a secreted protein provides evidence for an alternative antigen processing pathway: transport of precursors of major histocompatibility complex class I-restricted peptides from the endoplasmic reticulum to the cytosol. *J Exp. Med.* 186:479–487.

48 Mosse, C. A., L. Meadows, C. J. Luckey, D. J. Kittlesen, E. L. Huczko, C. L. Slingluff, J. Shabanowitz, D. F. Hunt, and V. H. Engelhard. 1998. The class I antigen-processing pathway for the membrane protein tyrosinase involves translation in the endoplasmic reticulum and processing in the cytosol. *J Exp. Med.* 187:37–48.

49 Reits, E. A., J. C. Vos, M. Gromme, and J. Neefjes. 2000. The major substrates for TAP in vivo are derived from newly synthesized proteins. *Nature* 404:774–778.

50 Orlowski, M. and S. Wilk. 2003. Ubiquitin-independent proteolytic functions of the proteasome. *Arch. Biochem. Biophys.* 415:1–5.

51 Yewdell, J. W., E. Reits, and J. Neefjes. 2003. Making sense of mass destruction: quantitating MHC class I antigen presentation. *Nat. Rev. Immunol.* 3:952–961.

52 Yewdell, J. W. and J. R. Bennink. 1999. Immunodominance in major histocompatibility complex class I-restricted T lymphocyte responses. *Annu. Rev. Immunol.* 17:51–88.

53 Wilson, N. S. and J. A. Villadangos. 2005. Regulation of antigen presentation and cross-presentation in the dendritic cell network: facts, hypothesis, and immunological implications. *Adv. Immunol.* 86:241–305.

III
Effect of the T Cell Repertoire on Dominance

Immunodominance: The Choice of the Immune System. Edited by Jeffrey A. Frelinger

ISBN: 3-527-31274-9

5
Regulation of Early T-Cell Development in the Thymus

Thomas M. Schmitt and Juan Carlos Zúñiga-Pflücker

5.1
Introduction

The thymus is a primary lymphoid organ that is essential for the generation of functional, self-tolerant T lymphocytes. The role of the thymus in immunity was not established until the early 1960s, when Jacques Miller discovered that neonatally thymectomized mice were highly susceptible to infection and could not reject allogeneic skin grafts [1]. Subsequent studies demonstrated that the thymus was essential for the generation of a new type of lymphocyte—the T lymphocyte, which was found to be essential for immunological tolerance and for most immune responses. [1]. Ever since these early studies, researchers have sought to elucidate how the thymus regulates T-cell development. In this chapter, we will describe the mechanisms involved in T-cell commitment and early T-cell differentiation in the thymus and discuss recent advances in our understanding of thymus-derived factors that regulate this complex process.

The thymus support structure is primarily composed of epithelial cells derived from the third pharyngeal pouch during embryogenesis and neural crest-derived mesenchymal cells [2, 3]. Histologically, the adult thymus can be separated into two distinct regions, the cortex and the medulla. Each of these regions contains distinct epithelial cell types, which mediate different aspects of thymocyte development. Consistent with this notion, cortical epithelial cells support early T-cell commitment and regulate the differentiation and ordered migration of immature thymocytes [4]. Cortical epithelial cells also appear to play an important role in positive selection [4, 5]. On the other hand, medullary epithelial cells are important for inducing self-tolerance through negative selection and support the maturation of CD4 and CD8 single-positive (SP) thymocytes [2, 6].

Mice that have genetic defects that disrupt the development of thymic epithelial cells are largely T cell–deficient. For example, the product of the *nude* gene locus, originally identified as Whn [7], and renamed Foxn1 [8], is known to play a critical role during thymus organogenesis [7, 9, 10]. In fact, the thymus of *nude* mice consists only of a cystic rudiment that is unable to support normal T-cell development, thus resulting in a near absence of peripheral T cells. Other factors that

Immunodominance: The Choice of the Immune System. Edited by Jeffrey A. Frelinger

ISBN: 3-527-31274-9

influence the development of thymic epithelial cells include fibroblast growth factors (FGFs) [11, 12] and the transcription factors HoxA3, Pax-1, and Pax-9 [13–15]. A loss of function in any of these factors interferes with the formation of a functional thymus microenvironment.

Although the precise nature of the hematopoietic progenitor cells (HPCs) that enter the thymus is not fully understood, it is clear that these cells undergo a well-defined program of differentiation. Thymocytes can be broadly characterized by the expression of CD4 and CD8, which have served to phenotypically define specific developmental stages. Early T-cell progenitors are CD4 and CD8 double negative (DN), while more mature T-cell precursors are CD4 and CD8 double positive (DP). Following T-cell receptor (TCR)-mediated selection events, thymocytes downregulate the expression of either CD4 or CD8 to become CD8 or CD4 SP T cells, respectively.

During the early DN stage of T-cell development, progenitor thymocytes commit to the T-cell lineage, proliferate, and undergo TCR gene rearrangements in response to developmental cues provided primarily by thymic epithelial cells. These early events occur prior to the expression of a mature $\alpha\beta$– or $\gamma\delta$ TCR and are therefore TCR independent. However, these early stages can be influenced by the presence of more mature TCR^{+} developmental intermediates [16]. During the DP and SP stages of thymocyte development, the T-cell repertoire is selected based on the antigen specificity of developing T-cell precursors. This process is regulated primarily by TCR-mediated signals. In this chapter, we will be focusing on the early, DN stages of T-cell development and the thymus-derived factors that regulate this process.

The development of functional T cells in the thymus from bone marrow (BM)-derived HPCs requires the coordinated activity of multiple regulatory factors. Migration of progenitor cells into the thymus requires chemokine-mediated signals, which are also important for the migration of T-cell progenitors within the thymus. Within the thymus environment, other soluble and membrane-bound factors regulate survival, proliferation, and differentiation of early progenitor thymocytes. These include growth factors, cytokines, and morphogens. One family of transmembrane receptors that is emerging as a key regulator of T-cell commitment and differentiation is Notch. Several important advances in this field have recently provided a unified picture of how T-cell development is regulated in the thymus (see Figure 5.1).

5.2
T-Cell Development in the Thymus

The HPCs that give rise to T cells in the thymus are generated in the adult BM or the fetal liver (FL) and migrate to the thymus through the blood. These progenitors have limited self-renewal potential; therefore, the thymus must be continuously seeded with progenitors from the BM throughout adult life [17]. The earliest population of progenitor thymocytes contains lineage-restricted progenitors that

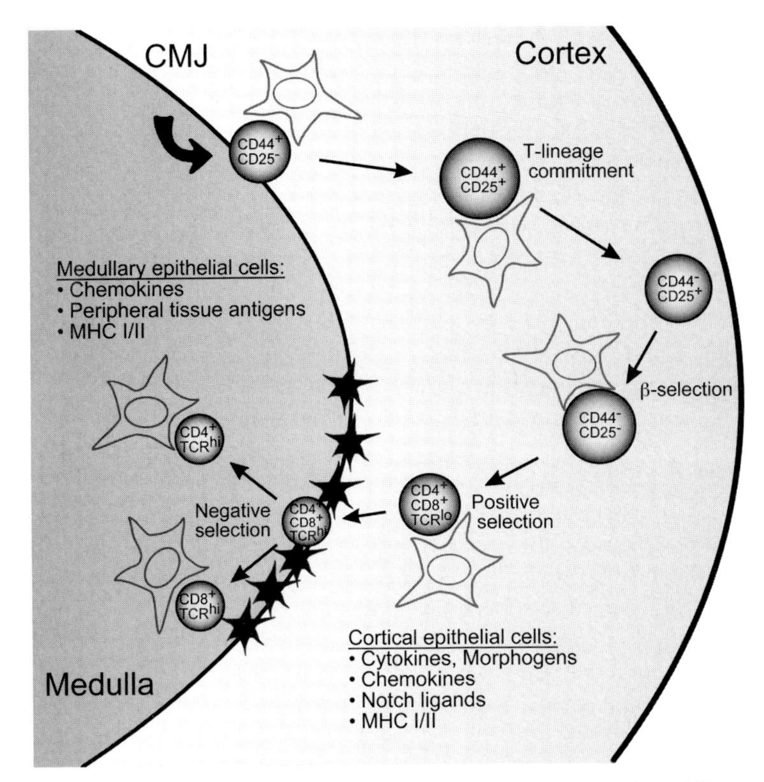

Figure 5.1 Schematic overview of T-cell development in the thymus. Progenitor thymocytes enter the adult thymus at the corticomedullary junction (CMJ) and then migrate towards the outer cortex during the early DN stages of T-cell development in response to several chemokines (see text). During this time, progenitor thymocytes undergo rearrangements at the Tcrb locus. Cells that have generated an in-frame Tcrb gene undergo β-selection by expressing the TCR-β> protein at the cell surface as part of the pre-TCR complex. Following β-selection, thymocytes migrate back towards the CMJ, where, following productive rearrangement of the Tcra gene, cells expressing an $\alpha\beta$ TCR with intermediate affinity for self-MHC–peptide expressed by cortical epithelial cells undergo positive selection. In response to chemokine signals, positively selected thymocytes enter the medulla and develop into CD4 or CD8 SP T cells. SP thymocytes that express a self-reactive $\alpha\beta$ TCR are eliminated by negative selection, which is mediated by DCs near the CMJ and medullary epithelial cells, which can express peripheral tissue antigens.

for the most part appear to be restricted to the T-cell and natural killer (NK)-cell lineages, although this population contains limited B-cell, dendritic cell (DC), and myeloid lineage potential as well [18]. This population is termed DN1 and is characterized by a high level of CD44 surface expression and the lack of lineage differentiation markers, including CD25. In the adult, this population can be further subdivided by CD117 and CD24 surface expression [19], as discussed in more detail below.

In response to thymus-derived factors, DN1 progenitors express CD25 to become DN2 thymocytes. DN2 thymocytes have long been thought to be T-lineage committed, as this population gives rise almost exclusively to T cells when cultured in fetal thymic organ culture (FTOC) or when injected into recipient animals [20, 21]. However, a number of studies have reported that DN2 cells have some potential to differentiate into alternate cell lineages. In particular, this population appears to retain some DC potential [22–24] as well as some NK-cell lineage potential [20, 25, 26]. Indeed, we have recently demonstrated that a substantial number of DN2 thymocytes retain NK-cell potential [26]. This potential is generally suppressed in the thymus; however, these studies demonstrate that many DN2 thymocytes likely retain some NK and perhaps DC potential if cultured under permissive conditions. At the DN2 stage, developing thymocytes express IL-7Rα and undergo a burst of IL-7-induced proliferation [27], followed by the onset of TCR gene rearrangements at the *Tcrb*, *Tcrg*, and *Tcrd* loci. TCR gene rearrangement coincides with the downregulation of CD44 expression, marking the progression to the DN3 stage.

DN3 thymocytes ($CD25^{+}CD44^{-}$) are committed to the T-cell lineage and therefore cannot adopt an alternate cell fate outside of the thymic microenvironment. At this stage of development, thymocytes complete TCR gene rearrangements at the *Tcrb*, *Tcrg*, and *Tcrd* loci. Cells that produce in-frame TCR-γ and TCR-δ gene rearrangements can adopt the $\gamma\delta$ T-cell fate at this point, through a process that remains not fully understood [28]. Cells that produce an in-frame rearrangement at the *Tcrb* locus express the TCR-β protein at the cell surface in a pre-TCR complex composed of CD3 chains and the invariant pTα protein. Expression of the pre-TCR at the cell surface of DN3 cells signals survival, proliferation, and developmental progression to the $CD25^{-}CD44^{-}$ DN4 stage. This process is termed β-selection and is dependent on a signaling competent pre-TCR complex [29]. Thymocytes from mice that are deficient for either of the two *Rag* genes [30, 31] or for the *pTα* gene [32] or that lack signaling components downstream of the TCR/pre-TCR (such as $p56^{lck}$ [33], Syk/ZAP-70 [34], LAT [35], or SLP76 [36]) are all developmentally blocked at the DN3 stage. Following β-selection, thymocytes begin to rearrange the *Tcra* locus and upregulate CD4 and CD8 to become DP T cells. Interestingly, no further exogenous signals are required for DN4 thymocytes to become DPs. Therefore, this population is also referred to as pre-DP [37].

DP thymocytes undergo selection events based on TCR antigen specificity, as discussed at length elsewhere in this volume. DP thymocytes that express an $\alpha\beta$ TCR capable of low-affinity interactions with self-peptide in the context of major histocompatibility complex (MHC) proteins on the surface of cortical thymic epithelial cells undergo positive selection, while those that express a TCR that cannot recognize self-peptide in the context of MHC are eliminated by a process termed "death by neglect." DP T cells that are capable of recognizing self-antigen in the context of MHC class I become CD8 single-positive (SP) cells, while those that recognize self-peptides in the context of MHC class II become CD4 SP cells. The mechanism by which a DP T cell distinguishes between a MHC class I–restricted TCR and a MHC class II–restricted TCR hinges on the duration and strength of TCR signaling at this stage of development [38–40]. Positively selected thymocytes then enter the thymic medulla, where they are subjected to negative selection—a process in which potentially autoreactive thymocytes, which express TCRs that have a high affinity for self-peptide/MHC complexes, are eliminated. T cells that successfully navigate negative selection undergo a poorly defined maturation process in the medulla before emigrating from the thymus as functional T cells.

5.2.1 Early T-Cell Progenitors

It is not clear whether hematopoietic stem cells (HSCs) in the BM or FL receive a specific signal that promotes the development of a lineage-restricted progenitor that then actively emigrates into the blood, and from the blood into the thymus. However, the notion that HSCs receive a direct signal, perhaps through Notch, that induces emigration from the BM is a compelling one. On the other hand, it is known that the thymic stroma plays an active role in the recruitment of progenitor cells from the blood into the thymus and that putative T-cell progenitors express surface receptors that make them susceptible to thymus-derived signals. For example, $\beta 1$ integrin and $\alpha 4$ integrin appear to play a role in progenitor homing, as HPCs deficient for either of these integrins failed to home to the adult thymus [41–43]. Likewise, antibodies specific for either $\alpha 6$ integrin or CD44 both blocked progenitor homing to the thymus when injected *in vivo* into mice [44, 45]. The active recruitment of HPCs from the blood into the thymus is likely mediated at least in part by chemokines. This notion is supported by the fact that blockade of pertussis toxin–sensitive heterotrimeric G protein signaling was shown to block the migration of progenitor cells towards fetal thymic lobes *in vitro* [46]. Indeed, the chemokines CCL21 and, perhaps, CCL25 appear to play a role in this process, as neutralizing antibodies against these molecules decreased fetal thymus colonization [47].

The nature of the progenitor cell that migrates from the BM or FL to the thymus also remains unresolved. As discussed above, the earliest population of progenitor thymocytes contains T-cell, B-cell, and NK-cell lineage potential [48]. Therefore, it has been suggested that the thymus is seeded by a common lymphoid progenitor (CLP) [49–51]. This CLP, or early thymocyte progenitor (ETP), is thought to lose B-lineage potential shortly after entering the thymus in response to thymic stroma–derived factors such as Notch [49]. It has also been argued that the early Lin^- $\mathrm{Sca\text{-}1}^+$ $\mathrm{CD117}^+$ progenitor is the BM-derived progenitor that seeds the thymus [52]. This notion is supported by the observation that migrant lymphoid precursors in the perithymic mesenchyme of day-12 embryos resemble putative lymphoid progenitors in the FL [49].

On the other hand, a number of studies have reported the identification of lineage-restricted progenitors in both the FL and fetal blood that lack B-cell potential but that could develop into T- or NK-cell lineages [53–57]. Furthermore, multiple studies from our lab and others have demonstrated that T/NK bipotent progenitors are prevalent within the earliest population of progenitor thymocytes [25, 26,

58, 59], suggesting that the progenitor that seeds the thymus is largely restricted to the T, NK, and perhaps thymic DC lineages. In the adult thymus, the DN1 population can be further subdivided by CD117 and CD24 [19]. In this study, only the CD117^{hi} CD24$^{-/lo}$ subset was capable of giving rise to T cells with the conventional kinetics and proliferation expected from canonical T-cell progenitors. In contrast, B-lineage potential was restricted to the CD117$^{-/lo}$ CD24^{+} subset. This population was not characteristic of multipotent progenitors, which express high levels of CD117, and lacked the potential for robust T-cell development. These data support the notion that the majority of T cells may not develop from a CLP in the thymus.

5.2.2
Thymocyte Migration

Blood-borne progenitor cells enter the thymus at the junction between the cortex and the medulla (the corticomedullary junction). Through the early DN stages, thymocytes migrate towards the outer cortex under the influence of several chemokines. In particular, CXCL12 (SDF-1), CCL19, and CCL21 appear to be essential for this process. Thymocytes deficient for CXCR4, the receptor for CXCL12 (SDF-1), have several defects in early thymocyte development [60, 61]. Interestingly, CXCR4-deficient progenitor thymocytes fail to migrate into the cortex from the corticomedullary junction and as a result are blocked at the DN1 stage of T-cell development [61]. Likewise, CCR7-deficient mice have been used to demonstrate a similar role for CXCL19 and CXCL21 at the early DN stages of T-cell development [62]. Furthermore, CCR9-dependant signals appear to be required for the localization of DN3 cells at the subcapsular zone prior to β-selection of DN3 thymocytes [63].

Following β-selection, DN thymocytes migrate back through the thymic cortex as they progress to the DN4 stage, rearrange *Tcra* genes, and develop into DP T cells. Early DP thymocytes undergo positive selection through interactions with self-peptide–MHC complexes expressed by cortical epithelial cells near the corticomedullary junction. The TCR signals associated with positive selection induce several changes in integrin and chemokine receptor gene expression, signaling positively selected thymocytes to migrate into the medulla. In particular, the chemokine receptors CCR7 [64] and CCR9 [65, 66] are upregulated by DP T cells following TCR stimulation. Importantly, the chemokine ligands for CCR7, CXCL19, and CXCL21 are expressed primarily by medullary epithelial cells in the thymus [67]. Utilizing CCR7-deficient mice, Takahama and colleagues made use of CCR7-deficient mice to demonstrate that signaling through CCR7 is essential for migration into the thymic medulla following positive selection [64]. This notion is supported by another study, in which ectopic expression of CCR7 in thymocytes prior to positive selection resulted in the localization of pre-selection DP thymocytes in the medulla of transgenic mice [68]. Taken together, these results make it clear that chemokine-directed movement within the thymus ensures that developing thymocytes are positioned at the appropriate microenvironments to support specific stages of T-cell differentiation.

5.2.3
Factors Regulating T-cell Development

The ordered migration of T-cell progenitors through distinct thymic niches suggests that a unique set of developmental cues is required at each successive stage of T-cell development. Indeed, many soluble and membrane-bound factors have been found to influence specific stages of T-cell development.

The thymus produces a large number of cytokines and growth factors that can influence T-cell development; yet, in most cases targeted deletion of these factors failed to reveal a non-redundant role in T lymphopoiesis [69]. Two cytokines that have been found to be essential for normal T-cell development are IL-7 and stem cell factor (SCF) [27, 70, 71]. Mice that are deficient for either SCF or its receptor, CD117 (c-kit), are anemic and have a severe reduction in most hematopoietic cell lineages. Although T cells developed in CD117-deficient animals, the DN1 population was reduced by 40-fold [72], indicating a progenitor defect in these mice. A similar defect was observed when SCF-deficient fetal thymic lobes were transplanted into wild-type recipient mice [73], indicating that SCF produced in the thymus is necessary for T-cell progenitor expansion.

A factor closely related to SCF, which has also been implicated in T-cell development and progenitor expansion, is FMS-related tyrosine kinase 3 ligand (Flt3-L) [74]. The ligand for Flt3-L is FMS-related tyrosine kinase 3 (Flt3), or, as it is more commonly referred to, fetal liver kinase-2 (Flk-2/CD135). CD135 is a receptor tyrosine kinase, similar to CD117 [75]. Flt3-L is produced by the thymic stroma, and it promotes the proliferation of both DN1 and DN2 thymocytes [76–78]. However, no overt defects in T-cell development are observed from mice that are deficient for either Flt3-L or CD135; suggesting that other factors, such as SCF or IL-7, can compensate in the thymus.

IL-7-mediated signaling is also required for the survival and proliferation of early progenitor thymocytes. Mice that have a targeted deletion of the gene that encodes IL-7, CD127 (IL-7Rα), or CD132 (the common cytokine receptor γ-chain) have a severe reduction in total thymocyte numbers and display a significant developmental block at the DN2 stage, in part because of a defect in IL-7-induced cell survival [27, 79]. Indeed, an important function of IL-7 signaling is to promote survival through the upregulation of Bcl-2. Normal $\alpha\beta$ T-cell development can be rescued in IL-7-deficient mice by the overexpression of Bcl-2 [80, 81], suggesting that this is the primary function of IL-7 signaling during $\alpha\beta$ T-cell development. Signals through IL-7 also play an essential role in the development of the $\gamma\delta$ T-cell lineage. Mice that lack any of the necessary components of IL-7 signaling fail to develop $\gamma\delta$ T cells [82–84]. Interestingly, Bcl-2 expression cannot rescue $\gamma\delta$ T-cell development in IL-7-deficient mice [80].

A group of molecules collectively referred to as morphogens have recently been implicated in T lymphopoiesis. Morphogens play an important role in embryonic patterning, where morphogen gradients influence cell fate decisions in a concentration-dependent manner [85]. This group of molecules includes bone morphogenetic proteins (BMPs), WNT proteins, and Hedgehog proteins. Several members of these different families of signaling molecules have been detected in the thymus, and the role of these proteins in T-cell development is starting to be understood.

BMPs and fibroblast growth factors (FGFs) are known to influence the development of the thymic stroma [11, 12, 86]. However, these molecules appear to regulate thymocyte development as well, by negatively impacting T lymphopoiesis. In particular, exogenous BMPs and FGFs both block T-cell development in FTOC prior to T-cell lineage commitment at the DN1 stage, and they inhibit further differentiation of committed T-cell progenitors [87–89]. However, it is not clear how these molecules act, and at least some of the observed effects may be the result of changes occurring in the thymic stroma [89].

The role of WNT proteins in T-cell development is more established. WNT proteins signal through the Frizzled family of transmembrane receptors, resulting in the stabilization of β-catenin, which then forms a complex with T-cell factor (TCF) family members that can activate transcription of target genes [90, 91]. Several WNT proteins are expressed by thymic stroma, including WNTs 4, 7a, 7b, 10a, and 10b [92]. WNT4 is expressed at the highest levels, which is consistent with reports that WNT4 can increase thymic cellularity [93, 94]. Two members of the TCF family, TCF-1 and lymphocyte enhancer factor-1 (LEF-1), are expressed in thymocytes and have been implicated at multiple stages of T-cell development [93, 95–100]. Mice deficient for TCF-1 have a cell autonomous defect in early DN thymocyte proliferation during fetal development [98]. However, adult animals exhibit a complete block at the DN1 stage [97], and TCF-1-deficient BM progenitors cannot reconstitute T cells in either irradiated recipient mice or FTOC [93]. These data indicate that TCF-1 is essential for the earliest stages of T-cell development. It has been proposed that the effect of TCF-1 deficiency is less severe during fetal development because of redundancy through LEF-1 [96]. TCF-1 also functions in DP thymocytes, where it is required for developmental progression following TCR stimulation [93, 99]. Interestingly, a constitutively active β-catenin transgene was found to allow thymocyte differentiation to the DP and SP stage in the absence of pre-TCR and TCR signals, respectively [95].

Another family of morphogens that has recently been implicated in T-cell development is the Hedgehog family. There are three mammalian Hedgehog family members: Sonic Hedgehog (Shh), Indian Hedgehog (Ihh), and Desert Hedgehog (Dhh) [101]. However, only Shh and Ihh are expressed in the thymus [102]. Shh-deficient embryos have a significant decrease in thymic cellularity. This appears to result from a defect in the proliferative capacity of DN1 and DN2 thymocytes [103]. Shh-deficient embryos also have a partial block at the DN-to-DP transition. However, the precise role of Shh at this stage of development is unclear, because a high level of exogenous Shh also results in a developmental block at this stage [102, 103].

5.2.4 Notch and T-cell Development

A number of seminal papers have been published in the last five years that address the role of Notch signaling during T-cell development. The Notch pathway plays an important role in regulating cell fate decisions in a surprisingly large number of cell types during ontogeny. However, the fundamental role that Notch plays in T-cell development has only recently been appreciated. It is now clear that the Notch pathway is essential for the commitment of HPCs to the T-cell lineage and that it plays a critical role in the subsequent differentiation of committed T-cell progenitors within the thymus.

In mammals, there are four Notch receptors and five Notch ligands. The ligands consist of three delta-like ligands, Delta-like 1 (Dll1), Delta-like 3 (Dll3), and Delta-like 4 (Dll4), and two jagged ligands, Jagged1 and Jagged2. Notch receptor–ligand interactions initiate a series of proteolytic cleavage events within the transmembrane and intracellular regions of the Notch receptor, which results in the release of the intracellular domain of Notch (NotchIC) [104–107]. The soluble NotchIC then translocates to the nucleus, where it interacts with the transcriptional repressor CBF-1 (CSL/RBP-J) [108], which is bound to the regulatory regions of Notch target genes. NotchIC converts CBF-1 from a repressor to a transcriptional activator [109], thus inducing the expression of target genes.

The first indication that Notch might play a role in T-cell development came in the early 1990s, when it was discovered that human T lymphoblastic leukemias contain a t(7;9)(q34;q34.3) chromosomal translocation, which juxtaposes the *Tcrb* gene promoter with the C-terminal region of the intracellular domain of Notch1 [110, 111], resulting in the high level of expression of a constitutively active Notch molecule in T cells. In 1996, a report from Robey et al. demonstrated that the constitutively active NotchIC could influence CD4 versus CD8 T-cell lineage commitment, favoring the differentiation of DP T cells towards the CD8 T-cell lineage [112]. This finding was followed by several papers arguing either for or against a role for Notch signaling at this stage of T-cell development [113–117]. In particular, Radtke et al. did not observe any change in the numbers of CD4 versus CD8 T cells in mice containing a conditional deletion of the *Notch1* gene [117], although other Notch family members may be involved. To date, the physiological role of Notch signaling at the DP stage remains unresolved.

In 1999, a report by Radtke et al. described a conditional knockout of Notch1 using an interferon-inducible Cre-lox system [118]. Surprisingly, Notch1-deficient progenitors failed to give rise to T cells and instead developed into B cells in the thymus of recipient mice. Later that year, Pear's group reported that transduced BM cells with NotchIC, which were then injected into mice, failed to differentiate into B cells and instead gave rise to DP T cells in the BM of recipient mice [119]. Taken together, these results established that Notch could influence T-cell versus B-cell fate determination and led to a closer analysis of the role played by Notch during T-cell commitment and early differentiation.

The findings of these initial reports suggest that Notch signals regulate a binary cell fate choice between the T-cell and B-cell lineages. However, subsequent reports demonstrated that Notch signaling acts continuously during early thymocyte development to induce progenitor cells to adopt the T-cell lineage. Several studies have shown that low levels of Notch signaling can extinguish B-lineage potential, either at the DN1 stage or prior to entry into the thymus [26, 120–122].

However, early progenitor thymocytes that do not receive continuous Notch signals cannot commit to the T-cell lineage and retain the ability to adopt alternate cell fates [26, 123]. Indeed, fetal DN2 thymocytes have been shown to possess substantial NK-lineage potential; however, NK cells develop only when DN2 cells are cultured in the absence of Notch signals [26]. These findings suggest that continuous Notch signaling maintains T-lineage specification at the DN2 stage of T-cell development and indicate that the Notch-induced loss of B-lineage potential is a temporally separate event from Notch-induced commitment to the T-cell lineage.

Notch signaling is essential for subsequent stages of early T-cell development in addition to the critical role it plays in T-lineage commitment. For example, Cre-mediated deletion of Notch1 at the DN2 stage revealed an essential role for Notch in *Tcrb* gene rearrangement [124]. Furthermore, our lab has demonstrated a requirement for Notch at the β-selection checkpoint, which is independent of the role that Notch plays in functional *Tcrb* gene rearrangement [125]. Indeed, we showed that Notch receptor–ligand interactions act in concert with pre-TCR-mediated signals to induce cell proliferation, differentiation, and survival at this stage of development. Furthermore, a report by Bevan et al. demonstrated that Notch1-IC could augment the progression of $Rag^{-/-}$ DN3 cells to the DP stage in response to CD3 engagement [126]. These results are consistent with the finding that Notch-dependent cellular transformation is dependent on the formation of a functional pre-TCR complex [127].

Several reports have focused on the role of Notch in regulating the $\alpha\beta$- versus $\gamma\delta$ T-cell fate decision. The first paper to implicate Notch signaling at this branch point in T-cell development was from Washburn et al. in 1997. In this report, mice that were heterozygous for Notch1 were found to have an increase in $\gamma\delta$ T-cell development, while ectopic expression of NotchIC led to a decrease in $\gamma\delta$ T cells and an increase in $\alpha\beta$T-cell development [128]. Likewise, mice bearing a conditional deletion of CBF-1 in T cells were found to have an increase in $\gamma\delta$ T-cell development [129]. However, in some experimental systems Notch appears to have the opposite effect. When human T-cell progenitors expressing Notch1-IC were cultured in FTOC, $\gamma\delta$ T cells developed at the expense of $\alpha\beta$ T cells [130]. In addition, Jagged2-deficient mice have a defect in $\gamma\delta$ T-cell development [131]. Thus, the precise role of Notch in regulating $\gamma\delta$ T-cell commitment remains to be elucidated.

5.3
In Vitro *T-cell Development*

Interestingly, despite our extensive knowledge of the factors that influence T lymphopoiesis, attempts to recapitulate T-cell development outside of an intact thymus have failed until recently. Thus, the prevailing dogma had long been that T-cell development requires multiple factors acting in the context of a three-dimensional thymic microenvironment [4]. As discussed above, many recent studies demonstrated an important role for Notch during T-lineage commitment and at multiple stages of T-cell development. Therefore, we hypothesized that the ability to induce strong Notch signals through abundantly expressed Notch ligands might be an important characteristic of the thymus that sets it apart from other hematopoietic environments, such as the BM. In order to test this hypothesis, the bone marrow stromal cell line OP9 was transduced to express the Notch ligand Dll1 [132]. The resulting OP9–DL1 cell line lost the ability to support B-cell development and gained the unique capacity to induce T-lineage commitment and differentiation from HSCs on a simple cell monolayer. Importantly, the OP9–DL1 system fully recapitulated normal T-cell differentiation in the thymus and supported the development of a diverse repertoire of T cells possessing unique antigen receptor specificities. This system was also adopted to support the *in vitro* generation of T cells from embryonic stem cells [133], which were shown to express multiple T lineage–associated genes when cultured in the OP9–DL1 microenvironment. Embryonic stem cell–derived T-cell progenitors were shown to effectively reconstitute the T-cell compartment of immunodeficient mice and were capable of generating an antigen-specific immune response to a viral challenge.

Taken together, these results demonstrate that a single gene, Delta-like 1, can transform a bone marrow microenvironment that strongly supports B-cell development into a thymus-like environment that strongly supports T-cell development. This finding suggests that the expression of Delta-like ligands underpins the unique ability of the thymus to promote T-cell lineage commitment and differentiation.

5.4 Concluding Remarks

The process of T-cell development is tightly regulated by multiple factors that likely act within distinct regions of the thymus. Recent studies implicating the WNT pathway, BMPs, Hedgehog proteins, and other soluble factors in T-cell development have clarified how the thymus regulates T-cell proliferation and differentiation. Furthermore, it has become clear that the Notch pathway plays a fundamental role in orchestrating T-cell commitment and differentiation. However, the interplay between Notch signals and those derived from the other factors produced in the thymus remains to be addressed. Indeed, many aspects of how the thymus controls the generation of the various lineages of T lymphocytes remain elusive. We remain hopeful that advances such as our *in vitro* system for T-cell development will facilitate future studies in this field and help to elucidate the mysteries of the thymic microenvironment.

References

1 Miller, J.F. 2002. The discovery of thymus function and of thymus-derived lymphocytes. *Immunol Rev* 185:7–14.

2 Anderson, G., and E.J. Jenkinson. 2001. Lymphostromal interactions in thymic development and function. *Nat Rev Immunol* 1:31–40.

3 Gordon, J., V.A. Wilson, N.F. Blair, J. Sheridan, A. Farley, L. Wilson, N.R. Manley, and C.C. Blackburn. 2004. Functional evidence for a single endodermal origin for the thymic epithelium. *Nat Immunol* 5:546–553.

4 Lind, E.F., S.E. Prockop, H.E. Porritt, and H.T. Petrie. 2001. Mapping precursor movement through the postnatal thymus reveals specific microenvironments supporting defined stages of early lymphoid development. *J Exp Med* 194:127–134.

5 Anderson, G., J.J. Owen, N.C. Moore, and E.J. Jenkinson. 1994. Thymic epithelial cells provide unique signals for positive selection of CD4+CD8+ thymocytes in vitro. *J Exp Med* 179:2027–2031.

6 Derbinski, J., A. Schulte, B. Kyewski, and L. Klein. 2001. Promiscuous gene expression in medullary thymic epithelial cells mirrors the peripheral self. *Nat Immunol* 2:1032–1039.

7 Nehls, M., D. Pfeifer, M. Schorpp, H. Hedrich, and T. Boehm. 1994. New member of the winged-helix protein family disrupted in mouse and rat nude mutations. *Nature* 372:103–107.

8 Kaestner, K.H., W. Knochel, and D.E. Martinez. 2000. Unified nomenclature for the winged helix/forkhead transcription factors. *Genes Dev* 14:142–146.

9 Holub, M., P. Rossmann, H. Tlaskalova, and H. Vidmarova. 1975. Thymus rudiment of the athymic nude mouse. *Nature* 256:491–493.

10 Wortis, H.H. 1971. Immunological responses of 'nude' mice. *Clin Exp Immunol* 8:305–317.

11 Celli, G., W.J. LaRochelle, S. Mackem, R. Sharp, and G. Merlino. 1998. Soluble dominant-negative receptor uncovers essential roles for fibroblast growth factors in multi-organ induction and patterning. *Embo J* 17:1642–1655.

12 Revest, J.M., R.K. Suniara, K. Kerr, J.J. Owen, and C. Dickson. 2001. Development of the thymus requires signaling through the fibroblast growth factor receptor R2-IIIb. *J Immunol* 167:1954–1961.

13 Manley, N.R., and M.R. Capecchi. 1995. The role of Hoxa-3 in mouse thymus and thyroid development. *Development* 121:1989–2003.

14 Peters, H., A. Neubuser, K. Kratochwil, and R. Balling. 1998. Pax9-deficient mice lack pharyngeal pouch derivatives and teeth and exhibit craniofacial and limb abnormalities. *Genes Dev* 12:2735–2747.

15 Wallin, J., H. Eibel, A. Neubuser, J. Wilting, H. Koseki, and R. Balling. 1996. Pax1 is expressed during development of the thymus epithelium and is required for normal T-cell maturation. *Development* 122:23–30.

16 Silva-Santos, B., D.J. Pennington, and A.C. Hayday. 2005. Lymphotoxin-mediated regulation of gammadelta cell differentiation by alphabeta T cell progenitors. *Science* 307:925–928.

17 Scollay, R., J. Smith, and V. Stauffer. 1986. Dynamics of early T cells: prothymocyte migration and proliferation in the adult mouse thymus. *Immunol Rev* 91:129–157.

18 Katsura, Y. 2002. Redefinition of lymphoid progenitors. *Nat Rev Immunol* 2:127–132.

19 Porritt, H.E., L.L. Rumfelt, S. Tabrizifard, T.M. Schmitt, J.C. Zúñiga-Pflücker, and H.T. Petrie. 2004. Heterogeneity among DN1 prothymocytes reveals multiple progenitors with different capacities to generate T cell and non T cell lineages. *Immunity* 20:735–745.

20 Moore, T.A., and A. Zlotnik. 1995. T-cell lineage commitment and cytokine response of thymic progenitors. *Blood* 86:1850–1860.

21 Zúñiga-Pflücker, J.C., J. Di, and M.J. Lenardo. 1995. Requirement for TNF-alpha and IL-1 alpha in fetal thymocyte commitment and differentiation. *Science* 268:1906–1909.

22 Ardavin, C., L. Wu, C.L. Li, and K. Shortman. 1993. Thymic dendritic cells and T cells develop simultaneously in the thymus from a common precursor population. *Nature* 362:761–763.

23 Shen, H.Q., M. Lu, T. Ikawa, K. Masuda, K. Ohmura, N. Minato, Y. Katsura, and H. Kawamoto. 2003. T/NK bipotent progenitors in the thymus retain the potential to generate dendritic cells. *J Immunol* 171:3401–3406.

24 Wu, L., C.L. Li, and K. Shortman. 1996. Thymic dendritic cell precursors: relationship to the T lymphocyte lineage and phenotype of the dendritic cell progeny. *J Exp Med* 184:903–911.

25 Ikawa, T., H. Kawamoto, S. Fujimoto, and Y. Katsura. 1999. Commitment of common T/Natural killer (NK) progenitors to unipotent T and NK progenitors in the murine fetal thymus revealed by a single progenitor assay. *J Exp Med* 190:1617–1626.

26 Schmitt, T.M., M. Ciofani, H.T. Petrie, and J.C. Zúñiga-Pflücker. 2004. Maintenance of T cell specification and differentiation require recurrent Notch receptor-ligand interactions. *J Exp Med* 200:469–479.

27 von Freeden-Jeffry, U., P. Vieira, L.A. Lucian, T. McNeil, S.E. Burdach, and R. Murray. 1995. Lymphopenia in interleukin (IL)-7 gene-deleted mice identifies IL-7 as a nonredundant cytokine. *J Exp Med* 181:1519–1526.

28 Hayday, A.C., D.F. Barber, N. Douglas, and E.S. Hoffman. 1999. Signals involved in gamma/delta T cell versus alpha/beta T cell lineage commitment. *Semin Immunol* 11:239–249.

29 Michie, A.M., and J.C. Zúñiga-Pflücker. 2002. Regulation of thymocyte differentiation: pre-TCR signals and beta-selection. *Semin Immunol* 14:311–323.

30 Mombaerts, P., J. Iacomini, R.S. Johnson, K. Herrup, S. Tonegawa, and V.E. Papaioannou. 1992. RAG-1-deficient mice have no mature B and T lymphocytes. *Cell* 68:869–877.

31 Shinkai, Y., G. Rathbun, K.-P. Lam, E.M. Oltz, V. Stewart, M. Mendelsohn, J. Charron, M. Datta, F. Young, A.M. Stall, and F.W. Alt. 1992. RAG-2-deficient mice lack mature lymphocytes owing to inability to initiate V(D)J rearrangement. *Cell* 68:855–867.

32 Fehling, H.J., A. Krotkova, C. Saint-Ruf, and H. von Boehmer. 1995. Crucial role of the pre-T-cell receptor alpha gene in development of alpha beta but not gamma delta T cells. *Nature* 375:795–798.

33 Molina, T.J., K. Kishihara, D.P. Siderovski, W. van Ewijk, A. Narendran, E. Timms, A. Wakeham, C.J. Paige, K.U. Hartmann, A. Veillette, D. Davidson, and T.W. Mak. 1992. Profound block in thymocyte development in mice lacking p56lck. *Nature* 357:161–164.

34 Cheng, A.M., I. Negishi, S.J. Anderson, A.C. Chan, J. Bolen, D.Y. Loh, and T. Pawson. 1997. The Syk and ZAP-70 SH2-containing tyrosine kinases are implicated in pre T cell receptor signaling. *Proc Natl Acad Sci U S A* 94:9797–9801.

35 Zhang, W., C.L. Sommers, D.N. Burshtyn, C.C. Stebbins, J.B. DeJarnette, R.P. Trible, A. Grinberg, H.C. Tsay, H.M. Jacobs, C.M. Kessler, E.O. Long, P.E. Love, and L.E. Samelson. 1999. Essential role of LAT in T cell development. *Immunity* 10:323–332.

36 Pivniouk, V., E. Tsitsikov, P. Swinton, G. Rathbun, F.W. Alt, and R.S. Geha. 1998. Impaired viability and profound block in thymocyte development in mice lacking the adaptor protein SLP-76. *Cell* 94:229–238.

37 Petrie, H.T., P. Hugo, R. Scollay, and K. Shortman. 1990. Lineage relationships and developmental kinetics of immature thymocytes: CD3, CD4, and CD8 acquisition in vivo and in vitro. *J. Exp. Med.* 172:1583–1588.

38 Hernandez-Hoyos, G., S.J. Sohn, E.V. Rothenberg, and J. Alberola-Ila. 2000. Lck activity controls CD4/CD8 T cell lineage commitment. *Immunity* 12:313–322.

39 Liu, X., and R. Bosselut. 2004. Duration of TCR signaling controls CD4-CD8 lineage differentiation in vivo. *Nat Immunol* 5:280–288.

40 Yasutomo, K., C. Doyle, L. Miele, C. Fuchs, and R.N. Germain. 2000. The duration of antigen receptor signalling determines CD4+ versus CD8+ T-cell lineage fate. *Nature* 404:506–510.

41 Arroyo, A.G., D. Taverna, C.A. Whittaker, U.G. Strauch, B.L. Bader, H. Rayburn, D. Crowley, C.M. Parker, and R.O. Hynes. 2000. In vivo roles of integrins during leukocyte development and traffic: insights from the analysis of mice chimeric for alpha 5, alpha v, and alpha 4 integrins. *J Immunol* 165:4667–4675.

42 Arroyo, A.G., J.T. Yang, H. Rayburn, and R.O. Hynes. 1996. Differential requirements for alpha4 integrins during fetal and adult hematopoiesis. *Cell* 85:997–1008.

43 Hirsch, E., A. Iglesias, A.J. Potocnik, U. Hartmann, and R. Fassler. 1996.

Impaired migration but not differentiation of haematopoietic stem cells in the absence of beta1 integrins. *Nature* 380:171–175.

44 Ruiz, P., M.V. Wiles, and B.A. Imhof. 1995. Alpha 6 integrins participate in pro T cell homing to the thymus. *Eur J Immunol* 25:2034–2041.

45 Wu, L., P.W. Kincade, and K. Shortman. 1993. The CD44 expressed on the earliest intrathymic precursor population functions as a thymus homing molecule but does not bind to hyaluronate. *Immunol Lett* 38:69–75.

46 Wilkinson, B., J.J. Owen, and E.J. Jenkinson. 1999. Factors regulating stem cell recruitment to the fetal thymus. *J Immunol* 162:3873–3881.

47 Liu, C., T. Ueno, S. Kuse, F. Saito, T. Nitta, L. Piali, H. Nakano, T. Kakiuchi, M. Lipp, G.A. Hollander, and Y. Takahama. 2004. Role of CCL21 in recruitment of T precursor cells to fetal thymus. *Blood* 105:31–39.

48 Wu, L., M. Antica, G.R. Johnson, R. Scollay, and K. Shortman. 1991. Developmental potential of the earliest precursor cells from the adult mouse thymus. *J Exp Med* 174:1617–1627.

49 Harman, B.C., E.J. Jenkinson, and G. Anderson. 2003. Entry into the thymic microenvironment triggers Notch activation in the earliest migrant T cell progenitors. *J Immunol* 170:1299–1303.

50 Kondo, M., I.L. Weissman, and K. Akashi. 1997. Identification of clonogenic common lymphoid progenitors in mouse bone marrow. *Cell* 91:661–672.

51 Martin, C.H., I. Aifantis, M.L. Scimone, U.H. von Andrian, B. Reizis, H. von Boehmer, and F. Gounari. 2003. Efficient thymic immigration of B220+ lymphoid-restricted bone marrow cells with T precursor potential. *Nat Immunol* 4:866–873.

52 Allman, D., A. Sambandam, S. Kim, J.P. Miller, A. Pagan, D. Well, A. Meraz, and A. Bhandoola. 2003. Thymopoiesis independent of common lymphoid progenitors. *Nat Immunol* 4:168–174.

53 Carlyle, J.R., and J.C. Zúñiga-Pflücker. 1998. Requirement for the thymus in $\alpha\beta$ T lymphocyte lineage commitment. *Immunity* 9:187–197.

54 Douagi, I., F. Colucci, J.P. Di Santo, and A. Cumano. 2002. Identification of the earliest prethymic bipotent T/NK progenitor in murine fetal liver. *Blood* 99:463–471.

55 Ikawa, T., K. Masuda, M. Lu, N. Minato, Y. Katsura, and H. Kawamoto. 2003. Identification of the earliest prethymic T-cell progenitors in murine fetal blood. *Blood* 103:530–537.

56 Kawamoto, H., T. Ikawa, K. Ohmura, S. Fujimoto, and Y. Katsura. 2000. T cell progenitors emerge earlier than B cell progenitors in the murine fetal liver. *Immunity* 12:441–450.

57 Rodewald, H.-R., K. Kretzschmar, S. Takeda, C. Hohl, and M. Dessing. 1994. Identification of pro-thymocytes in murine fetal blood: T lineage commitment can precede thymus colonization. *EMBO. J.* 13:4229–4240.

58 Carlyle, J.R., A.M. Michie, C. Furlonger, T. Nakano, M.J. Lenardo, C.J. Paige, and J.C. Zúñiga-Pflücker. 1997. Identification of a novel developmental stage marking lineage commitment of progenitor thymocytes. *J Exp Med* 186:173–182.

59 Michie, A.M., J.R. Carlyle, T.M. Schmitt, B. Ljutic, S.K. Cho, Q. Fong, and J.C. Zúñiga-Pflücker. 2000. Clonal characterization of a bipotent T cell and NK cell progenitor in the mouse fetal thymus. *J. Immunol.* 164:1730–1733.

60 Ara, T., M. Itoi, K. Kawabata, T. Egawa, K. Tokoyoda, T. Sugiyama, N. Fujii, T. Amagai, and T. Nagasawa. 2003. A role of CXC chemokine ligand 12/stromal cell-derived factor-1/pre-B cell growth stimulating factor and its receptor CXCR4 in fetal and adult T cell development in vivo. *J Immunol* 170:4649–4655.

61 Plotkin, J., S.E. Prockop, A. Lepique, and H.T. Petrie. 2003. Critical role for CXCR4 signaling in progenitor localization and T cell differentiation in the postnatal thymus. *J Immunol* 171:4521–4527.

62 Misslitz, A., O. Pabst, G. Hintzen, L. Ohl, E. Kremmer, H.T. Petrie, and R. Forster. 2004. Thymic T cell development and progenitor localization depend on CCR7. *J Exp Med* 200:481–491.

63 Benz, C., K. Heinzel, and C.C. Bleul. 2004. Homing of immature thymocytes to the subcapsular microenvironment within the thymus is not an absolute requirement for T cell development. *Eur J Immunol* 34:3652–3663.
64 Ueno, T., F. Saito, D.H. Gray, S. Kuse, K. Hieshima, H. Nakano, T. Kakiuchi, M. Lipp, R.L. Boyd, and Y. Takahama. 2004. CCR7 signals are essential for cortex-medulla migration of developing thymocytes. *J Exp Med* 200:493–505.
65 Norment, A.M., L.Y. Bogatzki, B.N. Gantner, and M.J. Bevan. 2000. Murine CCR9, a chemokine receptor for thymus-expressed chemokine that is up-regulated following pre-TCR signaling. *J Immunol* 164:639–648.
66 Uehara, S., K. Song, J.M. Farber, and P.E. Love. 2002. Characterization of CCR9 expression and CCL25/thymus-expressed chemokine responsiveness during T cell development: CD3(high)CD69+ thymocytes and gammadeltaTCR+ thymocytes preferentially respond to CCL25. *J Immunol* 168:134–142.
67 Ueno, T., K. Hara, M.S. Willis, M.A. Malin, U.E. Hopken, D.H. Gray, K. Matsushima, M. Lipp, T.A. Springer, R.L. Boyd, O. Yoshie, and Y. Takahama. 2002. Role for CCR7 ligands in the emigration of newly generated T lymphocytes from the neonatal thymus. *Immunity* 16:205–218.
68 Kwan, J., and N. Killeen. 2004. CCR7 directs the migration of thymocytes into the thymic medulla. *J Immunol* 172:3999–4007.
69 Zlotnik, A., and T.A. Moore. 1995. Cytokine production and requirements during T-cell development. *Curr Opin Immunol* 7:206–213.
70 Asamoto, H., and T.E. Mandel. 1981. Thymus in mice bearing the Steel mutation. Morphologic studies on fetal, neonatal, organ-cultured, and grafted fetal thymus. *Lab Invest* 45:418–426.
71 Godfrey, D.I., A. Zlotnik, and T. Suda. 1992. Phenotypic and functional characterization of c-kit expression during intrathymic T cell development. *J Immunol* 149:2281–2285.
72 Di Santo, J.P., and H.R. Rodewald. 1998. In vivo roles of receptor tyrosine kinases and cytokine receptors in early thymocyte development. *Curr Opin Immunol* 10:196–207.
73 Rodewald, H.R., K. Kretzschmar, W. Swat, and S. Takeda. 1995. Intrathymically expressed c-kit ligand (stem cell factor) is a major factor driving expansion of very immature thymocytes in vivo. *Immunity* 3:313–319.
74 Lyman, S.D., L. James, T. Vanden Bos, P. de Vries, K. Brasel, B. Gliniak, L.T. Hollingsworth, K.S. Picha, H.J. McKenna, R.R. Splett, F.A. Fletcher, E. Maraskovsky, T. Farrah, D. Foxworthe, D.E. Williams, and M.P. Beckmann. 1993. Molecular cloning of a ligand for the flt3/flk-2 tyrosine kinase receptor: a proliferative factor for primitive hematopoietic cells. *Cell* 75:1157–1167.
75 Matthews, W., C.T. Jordan, G.W. Wiegand, D. Pardoll, and I.R. Lemischka. 1991. A receptor tyrosine kinase specific to hematopoietic stem and progenitor cell-enriched populations. *Cell* 65:1143–1152.
76 Mackarehtschian, K., J.D. Hardin, K.A. Moore, S. Boast, S.P. Goff, and I.R. Lemischka. 1995. Targeted disruption of the flk2/flt3 gene leads to deficiencies in primitive hematopoietic progenitors. *Immunity* 3:147–161.
77 McKenna, H.J., K.L. Stocking, R.E. Miller, K. Brasel, T. De Smedt, E. Maraskovsky, C.R. Maliszewski, D.H. Lynch, J. Smith, B. Pulendran, E.R. Roux, M. Teepe, S.D. Lyman, and J.J. Peschon. 2000. Mice lacking flt3 ligand have deficient hematopoiesis affecting hematopoietic progenitor cells, dendritic cells, and natural killer cells. *Blood* 95:3489–3497.
78 Moore, T.A., and A. Zlotnik. 1997. Differential effects of Flk-2/Flt-3 ligand and stem cell factor on murine thymic progenitor cells. *J Immunol* 158:4187–4192.
79 Maraskovsky, E., L.A. O'Reilly, M. Teepe, L.M. Corcoran, J.J. Peschon, and A. Strasser. 1997. Bcl-2 can rescue T lymphocyte development in interleukin-7 receptor-deficient mice but not in mutant rag-1-/- mice. *Cell* 89:1011–1019.

80 Akashi, K., M. Kondo, U. von Freeden-Jeffry, R. Murray, and I.L. Weissman. 1997. Bcl-2 rescues T lymphopoiesis in interleukin-7 receptor-deficient mice. *Cell* 89:1033–1041.

81 von Freeden-Jeffry, U., N. Solvason, M. Howard, and R. Murray. 1997. The earliest T lineage-committed cells depend on IL-7 for Bcl-2 expression and normal cell cycle progression. *Immunity* 7:147–154.

82 Maki, K., S. Sunaga, Y. Komagata, Y. Kodaira, A. Mabuchi, H. Karasuyama, K. Yokomuro, J.I. Miyazaki, and K. Ikuta. 1996. Interleukin 7 receptor-deficient mice lack gammadelta T cells. *Proc Natl Acad Sci U S A* 93:7172–7177.

83 Malissen, M., P. Pereira, D.J. Gerber, B. Malissen, and J.P. DiSanto. 1997. The common cytokine receptor gamma chain controls survival of gamma/delta T cells. *J Exp Med* 186:1277–1285.

84 Moore, T.A., U. von Freeden-Jeffry, R. Murray, and A. Zlotnik. 1996. Inhibition of gamma delta T cell development and early thymocyte maturation in IL-7 –/– mice. *J Immunol* 157:2366–2373.

85 Tabata, T., and Y. Takei. 2004. Morphogens, their identification and regulation. *Development* 131:703–712.

86 Nosaka, T., S. Morita, H. Kitamura, H. Nakajima, F. Shibata, Y. Morikawa, Y. Kataoka, Y. Ebihara, T. Kawashima, T. Itoh, K. Ozaki, E. Senba, K. Tsuji, F. Makishima, N. Yoshida, and T. Kitamura. 2003. Mammalian twisted gastrulation is essential for skeleto-lymphogenesis. *Mol Cell Biol* 23:2969–2980.

87 Graf, D., S. Nethisinghe, D.B. Palmer, A.G. Fisher, and M. Merkenschlager. 2002. The developmentally regulated expression of Twisted gastrulation reveals a role for bone morphogenetic proteins in the control of T cell development. *J Exp Med* 196:163–171.

88 Hager-Theodorides, A.L., S.V. Outram, D.K. Shah, R. Sacedon, R.E. Shrimpton, A. Vicente, A. Varas, and T. Crompton. 2002. Bone morphogenetic protein 2/4 signaling regulates early thymocyte differentiation. *J Immunol* 169:5496–5504.

89 Tsai, P.T., R.A. Lee, and H. Wu. 2003. BMP4 acts upstream of FGF in modulating thymic stroma and regulating thymopoiesis. *Blood* 102:3947–3953.

90 Behrens, J., J.P. von Kries, M. Kuhl, L. Bruhn, D. Wedlich, R. Grosschedl, and W. Birchmeier. 1996. Functional interaction of beta-catenin with the transcription factor LEF-1. *Nature* 382:638–642.

91 Molenaar, M., M. van de Wetering, M. Oosterwegel, J. Peterson-Maduro, S. Godsave, V. Korinek, J. Roose, O. Destree, and H. Clevers. 1996. XTcf-3 transcription factor mediates beta-catenin-induced axis formation in Xenopus embryos. *Cell* 86:391–399.

92 Pongracz, J., K. Hare, B. Harman, G. Anderson, and E.J. Jenkinson. 2003. Thymic epithelial cells provide WNT signals to developing thymocytes. *Eur J Immunol* 33:1949–1956.

93 Staal, F.J., J. Meeldijk, P. Moerer, P. Jay, B.C. van de Weerdt, S. Vainio, G.P. Nolan, and H. Clevers. 2001. Wnt signaling is required for thymocyte development and activates Tcf-1 mediated transcription. *Eur J Immunol* 31:285–293.

94 Mulroy, T., J.A. McMahon, S.J. Burakoff, A.P. McMahon, and J. Sen. 2002. Wnt-1 and Wnt-4 regulate thymic cellularity. *Eur J Immunol* 32:967–971.

95 Gounari, F., I. Aifantis, K. Khazaie, S. Hoeflinger, N. Harada, M.M. Taketo, and H. von Boehmer. 2001. Somatic activation of beta-catenin bypasses pre-TCR signaling and TCR selection in thymocyte development. *Nat Immunol* 2:863–869.

96 Okamura, R.M., M. Sigvardsson, J. Galceran, S. Verbeek, H. Clevers, and R. Grosschedl. 1998. Redundant regulation of T cell differentiation and TCRalpha gene expression by the transcription factors LEF-1 and TCF-1. *Immunity* 8:11–20.

97 Schilham, M.W., A. Wilson, P. Moerer, B.J. Benaissa-Trouw, A. Cumano, and H.C. Clevers. 1998. Critical involvement of Tcf-1 in expansion of thymocytes. *J Immunol* 161:3984–3991.

98 Verbeek, S., D. Izon, F. Hofhuis, E. Robanus-Maandag, H. te Riele, M. van de Wetering, M. Oosterwegel, A. Wilson, H.R. MacDonald, and H.

Clevers. 1995. An HMG-box-containing T-cell factor required for thymocyte differentiation. *Nature* 374:70–74.

99 Xu, Y., D. Banerjee, J. Huelsken, W. Birchmeier, and J.M. Sen. 2003. Deletion of beta-catenin impairs T cell development. *Nat Immunol* 4:1177–1182.

100 Yu, Q., B. Erman, J.H. Park, L. Feigenbaum, and A. Singer. 2004. IL-7 receptor signals inhibit expression of transcription factors TCF-1, LEF-1, and ROR-gammat: impact on thymocyte development. *J Exp Med* 200:797–803.

101 Ingham, P.W., and A.P. McMahon. 2001. Hedgehog signaling in animal development: paradigms and principles. *Genes Dev* 15:3059–3087.

102 Outram, S.V., A. Varas, C.V. Pepicelli, and T. Crompton. 2000. Hedgehog signaling regulates differentiation from double-negative to double-positive thymocyte. *Immunity* 13:187–197.

103 Shah, D.K., A.L. Hager-Theodorides, S.V. Outram, S.E. Ross, A. Varas, and T. Crompton. 2004. Reduced thymocyte development in sonic hedgehog knockout embryos. *J Immunol* 172:2296–2306.

104 Brou, C., F. Logeat, N. Gupta, C. Bessia, O. LeBail, J.R. Doedens, A. Cumano, P. Roux, R.A. Black, and A. Israel. 2000. A novel proteolytic cleavage involved in Notch signaling: the role of the disintegrin-metalloprotease TACE. *Mol Cell* 5:207–216.

105 De Strooper, B., W. Annaert, P. Cupers, P. Saftig, K. Craessaerts, J.S. Mumm, E.H. Schroeter, V. Schrijvers, M.S. Wolfe, W.J. Ray, A. Goate, and R. Kopan. 1999. A presenilin-1-dependent gamma-secretase-like protease mediates release of Notch intracellular domain. *Nature* 398:518–522.

106 Mumm, J.S., E.H. Schroeter, M.T. Saxena, A. Griesemer, X. Tian, D.J. Pan, W.J. Ray, and R. Kopan. 2000. A ligand-induced extracellular cleavage regulates gamma-secretase-like proteolytic activation of Notch1. *Mol Cell* 5:197–206.

107 Taniguchi, Y., H. Karlstrom, J. Lundkvist, T. Mizutani, A. Otaka, M. Vestling, A. Bernstein, D. Donoviel, U. Lendahl, and T. Honjo. 2002. Notch receptor cleavage depends on but is not directly executed by presenilins. *Proc Natl Acad Sci U S A* 99:4014–4019.

108 Fortini, M.E., and S. Artavanis-Tsakonas. 1994. The suppressor of hairless protein participates in notch receptor signaling. *Cell* 79:273–282.

109 Jarriault, S., C. Brou, F. Logeat, E.H. Schroeter, R. Kopan, and A. Israel. 1995. Signalling downstream of activated mammalian Notch. *Nature* 377:355–358.

110 Ellisen, L.W., J. Bird, D.C. West, A.L. Soreng, T.C. Reynolds, S.D. Smith, and J. Sklar. 1991. TAN-1, the human homolog of the Drosophila notch gene, is broken by chromosomal translocations in T lymphoblastic neoplasms. *Cell* 66:649–661.

111 Reynolds, T.C., S.D. Smith, and J. Sklar. 1987. Analysis of DNA surrounding the breakpoints of chromosomal translocations involving the beta T cell receptor gene in human lymphoblastic neoplasms. *Cell* 50:107–117.

112 Robey, E., D. Chang, A. Itano, D. Cado, H. Alexander, D. Lans, G. Weinmaster, and P. Salmon. 1996. An activated form of Notch influences the choice between CD4 and CD8 T cell lineages. *Cell* 87:483–492.

113 Chang, D., P. Valdez, T. Ho, and E. Robey. 2000. MHC recognition in thymic development: distinct, parallel pathways for survival and lineage commitment. *J Immunol* 165:6710–6715.

114 Deftos, M.L., Y.W. He, E.W. Ojala, and M.J. Bevan. 1998. Correlating notch signaling with thymocyte maturation. *Immunity* 9:777–786.

115 Deftos, M.L., E. Huang, E.W. Ojala, K.A. Forbush, and M.J. Bevan. 2000. Notch1 signaling promotes the maturation of CD4 and CD8 SP thymocytes. *Immunity* 13:73–84.

116 Izon, D.J., J.A. Punt, L. Xu, F.G. Karnell, D. Allman, P.S. Myung, N.J. Boerth, J.C. Pui, G.A. Koretzky, and W.S. Pear. 2001. Notch1 regulates maturation of CD4+ and CD8+ thymocytes by modulating TCR signal strength. *Immunity* 14:253–264.

117 Wolfer, A., T. Bakker, A. Wilson, M. Nicolas, V. Ioannidis, D.R. Littman, C.B. Wilson, W. Held, H.R. MacDonald, and F. Radtke. 2001. Inactivation of Notch 1 in immature thymocytes does not perturb CD4 or CD8T cell development. *Nat Immunol* 2:235–241.
118 Radtke, F., A. Wilson, G. Stark, M. Bauer, J. van Meerwijk, H.R. MacDonald, and M. Aguet. 1999. Deficient T cell fate specification in mice with an induced inactivation of Notch1. *Immunity* 10:547–558.
119 Pui, J.C., D. Allman, L. Xu, S. DeRocco, F.G. Karnell, S. Bakkour, J.Y. Lee, T. Kadesch, R.R. Hardy, J.C. Aster, and W.S. Pear. 1999. Notch1 expression in early lymphopoiesis influences B versus T lineage determination. *Immunity* 11:299–308.
120 De Smedt, M., K. Reynvoet, T. Kerre, T. Taghon, B. Verhasselt, B. Vandekerckhove, G. Leclercq, and J. Plum. 2002. Active form of Notch imposes T cell fate in human progenitor cells. *J Immunol* 169:3021–3029.
121 Jaleco, A.C., H. Neves, E. Hooijberg, P. Gameiro, N. Clode, M. Haury, D. Henrique, and L. Parreira. 2001. Differential effects of Notch ligands Delta-1 and Jagged-1 in human lymphoid differentiation. *J Exp Med* 194:991–1002.
122 Lehar, S.M., J. Dooley, A.G. Farr, and M.J. Bevan. 2005. Notch ligands Delta 1 and Jagged1 transmit distinct signals to T-cell precursors. *Blood* 105:1440–1447.
123 van den Brandt, J., K. Voss, M. Schott, T. Hunig, M.S. Wolfe, and H.M. Reichardt. 2004. Inhibition of Notch signaling biases rat thymocyte development towards the NK cell lineage. *Eur J Immunol* 34:1405–1413.
124 Wolfer, A., A. Wilson, M. Nemir, H.R. MacDonald, and F. Radtke. 2002. Inactivation of Notch1 impairs VDJbeta rearrangement and allows pre-TCR-independent survival of early alpha/beta lineage thymocytes. *Immunity* 16:869–879.
125 Ciofani, M., T.M. Schmitt, A. Ciofani, A.M. Michie, N. Cuburu, A. Aublin, J.L. Maryanski, and J.C. Zúñiga-Pflücker. 2004. Obligatory role for cooperative signaling by pre-TCR and Notch during thymocyte differentiation. *J Immunol* 172:5230–5239.
126 Huang, E.Y., A.M. Gallegos, S.M. Richards, S.M. Lehar, and M.J. Bevan. 2003. Surface expression of Notch1 on thymocytes: correlation with the double-negative to double-positive transition. *J Immunol* 171:2296–2304.
127 Allman, D., F.G. Karnell, J.A. Punt, S. Bakkour, L. Xu, P. Myung, G.A. Koretzky, J.C. Pui, J.C. Aster, and W.S. Pear. 2001. Separation of Notch1 promoted lineage commitment and expansion/transformation in developing T cells. *J Exp Med* 194:99–106.
128 Washburn, T., E. Schweighoffer, T. Gridley, D. Chang, B.J. Fowlkes, D. Cado, and E. Robey. 1997. Notch activity influences the alphabeta versus gammadelta T cell lineage decision. *Cell* 88:833–843.
129 Tanigaki, K., M. Tsuji, N. Yamamoto, H. Han, J. Tsukada, H. Inoue, M. Kubo, and T. Honjo. 2004. Regulation of alphabeta/gammadelta T cell lineage commitment and peripheral T cell responses by Notch/RBP-J signaling. *Immunity* 20:611–622.
130 Garcia-Peydro, M., V.G. de Yebenes, and M.L. Toribio. 2003. Sustained Notch1 signaling instructs the earliest human intrathymic precursors to adopt a gammadelta T-cell fate in fetal thymus organ culture. *Blood* 102:2444–2451.
131 Jiang, R., Y. Lan, H.D. Chapman, C. Shawber, C.R. Norton, D.V. Serreze, G. Weinmaster, and T. Gridley. 1998. Defects in limb, craniofacial, and thymic development in Jagged2 mutant mice. *Genes Dev* 12:1046–1057.
132 Schmitt, T.M., and J.C. Zúñiga-Pflücker. 2002. Induction of T cell development from hematopoietic progenitor cells by delta-like-1 in vitro. *Immunity* 17:749–756.
133 Schmitt, T.M., R.F. de Pooter, M.A. Gronski, S.K. Cho, P.S. Ohashi, and J.C. Zúñiga-Pflücker. 2004. Induction of T cell development and establishment of T cell competence from embryonic stem cells differentiated in vitro. *Nat Immunol* 5:410–417.
134 Zúñiga-Pflücker, J.C. 2004. T-cell development made simple. *Nat Rev Immunol* 4:67–72.

6
CD8 T-cell Immunodominance, Repertoire, and Memory

Dalia E. Gaddis, Michael J. Fuller, and Allan J. Zajac

6.1
Introduction

In this chapter we focus on the properties of antigen-specific CD8 T cells and discuss how immunodominance within the overall ensemble of T cells that recognize any given antigen impacts immunological memory and recall responses.

A hallmark of adaptive immunity is antigen specificity, a trait that is well exemplified by CD8 T cells. CD8 T cells recognize short peptide epitopes, typically 8–10 amino acids in length, which are presented upon the surface of cells in association with major histocompatibility complex (MHC) class I molecules [1–4]. The MHC class I–peptide combinations that a CD8 T cell recognizes are determined by its expression of the clonally distributed, heterodimeric T-cell receptor (TCR) [5–9].

A diverse TCR repertoire is necessary to provide a pool of T cells that are capable of recognizing and responding to the numerous potential antigens that may be encountered during the lifespan of the host [5, 7, 9]. The particular TCR displayed by an individual CD8 T cell represents the end product of a series of random recombinatorial rearrangements of gene segments, which take place during T-cell ontogeny [5–9]. TCR genes are assembled by joining of the different V, D, and J segments to form the β chain, as well as combining V and J segments to form the α chain. Diversity is generated by the juxtaposition of these segments and by nucleotide additions or deletions at the junctions where the segments join. Further variation arises because of the heterodimerization of the α and β chains. This results in a huge potential repertoire of TCR combinations; however, an individual T cell usually expresses only one unique TCR clonotype.

As the chains of the TCR fold, they form three loop structures referred to as complementarity determining regions (CDR). CDR1 and CDR2 are encoded by the V segment, whereas CDR3 is the most diverse, spanning the junctions of the V, D, and J segments. Crystal structure analysis of the TCR and the peptide–MHC complex has shown that the CDR1 and CDR2 loops interact with MHC residues but that the CDR3, located at the center of the complex, interacts with amino acid side chains of the antigenic peptide that point away from the MHC [5, 7, 10–14]. Therefore, the sequence and length of the CDR3 are critical for determining the

Immunodominance: The Choice of the Immune System. Edited by Jeffrey A. Frelinger

ISBN: 3-527-31274-9

antigen specificity of a particular TCR and they set the T cell's recognition profile. Identical MHC–peptide complexes are usually recognized by several separate T-cell clones; nevertheless, the precise TCR sequence expressed by each T cell is unique. Therefore, an individual epitope usually activates an oligoclonal set of CD8 T cells. A central issue is how the repertoire and composition of the pool of T cells that recognize a particular epitope influence the development, functional competence, and longevity of the CD8 T-cell response.

Complex antigens, such as those encoded by viral, bacterial, and parasitic pathogens, typically contain a variety of distinct epitopes. Accordingly, the overall CD8 T-cell response, which is so critical for bringing about the control of intracellular pathogens, is not monoclonal or specific for only a single epitope. Because the magnitude and breadth of responses to each discrete epitope are not necessarily uniform, a hierarchy becomes established as certain specificities of T cells predominate, whereas other responses are less prominent [15–17]. Once the response is elaborated, the range of pathogen-derived epitopes that are detected, as well as the size and composition of the pools of T cells that recognize individual epitopes, may change over time. This can be influenced by a variety of parameters including virus persistence, epitope sequence variation, reexposure to the inducing antigen, cross-reactivity, and underlying immunodeficiency. In terms of understanding how CD8 T-cell responses contain intracellular pathogens, it is useful to delineate which epitope-specific responses are most effective at mediating rapid control of primary infections as well as which responses are subsequently capable of conferring long-lived immunity. Although focused immunodominant responses may be able to control particular pathogens, a more diverse overall response, which detects a broad array of epitopes, may be better suited for containing pathogens that are prone to mutating major epitopes.

In addition to antigen specificity, another distinguishing property of adaptive immune responses is memory. This feature enables the host to mount a bigger, better, and faster response if the initiating antigen is reencountered [18–25]. CD8 T-cell memory is brought about by the maintenance of an expanded pool of antigen-specific CD8 T cells and by the capacity of these antigen-experienced cells to mount accelerated responses following reactivation. By comparison with naïve (antigen-inexperienced) individuals, the frequency of antigen-specific cells is increased in immune hosts, and these memory cells are more tuned to respond to antigen [18–25].

Collectively, the network of CD8 T cells that is recruited following infections is remarkably heterogeneous (Figure 6.1). Not all T cells forming the pool that recognize a particular epitope necessarily react similarly, and this can result in a skewing of the response as particular clones are favored. Moreover, pathogens usually encode several distinct epitopes, and the responses to each of these epitopes will be clonally distinct and may differ in magnitude. Finally, the functional properties and phenotypic traits of the subsets of CD8 T cells, which constitute the overall response, are also diverse.

A major objective for dissecting the components of the immune system is to provide a better understanding of the factors that confer immunity to infections. Pertinent issues regarding immunodominance, repertoire, and memory include

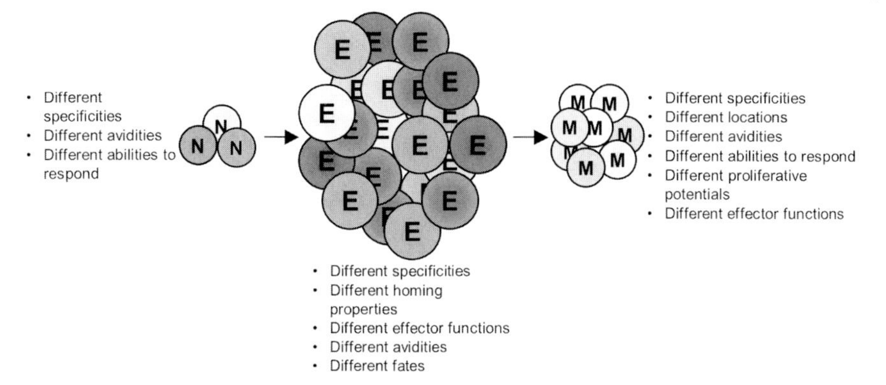

Figure 6.1 CD8 T-cell responses are remarkably heterogeneous. CD8 T-cell responses differ in many ways, including in their epitope specificity and in additional properties stated in this figure. Heterogeneity is observed in naïve (N) CD8 T cells in the periphery, through the expansion phase of the response as effector (E) cells develop, and in the memory (M) pool.

delineating the parameters that shape pathogen-specific responses, defining how the composition of the pool of antigen-specific CD8 T cells impacts the effectiveness of infection control, and elucidating how the response can be tailored to best confer long-lived, protective immunity.

6.2 CD8 T-Cell Responses and Memory

CD8 T cells play a vital role in controlling intracellular pathogens by bringing about the targeted removal of infected cells. The ability of CD8 T cells to identify pathogen-infected cells is mediated by the interaction between the TCR, expressed by the T cell, and the MHC class I complexes that present antigenic peptides upon the surface of infected cells [5, 7, 10–14]. Initial exposures to antigen, such as those that occur during the course of acute viral infections, result in marked activation and expansion of antigen-specific CD8 T cells [26–31]. Subsequently, the successful removal of the antigen (i.e., clearance of the pathogen) is associated with the development of memory T cells, which help confer long-lived immunological protection if the priming antigen or pathogen is reencountered. Before discussing how the T-cell repertoire impacts memory responses and immunodominance, it is relevant to first overview how memory CD8 T cells develop. More comprehensive information regarding this subject can be found in numerous excellent reviews [18–25].

CD8 T cells, that survive the selective process of thymic education emigrate into the periphery, where they circulate and migrate through secondary lymphoid organs [32, 33]. These naïve cells survey MHC class I–peptide complexes displayed on the surface of professional antigen-presenting cells (APCs) such as dendritic cells [34]. Although a vast array of peptide epitopes can potentially be presented at the cell surface, activation is triggered only if the CD8 T cell's rearranged TCR recognizes the specific MHC–peptide combination with sufficient avidity. TCR-dependent activation of the CD8 T cell launches a proliferation and differentiation program, which, once initiated, proceeds even if the antigenic stimulus is withdrawn [35, 36]. Parameters such as the duration and strength of antigenic stimulation, costimulatory interactions, presence of cytokines, and provision of CD4 T-cell help guide the ensuing response. Thus, early activation events play a critical role in shaping the subsequent development of CD8 T cells.

Typical CD8 T-cell responses can be divided into three distinct phases: expansion, contraction, and maintenance (Figure 6.2). As naïve CD8 T cells respond to their cognate antigens, patterns of gene expression change, promoting the expression of cytokines and cytotoxic effector molecules, as well as alterations in surface molecules, including cytokine receptors and adhesion molecules [37–41]. In the case of acute infection with the natural mouse pathogen lymphocytic choriomeningitis virus (LCMV), 50% or more of the CD8 T cells detectable at the peak of the response are antigen specific, and marked expansions of CD8 T cells are also apparent following other infections, including influenza, Epstein-Barr virus (EBV), *Listeria monocytogenes*, and human immunodeficiency virus (HIV) [28–31, 42–45]. Such overwhelming responses play a principle role in removing pathogen-infected cells. The effector pool is remarkably heterogeneous, as these cells differ in many ways, including in the arrays of effector molecules and patterns of activation markers that they express (Figure 6.1). Effector cells also differ in their ana-

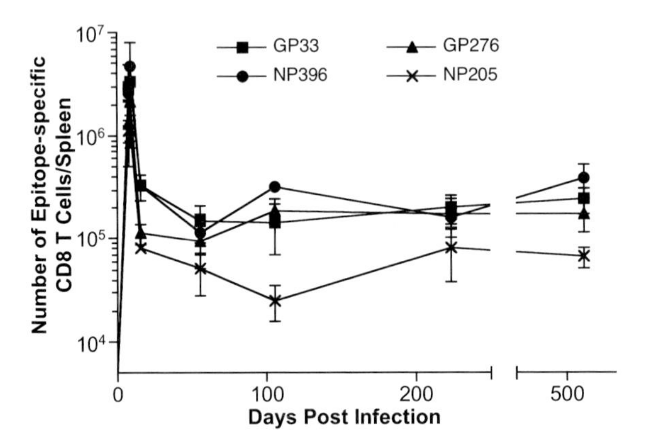

Figure 6.2 Immunodominance and kinetics of antiviral CD8 T-cell responses. Acute infection of mice with LCMV illustrates the three phases of a typical CD8 T-cell response: expansion, contraction, and memory. Following infection, naïve virus-specific CD8 T cells are activated and expand, vastly increasing in number to form a massive pool of effector cells. The resolution of the infection is associated with a marked contraction of the response when most, but not all, of the antigen-specific CD8 T cells die by apoptosis. The surviving T cells form the memory pool, which is usually stably maintained over time. Many infections elicit responses to multiple epitopes, as shown here for the LCMV GP33, GP276, NP396, and NP205 epitopes. These epitope-specific responses undergo different magnitudes of expansion and contract to different levels, creating a hierarchy of responses.

tomic location because, although naïve T cells are primed in secondary lymphoid organs, the effector cells that develop disperse throughout the host, trafficking to tertiary tissues [46–48]. These cells can be recovered from many sites, including the liver, lung, kidney, fat pad, and intestine, in addition to the spleen and lymph nodes. This diaspora allows the effector CD8 T cells to reach the sites of infection, where they can act locally to eliminate infected cells (Figure 6.3).

The heterogeneous populations of T cells that make up the effector pool succumb to different fates, as they do not remain constitutively activated. Instead, the response is downregulated as often over 90% of the T cells present undergo apoptosis [26, 27, 49]. The T cells that survive this contraction phase undergo further phenotypic and functional changes as they go on to form the memory pool [37–39, 41]. The population of memory CD8 T cells that emerges as homeostasis is restored following the contraction phase, and clearance of the inducing antigen is not uniform because phenotypic and functional diversity is apparent even within subsets of memory CD8 T cells that recognize the same epitope. This is well illustrated by the classification of memory T cells into broad categories termed effector memory and central memory T cells. These subsets have been defined based on their anatomical location, functional quality, proliferative potential, and expression of surface molecules [47, 50–52]. Effector memory CD8 T cells display a $CD62L^{lo}$, $CCR7^{-}$, $CD27^{lo}$ phenotype, express lower levels of IL-2 following stimulation, and may exhibit a diminished capacity to undergo antigen-driven proliferation. Central memory CD8 T cells display a contrasting phenotype and are $CD62L^{hi}$, $CCR7^{+}$, $CD27^{hi}$, express higher levels of IL-2, and may proliferate more

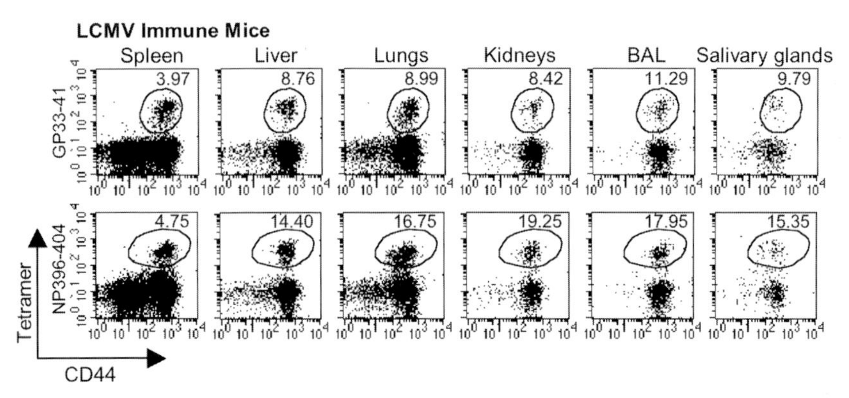

Figure 6.3 Distributions of memory CD8 T cells. An impressive feature of memory CD8 T cells is their disbursement to tissues throughout the host. This ensures that antigen-specific CD8 T cells are locally available to rapidly react if the inducing antigen is reencountered. These flow cytometry plots show the presence of virus-specific CD8 T cells in various tissues of C57BL/6 mice that were inoculated with LCMV-Armstrong >3 months previously and have resolved the infection. Antiviral CD8 T cells specific for two different epitopes were detected by MHC tetramer staining (GP33 responses, upper panels; NP396 responses, lower panels). The profiles of MHC tetramer and anti-CD44 antibody staining are shown for gated CD8 T cells. The values indicate the percentage of CD8 T cells recovered from each site that stain positively with the indicated MHC tetramer. BAL= bronchoalveolar lavage.

vigorously during recall responses. Although the overall numbers of memory CD8 T cells specific for any given epitope may remain stable over time, the relative abundances of effector memory and central memory CD8 T cells change; the proportion of effector memory cells tends to decline and the central memory population increases [51, 53, 54]. Although the segregation of memory CD8 T cells into central and effector memory subsets is useful, this definition is not always clear-cut and additional complexities arise; for example, $CD62L^-$ $CCR7^+$ memory cells are detectable that cannot be neatly classified as either central or effector memory CD8 T cells [55]. Thus, memory T-cell pools are complex, dynamic, and diverse and display a gradation of phenotypes and functions.

The establishment of memory CD8 T-cell responses provides the host with a beneficial mechanism for recalling past antigenic experiences, as an increased frequency of antigen-specific CD8 T cells are maintained that can more rapidly respond if the original antigen is reencountered. These memory T-cell populations are detectable in the mouse several years following priming [26, 27, 56, 57], and in humans CD8 T cells induced by the smallpox vaccine have been detected more than 60 years following inoculation [58, 59]. By maintaining memory CD8 T cells for these extended periods, the adaptive immune response is able to contain the dissemination of secondary infections, thereby reducing morbidity and mortality. Nevertheless, maintaining elevated numbers of antigen-specific T cells can be detrimental in certain circumstances because they may enhance immunopathology or diminish responses to altered epitopes that evolve as a result of the selective pressure of the immune response [60–62].

6.3 Analyzing the Memory Repertoire

Given the interconnections between TCR usage, clonal abundance, epitope immunodominance, and the phenotypic and functional properties of memory CD8 T cells, it is appropriate to briefly discuss some of the experimental approaches that are employed to probe the complexities of memory T-cell networks.

Multiparameter flow cytometry has proven to be a robust and powerful tool for characterizing T-cell responses. Much of our current appreciation of the heterogeneity of memory CD8 T-cell responses has been established by flow-based assays [63, 64]. If appropriate antibodies or staining reagents are available, the expression of cell-surface molecules can be easily assessed. In addition, functional properties can be assessed, including patterns of cytokine production, cytotoxicity, and proliferative potential. By simultaneously analyzing multiple phenotypic traits, comprehensive information can be acquired regarding the composition of T-cell pools.

TCR usage defines the diversity of the T-cell repertoire and determines antigen recognition patterns. Panels of monoclonal antibodies have been developed that are specific for different $V\alpha$ and $V\beta$ gene segments and can be used in conjunction with flow cytometry to reveal overall patterns of TCR usage by populations of T cells [45, 65–68]. By using this approach, shifts in the patterns of TCR V gene usage and the expansions of families of T cells, which express the same V genes, can be tracked during the course of an immune response and into the memory phase. This antibody-staining approach provides a gross overview of TCR V gene usage; nevertheless, it does not reveal the diversity of the variable CDR3 sequences within subsets of T cells that express identical TCR V segments. This region is critical for determining the antigen specificity of the TCR and varies in length and sequence as a result of recombinatorial processes that rearrange germ line–encoded TCR gene segments.

Molecular analyses of expressed TCR sequences have provided important information regarding the T-cell repertoire. Approaches have included cell sorting, single-cell PCR, and sequence analyses [45, 69, 70]. A powerful technique for determining the clonality of the TCR repertoire is CDR3 spectratyping, also known as immunoscope analysis [71, 72]. This is a quantitative, PCR-based method that reveals the spectrum of CDR3 lengths of all possible TCR Vβ–Cβ combinations (Figure 6.4). By amplifying complementary DNA with panels of different Vβ primers and a Cβ primer, a number of PCR products are generated that span the CDR3 region. Next, the profile of CDR3 lengths is revealed by performing runoff reactions using fluorescently labeled primers. A low-resolution spectratype can be obtained with Cβ-specific primers, but for more refined analyses, panels of Jβ primers can be used. The length and abundance of each runoff product are assayed using an automated DNA sequencer and specialized computer software. Typical spectratypes consist of 6–8 distinct peaks, which differ in size by multiples of three nucleotides. The profile of peaks portrays the range of amino acid lengths exhibited by the particular amplified CDR3 region. The area of each peak represents the relative abundance of T cells that express a particular Vβ segment (and possibly Jβ segments for high-resolution analyses) and contain a CDR3 of that specific length. This analysis can be further refined by sequencing the runoff reactions to give precise information regarding clonal diversity. Naïve T cells have a random distribution of CDR3 lengths, and thus the pattern of peaks forms a Gaussian (bell-shaped) distribution curve. However, as antigen-driven clonal expansions occur, T cells with particular CDR3 lengths and sequences preferentially respond, causing a deviation in the Gaussian curve as the pattern skews towards a CDR3 of a certain amino acid length indicative of the dominant response.

A significant step forward in our ability to scrutinize T-cell responses was the development of MHC tetramers by Altman and coworkers [73–75]. These reagents are typically comprised of recombinant MHC molecules complexed with a synthetic peptide epitope and are tetramerized via a biotin–avidin coupling. Other strategies for multimerizing MHC molecules have also been successfully employed, and both MHC class I and MHC class II multimers have been produced [76, 77]. These reagents can be used to identify, isolate, and characterize antigen-specific T cells directly *ex vivo* without necessity of functional readouts. Thus, T cells capable of recognizing a particular MHC–peptide combination can be enumerated and their phenotypic traits assessed. MHC tetramers have proven useful for studies of immunodominance. If the epitope sequences are known, then panels of these reagents can be generated and used to assess the abundance and heterogeneity of T-cell responses to multiple epitopes in the same host (see also Figure 6.3).

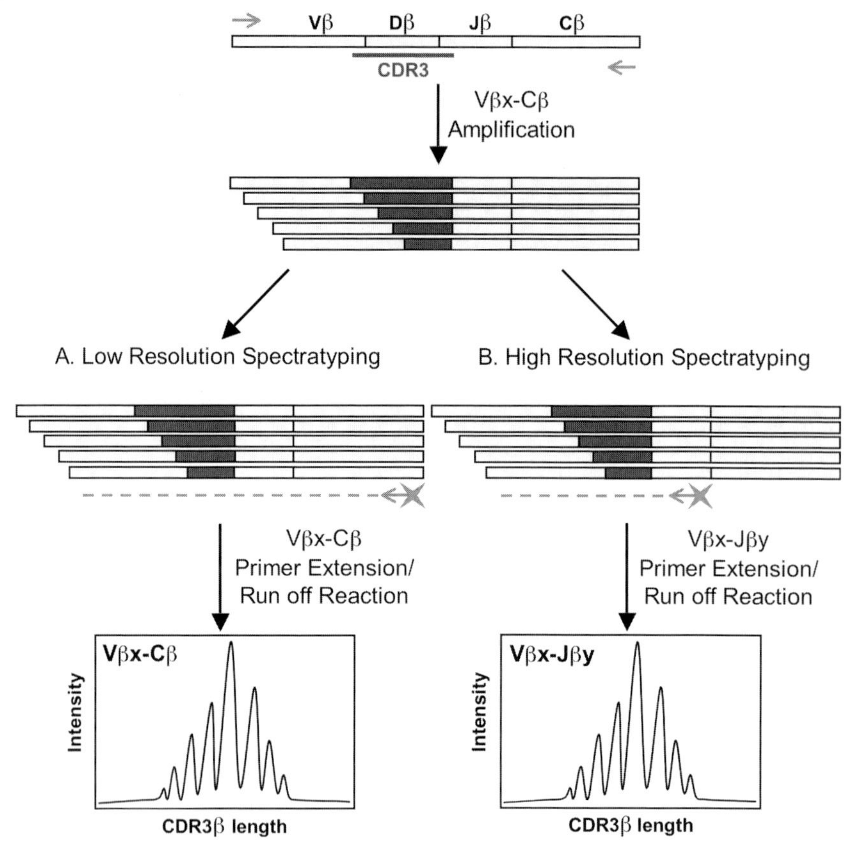

Figure 6.4 Spectratype analysis is an informative approach for defining CDR3 lengths and Vβ usage by populations of T cells. This technique, shown here for the TCR β chain, involves two major steps. First TCR cDNA are generated by reverse transcription and are amplified using C and various V region primers. Second, runoff reactions are performed using fluorescently labeled primers. Cβ primers are used to provide low-resolution spectratypes (A); more detailed, high-resolution information regarding CDR3 lengths and J gene usage are obtained using sets of Jβ-specific primers (B). Analysis of the products with an automated sequencer and specialized immunoscope software gives a series of distinct peaks. Each peak represents a one-amino-acid difference in CDR3 length, and the size of the peak indicates the relative frequency of that particular CDR3 length.

6.4 Immunodominance

Immunodominance is a widely recognized feature of CD8 T-cell responses [15–17]. Not all potential epitopes encoded by complex antigens elicit responses of equal magnitudes. Consequently, a stratified hierarchy emerges as T cells specific for certain epitopes preferentially expand, whereas subdominant responses are elicited to other epitopes. Patterns of immunodominance have been observed following many viral infections. In the case of LCMV, at least nine MHC class I–restricted epitopes have been defined in H-2^b mice [78]. Following acute infections, more dominant responses have been ascribed to glycoprotein (GP)33 and GP276 epitopes as well as to the nucleoprotein (NP)396 epitope; however, responses to NP205, GP118, and GP92 epitopes are less prominent (see Figure 6.2). In BALB/c (H-2^d) mice, viral epitope hierarchies are even more striking because acute LCMV infection results in a massive response specific for the NP118 epitope, whereas subdominant GP283- and GP99-specific responses are weakly detected [31, 79, 80]. This example illustrates the complexity of the overall T-cell response to a natural mouse pathogen. There are numerous examples of quantitative and qualitative discordances in the spectrum of epitope-specific responses to a single pathogen. It remains relatively ill defined why different patterns of responses arise and what the relative contributions of multiple key factors are in shaping the response [15, 17]. Because immunodominance influences the resolution of primary infections and also impacts the success of recall responses, understanding the underlying factors that shape the response profile is critical.

Overall, immunodominance is influenced by two major factors: the antigens that are presented for inspection by CD8 T cells and the CD8 T cells that respond to the antigen (Figure 6.5).

6.4.1 Antigen-related Factors

Every step that is involved in MHC class I antigen processing and presentation can influence immunodominance [15–17], including (1) the kinetics and temporal

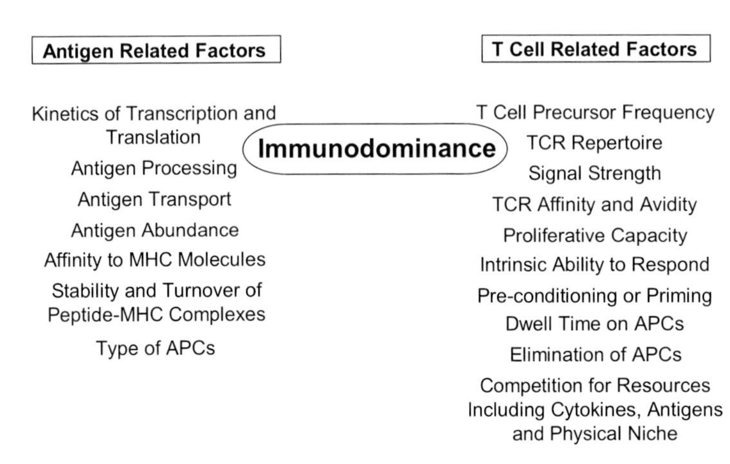

Figure 6.5 Immunodominance is influenced by multiple factors. Epitope-dependent differences in the magnitude and functional quality of CD8 T-cell responses are routinely observed but remain poorly understood. Both antigen-related factors, responsible for activating the T-cell response, and the properties of the T cells that respond to the antigens contribute to the structuring of immunodominance.

order of protein synthesis; (2) the degradation of the protein by the immunoproteasome; (3) transport of peptides into the endoplasmic reticulum; (4) the ability to associate with an MHC class I heavy chain and traffic to the cell surface; (5) the overall abundance of the peptide; (6) the affinity of the peptide for the MHC molecule; (7) the stability and turnover of the peptide–MHC complex; and (8) the cell type presenting the antigen. Therefore, when there is a choice between multiple epitopes, these factors, which govern antigen presentation, influence the hierarchy of the resulting T-cell response.

A prime determinant of immunogenicity is the ability of the antigenic peptide to bind to MHC molecules. Several studies have investigated the binding affinities of known and predicted epitopes to MHC class I molecules, and the general trend that emerges is that immunogenic epitopes are better MHC binders than are weakly or non-immunogenic peptides [79, 81–83]. This is a straightforward concept, but when multiple epitopes are present, the binding affinity of each individual epitope may not correlate with the resulting patterns of immunodominance.

The numbers of MHC complexes that present a particular antigenic peptide may impact immunodominance; however, this is not always the case. The hierarchy of CD8 T-cell responses to the intracellular bacterium *Listeria monocytogenes* in $H\text{-}2^d$ mice is (1) listeriolysin O (LLO) 91–99, (2) p60 217–225, and (3) p60 449–457 [84]. By contrast, determination of the number of peptide epitopes presented by infected cells revealed an opposing order, as the abundance of peptide was p60 449–457> p60 217–225> LLO 91–99 [85, 86]. Similarly, analyses of $H\text{-}2K^d$-restricted, influenza virus PR8–specific CD8 T-cell responses also revealed disparities between immunodominance and the abundance of peptide epitopes extracted from MHC class I molecules on infected cells [87]. In this case the most sizable response, in terms of IFN-γ production, was directed to the NP147 epitope followed by subdominant responses to HA518 and NP39 epitopes, whereas responses to the HA462 epitope were barely detectable. However, the HA462 peptide was the most abundant epitope presented at the cell surface by the K^d molecule. Both of these examples illustrate that the immunodominance of primary responses, which shapes the hierarchy of the memory pool, does not necessarily correspond with the levels of epitopes displayed at the cell surface. In other studies, elution of peptides from MC57 fibroblast cell lines at 48 hours following infection with LCMV estimated that the numbers of peptides recovered per cell were 1081, 162, and 92 for the GP33, NP396, and GP276 epitopes, respectively [88]. This correlates with the hierarchy of the antiviral CD8 T-cell response following primary infection with low doses of the WE strain of LCMV; however, as discussed later, more disseminated infections result in striking shifts in immunodominance. Interestingly, adoptive transfer studies suggest that LCMV NP396-specific CD8 T-cell lines are more efficacious at protecting the host than either GP276- or GP33-specific CTL lines [88]. Collectively, these observations underscore the complexity of the multifactorial process that influences whether an epitope elicits the most favorable CD8 T-cell response.

6.4.2 T Cell–related Factors

Although the presence of antigen is critical for initiating CD8 T-cell responses, intrinsic properties of the epitope-specific T cells also impact immunodominance. The process of thymic selection shapes the diversity of the naïve CD8 T-cell repertoire, setting the precursor frequency of T cells reactive against particular epitopes [6]. Limiting the complexity of the T-cell repertoire can, therefore, impact immunodominance, especially if the numbers of naïve cells that give rise to usually conserved (sometimes referred to as "public") responses are altered (see Section 6.8). The avidity and duration of the interaction between the T cell and the APC may also influence the response because these help to determine the strength and duration of activatory signals delivered through the TCR as well as the potential to perceive accessory signals from costimulatory molecules. Importantly, because the immunodominance of the effector pool usually parallels that of the memory pool, these initial activation events contribute to the structuring of both the primary response and the resulting set of memory T cells.

Upon encountering antigen, naïve T cells do not mount identical responses (Figure 6.1). This variability may reflect intrinsic differences in the responsiveness of the T cells. For example, certain cells may be pre-triggered to respond because they adopt a semi-memory-like phenotype resulting from cross-reactivity with endogenous or environmental antigens, or even as a consequence of altered homeostatic proliferation [89–92]. The hierarchy of the epitope-specific response is also molded by the relative kinetics and competitiveness of dominant and subdominant responses [15, 17]. Here the link between antigen presentation and the swiftness of response initiation may prove critical. Once dominant responses are launched, cells that present the priming antigen may be cleared, causing a temporal restriction in the availability of antigen and impeding the stimulation of subdominant responses [93].

Divergence between major and minor responses may be amplified by the ability of the leading responders to sequester resources and out-compete weaker ones. Several interesting studies have highlighted opposing roles for IFN-γ in this regard. By comparison with wild-type hosts, a greater abundance of antigen-specific CD8 T cells were recovered from IFN-γ knockout mice following infection with *Listeria monocytogenes*, suggesting that IFN-γ plays a role in downsizing primary responses and in this way contributes to the structuring of the epitope hierarchy [94]. By contrast, a series of studies of LCMV-specific responses by Whitton and coworkers suggest that the ability to quickly produce and perceive IFN-γ promotes dominant responses but may suppress subdominant T cells [95–97]. These divergent activatory verses inhibitory activities of IFN-γ may be a consequence of the activation state of the cell at the time of cytokine exposure. The key point is that epitope-specific T cells that are activated earlier and expand in number more quickly may create an immunological environment that is unfavorable to slower and smaller responses.

In summary, for an epitope to have the prime position, abundance of the peptide being presented alone is not sufficient to generate a prominent response. Precursor frequencies, avidity, intrinsic responsiveness, and competitive ability all work together to configure immunodominance. These factors shape both primary responses and the pool of memory T cells that emerge and serve as the starting set of cells for secondary responses.

6.5
Epitope-dependent Skewing of the Repertoire During Primary, Memory, and Recall Responses

Several studies have examined the changes in the T-cell repertoire that occur during the course of an immune response and have addressed whether naïve, effector, memory, and recall responses can be distinguished by TCR usage and repertoire diversity [45, 65–67, 84, 98, 99]. As naïve CD8 T cells are recruited during the course of a primary immune response, antigen-specific cells undergo clonal expansion [7, 26, 27, 29–31, 35, 36]. Because only a selective subset of T cells recognizes a particular antigen, the diversity of TCRs utilized by the responding cells is more limited than that displayed by the overall pool of naïve cells. As a consequence of these restricted patterns of TCR usage, a skewing of the TCR repertoire becomes apparent as particular CDR3 lengths and sequences are favored (Figure 6.6).

Epitope-dependent skewing of the repertoire has been documented during the acute phase of several experimental infections. Inspection of TCR usage by CD8

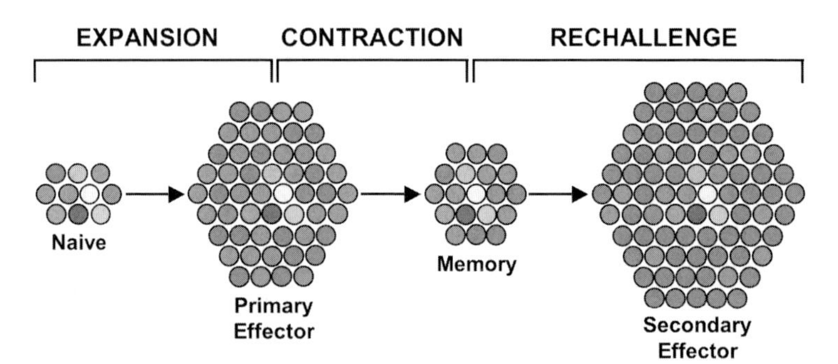

Figure 6.6 Shifts in repertoire and immunodominance following primary and secondary exposures to antigen. As primary immune responses are initiated, antigen-specific T cells become activated and expand in number. This results in a discernable shift in the T-cell repertoire as antigen-specific cells, indicated here in red, green, and blue, increase in frequency. A contraction phase ensues, following clearance of the inducing antigen; however, this downsizing is typically proportional. Consequently, the skewing of the repertoire and the hierarchy of immunodominance that develops during the expansion phase are imprinted on the memory pool. This phenomenon is sometimes referred to as immunological scarring. Although the repertoire and hierarchies of the primary effector and memory pools are usually similar, marked differences can arise following rechallenges. In this illustration, the red responders become most dominant.

T cells specific for the immunodominant K^d-restricted *Listeria monocytogenes* LLO 91–99 epitope revealed that these T cells preferentially use Vβ8, followed by Vβ2 and Vβ10 [65, 100, 101]. Use of the Vβ8 segment was prevalent in all mice examined, indicative of a "public" or common reactivity; however, there was slight variation in the use of alternative Vβ segments, suggesting "private" or unique, individual responses. Consequently, the overall repertoire of LLO 91–99-specific effector CD8 T cells appears somewhat diverse.

In the case of LCMV infection of BALB/c (H-2^d) mice, Vβ profiling by antibody and MHC tetramer costains, as well as spectratype analyses of sorted antigen-specific CD8 cells, showed that 30% of the L^d-restricted, NP118-specific cells utilized Vβ10 and that the preferred CDR3 length was nine amino acids [66]. Within the primary effector pool ~85% of all V$\beta10^+$ cells were NP118 specific. The next most common NP118-specific responses were V$\beta8.1^+$ and V$\beta8.2^+$, with CDR3 lengths of eight and six amino acids, respectively. By contrast with H-2^d mice, C57BL/6 (H-2^b) mice mount a more diverse response to acute LCMV infection as a broader array of epitope-specific responses develop. Comparison of the NP396-, GP33-, and GP34-specific responses in H-2^b mice showed a Vβ8.1/8.2 public response to all three epitopes; however, other Vβ segments were also utilized [67, 98]. In the case of NP396-specific CD8 T cells, Vβ6 and Vβ9 were also favored and spectratyping established that certain CDR3 lengths were commonly used by the responding T cells.

Discernable shifts in the T-cell repertoire as a result of the induction of dominant responses also manifests during influenza virus infection. During primary influenza infection, major H-$2D^b$-specific responses are elaborated to the NP366 and PA224 epitopes [45, 99, 102, 103]. Here, epitope-dependent clonal expansions become apparent as Vβ8.3 and Jβ2.2 TCRs with a CDR3 length of nine residues are preferentially used by NP366-specific cells [99, 103, 104]. By contrast with the more reproducible or public NP366-specific response, elaboration of PA224-specific responses also results in repertoire skewing, typically expanding V$\beta7^+$ CD8 T cells, but this response is broader and greater diversity is apparent between individual mice [45]. These more private or individualistic responses are associated with variable CDR3 lengths of between five and seven amino acids and with usage of Jβ1.1 or Jβ2.6. Structural analyses of these antigenic peptides in association with the D^b molecule showed that the NP366 peptide adopts a flatter, featureless conformation; however, the PA224 peptide forms a carboxy-terminal arch with the side chain of the arginine residue, at position 7 of the epitope, solvent exposed, pointing upwards towards a possible TCR encounter [105]. Reverse genetics elegantly revealed that the usually broader PA224-specific response becomes more restricted and limited in diversity if the P7 arginine residue is substituted with alanine. This amino acid change alters the topography of peptide binding, forming a featureless configuration as the shorter alanine side chain is less protrusive [105]. Thus, for these two influenza epitopes the presence or absence of prominent structural features, which may impact how the MHC–peptide complex is interrogated by TCRs, appears to be a determinant of the diversity of the T-cell response.

Differential epitope-specific TCR usage has also been reported for EBV-specific CD8 T cells. Masucci and coworkers analyzed *in vitro*–expanded T-cell clones specific for two HLA-A11 epitopes derived from the same latent protein, EBNA 4 (also know as EBNA 3B) [106]. RT-PCR and sequence analyses of both α and β chains of the TCR showed that the nucleotide sequence and length of the CDR3 region, as well as the recombination between different Vα and Jα segments or Vβ, Dβ, and Jβ segments, were more diverse for the immunodominant EBNA 4 416–424 (IVT)-specific clones than for T-cell clones specific for the subdominant EBNA 4 399–408 (AVF) epitope. In this case the immunodominant epitope was recognized by a broad repertoire of T cells, whereas responses to the subdominant epitope were more restricted in terms of TCR usage.

Taken together, all of these findings highlight that primary immune responses are associated with the selective expansion of epitope-specific T cells. This clonal expansion process results in the skewing of the overall T-cell repertoire as particular T cells become more prevalent. Responses to certain epitopes elicit more restricted or conserved responses; however, broader and more host-dependent responses are detectable and form the predominant responses to other epitopes. Thus, different epitopes imprint a distinctive signature of TCR usage, and this immunological scarring becomes visible during the initial phase of the response (Figure 6.6).

Comparative longitudinal analyses of epitope-specific TCR repertoires have addressed whether effector and memory T-cell subsets can be distinguished by TCR usage patterns and whether memory repertoires are stable or subject to variation. Collectively, the results of several independent studies indicate that the repertoires of effector and memory T cells are usually similar and that the repertoires of memory T cells that develop following acute antigenic exposures remain quite stable over time [30, 45, 65–67, 84, 98, 99, 102]. Thus, even though the antigen-specific T-cell repertoire is unique for any given epitope and may differ between individual hosts, within an individual the epitope-specific repertoire apparent at the peak of the response sets the repertoire of the memory pool.

Primary effector as well as memory CD8 T cells become distributed throughout the host and can be recovered from lymphoid as well as non-lymphoid organs [45–48] (see Figure 6.3). Studies by Doherty and coworkers addressed whether the TCR usage of CD8 T cells specific for the influenza virus D^b-restricted PA224 epitope differed depending upon the anatomical localization of the T cells [45]. They showed that the antigen-specific T cells randomly dispersed into tissues and that particular clones of antigen-specific cells did not selectively localize in particular tissues.

Memory CD8 T cells can be subdivided based upon their expression of the surface molecule CD62L. An interesting study used repertoire analyses to track the lineage relationships of central ($CD62L^{hi}$) and effector ($CD62L^{lo}$) male antigen–specific memory T cells [54]. Consistent with the results from the influenza system, there did not appear to be selective trafficking of distinct T-cell clones to particular tissues. Although central and effector memory T cells were clearly phenotypically and functionally distinct, the majority of these memory cells shared a common repertoire; however, approximately one-third of the clonotypes were not "shared" between the subsets. By performing hemisplenectomy and analyzing the repertoires of sorted CD62L^{hi} and CD62L^{lo} cells at two different time points, the interrelatedness of these memory subsets was determined. This analysis revealed that (1) CD62L^{lo} cells were not maintained over time and (2) a number of the CD62L^{lo} cells could convert to a CD62L^{hi} state.

Although the repertoire of primary effectors and memory cells typically remains stable, increases in the avidity of the T cells may occur as the memory pool emerges. This functional maturation results even though TCR usage remains fairly constant, suggesting that modifications in the intrinsic properties of the memory T cells occur that fine-tune their responsiveness independently of changes in TCR–MHC affinity [101, 107–110]. These types of enhancements in the ability of T cells to respond to antigenic stimulation may impact the likelihood of particular T cells becoming recruited during secondary or recall responses.

The biological role of memory T cells is to help protect the host during secondary antigenic encounters [18–25]. These secondary exposures differ from primary exposures in two principle ways. First, an existing ensemble of memory T cells is available to initiate secondary response. Therefore, by comparison with primary responses, an increased frequency of antigen-specific T cells that are capable of mounting accelerated responses is available. Second, because of the anamnestic recall response, the duration of antigenic stimulation is shorter during secondary responses because the antigen is cleared more quickly. Although the repertoires of effector and memory pools are generally similar, reconfigurations can occur as the memory T cells become reactivated during recall responses (Figure 6.6).

Analyses of secondary responses to *Listeria monocytogenes* have shown that the repertoire of LLO 91–99-specific CD8 T cells becomes more focused [65]. In particular, epitope-specific clones, which were infrequent in the memory pool, were disfavored during the secondary response, whereas more prevalent subsets were preferentially recruited. These observations indicate that, in this instance, the timing of recruitment and rate of T-cell expansion during the secondary response may result in a less diverse repertoire of epitope-specific T cells following secondary stimulations.

A contrasting pattern emerges following secondary infection of BALB/c (H-2^{d}) mice with LCMV [66]. Following re-infection, major expansions of Vβ10^{+} and Vβ8.1/2^{+} NP118-specific T cells occur; however, the TCR usage remains conserved between primary, memory, and recall responses. Inspection of more minor responses (Vβ7, Vβ13 and Vβ6) also indicated that there was no difference in TCR usage between primary and secondary effectors. This trend of conserved TCR usage patterns was further confirmed by spectratype analyses as well as functional fingerprinting using variant NP118 peptides.

In H-2^{b} (C57BL/6) mice the relative magnitudes of epitope-specific CD8 T cells are quite strikingly different between primary acute and secondary LCMV infections [67]. As illustrated in Figure 6.7, following rechallenge of immune mice with LCMV, the NP396-specific response is overwhelming and the GP276 response reproducibly becomes the second most prevalent set of cells. Despite the obvious differential expansions of epitope-specific T cells, no convincing shifts in the TCR repertoire develop during the secondary response [67, 98]. Several factors may contribute to the marked secondary NP396-specific response. The temporal order of viral protein synthesis may influence the hierarchy of the secondary response. LCMV utilizes an ambisense coding strategy in which viral NP mRNAs are transcribed prior to synthesis of complementary viral genomes and several hours before GP mRNAs [111–113]. Additionally, studies have shown that NP-derived epitopes are presented earlier than GP epitopes [114]. Thus, during secondary infection preexisting memory T cells may first encounter NP epitopes, which results in the marked NP396-specific recall response. In addition to the timing of viral protein synthesis and antigen presentation, other factors, including the fitness or intrinsic responsiveness of the antigen-specific T cells, are likely to impact immunodominance [96, 107–109]. It is plausible that such inherent properties become differentially imprinted during the primary infection and depend upon the viral epitope recognized. Other factors, including the type of cells that display the antigen for inspection by T cells, may also impact immunodominance during recall responses, and this is perhaps best illustrated by influenza virus infection.

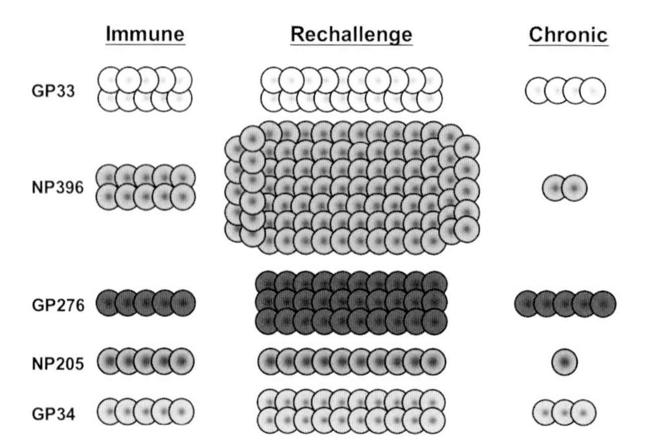

Figure 6.7 Contrasting profiles of immunodominance during protective secondary and non-protective chronic LCMV infections. Following acute LCMV infection of C57BL/6 mice, memory CD8 T cells are detected that are specific for various viral epitopes, including GP33, NP396, GP276, NP205, and GP34. These memory T cells are stably maintained but differ in abundance depending upon their epitope specificity, as depicted in the left column. The center column illustrates that if these LCMV-immune mice are rechallenged, all specificities of antiviral CD8 T cells expand to some extent, but the NP396-specific response is overwhelming. This overall anamnestic response rapidly clears the secondary infection. By contrast, as indicated in the right column, primary infections that are not controlled are associated with an altered epitope hierarchy and a marked diminution of the NP396 response. Thus, the specificity of antiviral T cells that respond most vigorously during protective secondary responses succumbs to deletion during chronic infections.

Studies with influenza virus have shown that both PA224- and NP366-specific responses are codominant during the primary infection and in the memory pool; however, secondary infections are characterized by more dominant NP366-specific responses [115–118]. Although the PA224-specific CD8 T cells do not expand as vigorously as NP366-specific cells following re-infection, they displayed superior functional qualities in that they appeared more capable of producing IL-2 and TNF-α [118]. This may reflect differences in the duration and magnitude of antigenic stimulation during the secondary response. Furthermore, assessment of TCR avidity by using MHC class I tetramers showed that PA224-specific CD8 T cells interacted more strongly than NP366-specific cells [118]. So why do NP366-specific T cells dominate the secondary response to influenza virus? Crowe et al. reported that the NP366 epitope is presented by both dendritic cells and non-dendritic cells, but only dendritic cells present the PA224 epitope [116]. Thus, during the primary response the presentation of both epitopes by dendritic cells drives comparable NP366- and PA224-specific responses. During secondary responses the preexisting memory cells will respond to antigen presented by both dendritic and non-dendritic cells. Since the NP366 epitope is more promiscuously presented, this is likely to drive a more vigorous response following secondary infections. Although differential antigen presentation may shape the pattern of secondary responses, these findings also suggest that certain specificities of T cells are likely to be more effective at clearing particular types of infected cells *in vivo*.

A common theme is that the memory pool usually portrays the shifts in the T-cell repertoire that become apparent as epitope-specific T cells expand during primary responses and successfully clear the antigen from the host. The downsizing or contraction of the primary response is proportional, and therefore the repertoire of the primary effector pool is transmitted to the memory pool. Hence, developing the most appropriate repertoire during the primary response is critical for shaping the memory pool. Although the primary and memory pools share similar repertoires, marked shifts in immunodominance, as well as further repertoire refinements, have been documented during certain secondary responses.

6.6 Heterologous Infections and Immunodominance

Experimentally controlled exposures to antigens have provided an informative means of carefully dissecting the dynamics of CD8 T-cell responses. As discussed above, such studies have shown that the composition of the memory CD8 T-cell pool and hierarchies of immunodominance usually remain remarkably stable following clearance of the priming antigen. In natural settings, however, contact with antigens is not tightly regulated. Therefore, hosts successively encounter, and perhaps reencounter, a plethora of immunogens over time.

Pioneering studies from the laboratories of Selin and Welsh have begun to address how serial infections with antigenically distinct viruses alter the abundance and traits of preexisting memory CD8 T cells [23, 92, 119]. The major finding is that heterologous infections are associated with quantitative and qualitative changes in the memory T-cell compartment. In mice that had been previously primed by acute LCMV infection, subsequent inoculation with Pichinde virus (PV), murine cytomegalovirus (MCMV), vaccinia virus (VV), or vesicular stomatitis virus (VSV) caused selective reductions in the numbers of LCMV-specific CD8 T cells [120, 121]. Single or multiple series of heterologous infections resulted in varying degrees of attrition. This trend was not restricted to LCMV-specific memory cells but was also observed in subsets of PV- as well as VV-specific memory T cells following subsequent heterologous challenge. Interestingly, epitope-specific memory T cells differed in their susceptibility to loss following a second infection. The removal of existing memory CD8 T cells during new primary infections may make available the necessary resources, such as geographic niches and cytokines, that promote the induction of responses to the current infection. This plasticity results in the downsizing of T cells specific to previously cleared infections, causing a reconfiguration of the memory pool.

Heterologous infections do not always result in reductions in the numbers of memory CD8 T cells, as cross-reactive specificities may be recruited [122, 123]. Both LCMV and PV share an epitope (NP205) with amino acid sequence similarity. The only differences are at residues 5 and 8 of the antigenic peptide. In both PV- and LCMV-immune mice, the response to this epitope is subdominant and does not exceed more than 1% of the memory CD8 T-cell population. Nevertheless, CD8 T cells specific for the NP205 epitope expand when PV-immune mice are infected with LCMV. This leads to a dramatic shift in immunodominance as the usually trailing response becomes preeminent [123]. Conversely, infection of LCMV immune mice with PV results in reduced responses to normally immunodominant PV epitopes but an enhancement of the NP205 response. This provides a rather clear-cut case of epitope cross-reactivities and demonstrates that the profiles of immunodominance can be modified by subsequent heterologous viral infections.

Cross-reactive CD8 T-cell specificities can be beneficial to the host because they can provide a degree of immunological protection. Prior infections with LCMV or PV are associated with lower viral titers during primary infection with VV [124]. Moreover, LCMV-immune mice are more resistant to death following exposure to a usually lethal dose of VV. This decreased mortality is associated with contrasting pathology in lung tissue and the formation of bronchus-associated lymphoid tissue (BALT). Evidently, LCMV and VV elicit some level of cross-reactive immunity.

In follow-up studies it was shown that whereas sequential PV and LCMV infections reproducibly augment NP205-specific CD8 T cells, the patterns of cross-reactivity are less predictable following VV infections of LCMV-immune mice [125]. The profiles of responses differ depending upon the individual mouse analyzed. Thus, LCMV NP205-specific cells were prominent in certain mice, but in others different sets of LCMV epitope–specific T cells preferentially responded during VV infection. These discrepancies suggest that variations in the T-cell repertoires between individual mice govern the patterns of cross-reactivities. Accordingly, the attributes of the host's private T-cell responses may modulate the ability to contain unrelated infections and account for intrinsic variability in cellular immune responses.

Cross-reactivity between unrelated viruses is not a phenomenon that is restricted to experimental models of infection. For example, CD8 T-cell responses to a hepatitis C virus (HCV) NS3 epitope were detectable in 60% of HCV seronegative donors [126]. This reactivity manifested because of prior infection with influenza virus, which primes CD8 T cells specific for a neuraminidase epitope with a high degree of sequence similarity with the NS3 peptide.

Heterologous viral infections illustrate the adaptability of CD8 T-cell responses. The effect of cross-reactivities on attrition or expansion of preexisting sets of the memory T cells is not necessarily predictable or reciprocal. It is molded by the antigenicity of the pathogen, the sequential order of infection, and the T-cell repertoire of the host. It should be noted that cross-reactivity may not always be advantageous [23]. The severity of several infections including EBV and varicella-zoster virus increases with age. It is tempting to speculate that the more extensive networks of memory T cells that form over time, as a result of exposures to various antigens, may encompass cross-reactive T-cell clones that enhance the pathology of particular infections.

6.7 Chronic Infections and T-cell Heterogeneity

Ideally, exposures to intracellular pathogens elicit functionally competent arrays of CD8 T cells that swiftly eradicate the infection as well as establish a pool of memory T cells, which are maintained following clearance and help to confer long-lived immunological protection. Numerous pathogens are, however, not rapidly controlled by the actions of the host's immune response. In these instances a protracted infection may ensue that is only slowly resolved, or the pathogen may never be eliminated, resulting in a lifelong persistent infection [24, 127–130]. Both of these scenarios may be associated with a more severe prognosis as well as pose public health problems because of increased risks of transmission resulting from shedding of the infectious agent.

Various defects in CD8 T-cell responses have been documented during persistent infections. The failure to rapidly control infections has been associated with lower magnitudes of T-cell expansion as well as with the arrested maturation of functionally competent and long-lived responses, which impedes the ability to elaborate or sustain critical T-cell effector functions. In persistently infected hosts, antigen-specific CD8 T cells may also succumb to deletion, leading to changes in repertoire diversity and altered epitope hierarchies (Figure 6.8) [57, 68, 131–134]. Other components of the immune system may also be either directly or indirectly suppressed by persistent infections. Notably, reduced CD4 T-cell activities may further diminish the efficacy of CD8 T-cell responses [135–137].

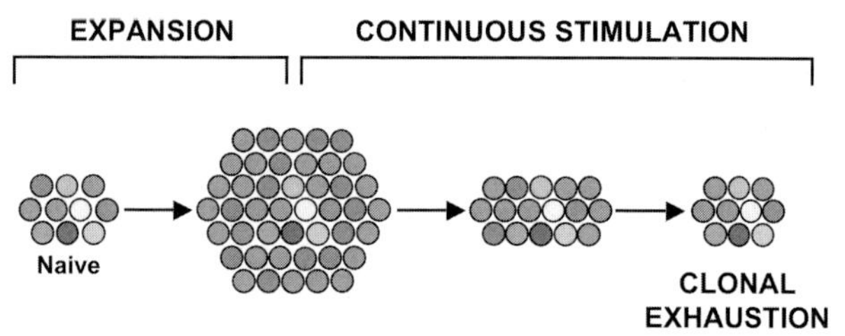

Figure 6.8 Continuous antigenic stimulation can drive responding CD8 T cells to deletion. Certain foreign antigens are not rapidly removed by the actions of the host's immune response. This is perhaps best exemplified by persistent viral infections. In these instances an initial response becomes detectable, resulting in repertoire shifts and the development of immunodominance. If the infection is not cleared by the overall immune response, then the responding cells may be subject to repetitive antigenic stimulation. Under these conditions certain clones and specificities of CD8 T cells may succumb to deletion, resulting in further changes in repertoire and epitope hierarchies. Notably, the deletion of CD8 T cells is exacerbated by the absence of CD4 T-cell help. (This figure also appears with the color plates.)

LCMV infection provides an informative system for comparing the induction and functions of antiviral CD8 T-cell responses during acute, prolonged, and chronic infections. A lifelong virus carrier state develops if infection occurs *in utero* or within one day of birth [138–140]. This inability to clear the infection is attributed to the presence of viral antigen in the thymus, which results in the negative selection of virus-specific T cells. This virus-induced "hole" in the T-cell repertoire ablates the elaboration of a high-affinity T-cell response capable of clearing the infection from the carrier mouse. More recently, studies of transgenic mice expressing the LCMV NP in the thymus have further highlighted the effect of thymic selection on peripheral antiviral T-cell responses. In these transgenic mice, no gross shifts in the TCR usage were detectable; nevertheless, the usually immunodominant NP118-specific response was reduced, as T cells with high affinity for this epitope failed to egress from the thymus [141, 142]. Interestingly, T cells specific for subdominant epitopes did emerge and were represented in the periphery. Therefore, the presence of viral antigen in the thymus may alter reactivities in the periphery, but if the repertoire remains sufficiently diverse, antiviral responses can be elicited.

Acute LCMV infection of adult mice is one of the best examples of the induction of massive and effective CD8 T-cell responses [29, 31]. By contrast, certain virulent LCMV variants cause a more potent infection that is only slowly or never brought under control. Similarly, a chronic infection also ensues following infection of certain immunodeficient strains of mice [138–140]. During the course of these types of prolonged or persistent LCMV infections, marked changes are observed in the hierarchy of CD8 T-cell responses. Although a multi-epitope-specific response is initially induced in C57BL/6 mice, the failure to resolve LCMV infection is associated with the deletion of T cells specific for the NP396 and NP205 epitopes. GP276-specific T cells emerge as the preeminent response, followed by T cells that recognize the GP33 epitope (Figure 6.7) [57, 68, 134]. If viral loads remain high, and if CD4 T-cell responses are absent, the numbers of GP-specific T cells also gradually decline [57, 143]. The rapid loss of the NP396-specific T cells during chronic LCMV infections is particularly noteworthy because this is a major immunodominant epitope during acute infections and these cells respond most vigorously following secondary challenges of immune mice (Figure 6.7) [53, 88]. These findings are consistent with the concept that epitopes ascribed as immunodominant in chronically infected hosts may not represent the specificities of T cells that are most effective at containing the infection.

Changes in immunodominance have also been described during the course of other persistent viral infections. During HCV infection, responses to the viral NS3 initially predominated, but as viremia declines, CD8 T cells specific for an NS5-derived peptide emerge [144]. Restructuring of immunodominance is also detectable during EBV infections. The acute phase of EBV infection is associated with a broad initial CD8 T-cell response directed towards lytic cycle antigens, although weaker reactivities against latent proteins are evident. These responses can be further subdivided among immediate early, early, and late viral proteins [145]. Interestingly, the hierarchy of the primary response appears to correlate with the temporal order of viral protein synthesis. The most prevalent initial responses are directed to immediate early and early gene products, but late lytic viral proteins appear less immunogenic. With the onset of latency, changes in the patterns of immunodominance arise as responses to lytic antigens dramatically decline, while latent antigen-specific CD8 T cells persist [146].

Analyses of TCR usage has been used to track changes in T-cell repertoires during persistent infections. Perforin-deficient mice cannot control LCMV infection because of defective CTL function [132, 147, 148]. In these chronically infected mice, the GP276 response becomes immunodominant, but, at the level of anti-Vβ antibody profiling, TCR usage by these cells is indistinguishable from that of primary effector and memory GP276-specific T cells in acutely infected mice [68]. Spectratype analysis of CDR3 lengths in Vβ8.1$^+$ CD8 T cells demonstrated that the T-cell response evolves during the course of protracted LCMV clone 13 infection of C57BL/6 mice. The distribution of CDR3 lengths is not fixed and continues to vary between days 7 and 13. Further longitudinal analyses reveal that the spectrum of CDR3 lengths regains a Gaussian distribution over time, and the patterns eventually resembled those observed in naïve mice [98].

Studies of individuals infected with HIV have demonstrated that a constantly changing repertoire of antiviral responses may help to slow the progression of disease. CD4 T-cell counts have been reported to decline more slowly in individuals who display more perturbations in Vβ usage by CD8 T cells [149]. Better infection control by a more diversified repertoire is further indicated by other studies of HIV-infected patients, because focused responses, resulting in massive expansions of T cells that use only a single Vβ gene, are associated with rapid progression to AIDS [42, 150].

In chimpanzees chronically infected with HCV, liver- and blood-derived CD8 T-cell clones that share highly conserved TCR CDR3 sequences are associated with the selection of epitope escape mutants [151]. By comparison, more diverse responses are associated with fewer amino acid substitutions in epitope sequences. Thus, the development of a broad and less focused antiviral T-cell response may be more effective at controlling chronic viral infections. Alternatively, it is possible that constant changes in the CD8 T-cell repertoire signify that responses which initially expand to combat the infection are prematurely terminated [98, 131–134, 152]. Hence, the constant evolution of T-cell responses to chronic viral infections may have opposing effects, providing a compensatory mechanism for controlling pathogens that are attempting to evade the immune response but also leading to the loss of clones that may be effective at eliminating virus-infected cells.

Although shifts in immunodominance are a common feature of persistent infections, alterations in the phenotypic and functional properties of the responding T cells are also observed [24, 57, 68, 128, 133–136, 153–159]. Persistent infections are associated with the aberrant development of memory T cells in the periphery. This failure of antiviral T cells to acquire a memory phenotype may contribute to changes in epitope hierarchies, as certain specificities are unable to survive over time. In addition, this altered maturation of virus-specific T cells may account for the loss of T-cell effector activities that has been shown to occur during persistent infections. The inability to elaborate necessary functional activities may in turn further promote virus persistence.

A stepwise and progressive functional inactivation of virus-specific CD8 T cells occurs during protracted and chronic LCMV infection of adult mice (Figure 6.9) [24, 57, 68, 132, 134]. Intracellular cytokine staining and multicolor flow cytometric analysis has shown that the generation of CD8 T cells that co-produce IL-2, TNF-α, and IFN-γ is markedly reduced early after infection [57, 68, 134]. As the infection continues, the capacity to produce TNF-α and IFN-γ declines further. The rate of functional impairment differs depending upon the epitope specificity of the responding T cells and precedes the physical deletion of the T cells. NP396- and NP205-specific cells succumb to the most rapid functional exhaustion, whereas the functional activities of GP33 and GP276 responses are extinguished more slowly. Under these conditions the degree of exhaustion may be proportional to the extent and repetitiveness of epitope-specific antigenic stimulation.

The functional inactivation of virus-specific CD8 T cells has also been observed following chronic viral infection of humans and non-human primates. During HIV infection, defects such as decreased production of IL-2 and TNF-α as well as reduced cytotoxic activities have been reported [136, 155, 160, 161]. In late-stage patients, diminished IFN-γ production has also been documented. During HCV infection, a transient "stunned" phenotype has been described, as IFN-γ production depreciates while viremia remains high but recovers as viral loads drop [162].

The notion that resolution of the infection can promote the recovery of antiviral T-cell functions is further suggested by studies at later time points following LCMV clone 13 infection of mice. In C57BL/6 mice, clone 13 infection is gradually resolved over a period of months. As the levels of virus decline, a partial restoration of cytokine production is observed, with IFN-γ, TNF-α, and IL-2 reemerging over time [57, 143, 163]. This is in striking contrast with chronic LCMV infection in the absence of CD4 T-cell help, because under these conditions, effector functions are abolished and all specificities of antiviral T cells are eventually deleted (Figure 6.9) [57, 143].

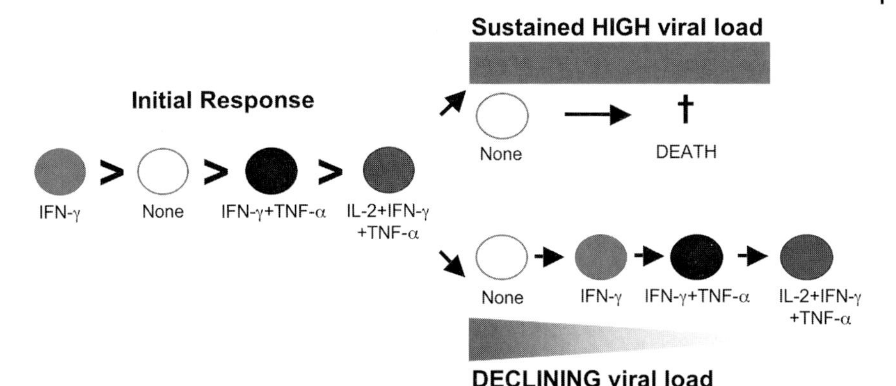

Figure 6.9 Decreased functional quality of antiviral T cells during chronic infections. The failure of adult mice to rapidly resolve LCMV infections is associated with a functionally impaired antiviral CD8 T-cell response. During the initial phase of the infection, virus-specific CD8 T cells can be subdivided based upon their patterns of cytokine production. CD8 T cells that produce only IFN-γ; co-produce IFN-γ and TNF-α; co-produce IFN-γ, TNF-α, and IL-2; or do not produce any of these cytokines can be identified and are present in different abundances. If viral loads remain high and CD4 T-cell help is absent, then cytokine-producing activities continue to deteriorate and this culminates in the physical loss of the virus-specific CD8 T cells. If, however, the initial viral loads decline within a sufficient time frame, at least a partial restoration of effector activities is detectable. Notably, the rates of functional impairment, deletion, and rebound differ depending upon the epitope specificity of the antiviral CD8 T cells.

Taken together, studies of persistent infections have illustrated the heterogeneity of peripheral CD8 T-cell responses. During these types of infections, shifts in immunodominance and repertoire usage may arise as more protective subsets of T cells succumb to deletion. Notably, more diversified responses may provide better long-term infection control. Persistent antigenic stimulation also prevents the transition of responding T cells into resting memory cells. This arrested development of antiviral T cells is consistent with dysregulated cytokine and cytotoxic capabilities and decreased proliferative potential. Moreover, findings in both murine and primate systems indicate that sustained high viral loads may result in a severe ablation of antigen-specific CD8 T-cell responses. Nevertheless, encouraging reports suggest that if successful infection control measures can be applied within a sufficient time frame, then at least a partial restoration of responsiveness may occur.

6.8 Repertoire Limitation and Immunodominance

As discussed above, T-cell development generates a massive number of CD8 T-cell clones that express unique TCRs. Consequently, restricting the diversity of the T-cell repertoire may alter immunodominance and affect the outcomes of infection. Limitation of the TCR repertoire can be experimentally achieved by generating mice transgenic for rearranged TCR α or β chains, by antibody depletion, or by superantigen treatment to delete particular Vβ families [104, 164–167]. Constrictions of the repertoire also develop naturally as a consequence of aging, which is associated with variable clonal expansions of subsets of T cells but a decrease in overall diversity [168–171]. Studies addressing the impact of repertoire limitation have given contrasting results, which most likely reflect the severity of the restriction as well as the plasticity of the response to the chosen epitope.

Adult immunocompetent mice infected by intracranial injection with LCMV typically die within one week as a result of lethal CD8 T cell–mediated immunopathology [138–140]. Investigations using transgenic mice, which express a Vβ8.1, Dβ2, Jβ2.3, Cβ2 rearranged TCR β chain (TG8.1), demonstrated that restricting TCR usage can modify the course of disease [164]. In this case the repertoire is limited because of exclusion of endogenous TCRβ rearrangements that are due to expression of the rearranged transgenic β chain; however, TCR α chain recombination may generate some diversity. TG8.1 mice do not mount a CD8 T-cell response that is capable of killing the host following intracranial LCMV infection, but antiviral CD8 T cells are functionally detectable [164]. Thus, limiting the repertoire may not completely abrogate the immune response but may profoundly alter the course of infection.

Restricting the diversity of the T-cell repertoire may also modify immunodominance [165]. Analyses of the response profiles following infection with influenza A virus (H3N2 strain) showed that both wild-type and TG8.1 mice elaborated an antiviral T-cell response, but the patterns of epitope recognition were markedly different. T-cell hybridomas generated from wild-type mice recognize a K^k-restricted NP or D^k-restricted PB-1 epitope. By contrast, TG8.1 mice mounted responses specific for a usually subdominant K^k-restricted NS-1 epitope and a D^k-restricted M-1 epitope. This rather striking alteration of immunodominance as a result of repertoire restriction may be due to changes in the avidity or in the precursor frequencies of T cells, which usually detect the arrays of epitopes presented following influenza virus infection.

Investigations using the superantigen staphylococcal enterotoxin B (SEB) to ablate the influenza virus–specific response revealed different effects depending upon the timing of administration [166, 167]. In C57BL/6 mice Vβ8.3$^+$ T cells respond prominently to the influenza virus D^b-restricted NP366 epitope [99, 104]. Treatment with SEB activates Vβ8$^+$ T cells and drives them to deletion. Transient treatment with SEB during the acute phase of the response resulted in a modest delay in influenza clearance but little overall impact on the D^b NP366 response [166]. This result is consistent with earlier antibody depletion studies designed to remove Vβ8$^+$ T cells during primary and secondary influenza virus responses [104]. In follow-up experiments SEB was administered to influenza virus–immune mice during the memory phase of the response via an implanted Alzet pump [167]. This treatment regimen resulted in a marked reduction in the percentage of NP366-specific Vβ8.3$^+$ CD8 T cells following secondary viral challenge, but there were no differences in the overall numbers of NP366-specific CD8

T cells. Instead, superantigen-mediated deletion of influenza virus–specific CD8 T cells resulted in a compensatory repertoire shift during the secondary response as $V\beta4^+$, $V\beta11^+$, and $V\beta13^+$ NP366 T cells expanded [167]. These findings illustrate the plasticity of the memory repertoire and show that even limited repertoires can combat pathogens.

The T-cell repertoire becomes progressively focused over the lifespan of the host. As individuals age the overall numbers of peripheral T cells do not decline even though thymic output is reduced. Studies of both humans and mice have shown selective expansions of particular T-cell clones in aged individuals, leading to a more restricted CD8 T-cell pool. This age-associated constriction of the T-cell repertoire is variable and is not always detectable [168–171]. The mechanisms responsible for this skewing are ill defined but are likely to be governed by at least two factors. First, continuous low-grade activation of T-cell clones, which are specific for environmental or cross-reactive endogenous antigens, may result in their preferred accumulation over time. Second, altered homeostatic proliferation may drive age-related clonal expansions.

The limitation of the TCR repertoire due to aging impacts the performance of the immune system. Treatment of influenza-immune mice with SEB had more dramatic effects on the ability to control secondary infections in aged mice than in younger counterparts [167]. This result is consistent with the notion that superantigen-mediated limitations in T-cell diversity are compounded by aging and this impairs the host's capacity to mount compensatory responses. The efficacy of herpes simplex virus (HSV)-1-specific antiviral CD8 T-cell responses also deteriorates with age [172]. In C57BL/6 mice, infection with HSV-1 elicits an immunodominant response towards the glycoprotein B (gB) 498–505 epitope that is characterized by usage of $V\beta10$ or $V\beta8$ by >90% of the CD8 T cells. Although young mice mounted potent HSV-specific responses, older mice mounted weaker responses. Antiviral CD8 T-cell activities are most severely reduced in older mice that exhibit age-related clonal expansion of $V\beta10^+$ or $V\beta8^+$ T cells. These diminished responses demonstrate that reduced repertoire diversity caused by aging can negatively impact the capacity of the host to mount vigorous primary responses.

6.9 Impact of Epitope Variation

Because immunodominance is in part determined by the sequence and availability of peptide antigens, a relevant question is how antigenic variation affects epitope hierarchies. This is of particular importance when considering responses to pathogens that are prone to sequence variation and can alter antigenicity in order to potentially escape immunological surveillance.

Amino acid substitutions within epitopes or in nearby flanking regions can have several consequences [173–175]. Mutations of residues that anchor the epitope in the peptide-binding groove of the MHC molecule can reduce the affinity, lower the complex stability, and decrease the presentation of the antigen. Alterations of flanking residues can modify the interaction of the newly synthesized proteins with the antigen-processing machinery and thus ablate presentation of the epitope. In both these instances the display of the peptide antigen to reactive T cells may be diminished. Mutations can also occur in peptide residues whose side chains structurally point away from the MHC molecule and serve as TCR contact residues. These types of changes can result in a reshaping of the repertoire as different sets of T-cell clones respond to the new epitope sequences [105]. Also, changes in TCR contact residues can modify the types of effector functions elaborated by preexisting T cells by operating as weaker agonists or antagonists [61, 62]. Consequently, changes in the epitope sequence may not result in the complete abolition of responsiveness but instead may activate suboptimal arrays of effector functions.

Pathogens may not necessarily tolerate all potential mutations within epitopes. Variation of regions that are functionally or structurally critical may decrease fitness, reducing replication rates or transmissibility. In these instances the escaped variant quasispecies may be underrepresented and out-competed. Epitope alterations can both directly impact T-cell responses and indirectly affect responses, for example, by altering viral loads or dissemination patterns. Thus, the host may be exposed to multiple variant viruses, which may coexist in a persistently infected individual. This renders analyses of responses more complex but elucidation of the biological impact of epitope alterations more significant.

The selection and maintenance of epitope escape mutants are driven by ongoing immune responses. This is perhaps best illustrated by studies using TCR-transgenic mice, where the majority of circulating T cells express rearranged TCRs that recognize a particular MHC–peptide combination. In P14 TCR-transgenic mice, the majority of CD8 T cells are specific for the H-2D^b-restricted LCMV GP33 epitope. Infection of these mice with LCMV selects for variant viruses, which escape recognition by the monoclonal TCR-transgenic CD8 T cells [176]. Similarly, influenza virus infection of mice transgenic for the F5 TCR, which recognizes the NP366 epitope, also results in the emergence of viral variants that escape immunosurveillance [177]. Both of these examples illustrate that in experimental settings a highly focused response may be detrimental, as this may select for variants that escape the actions of antigen-specific CD8 T cells.

Seminal studies from the laboratory of Simon Mallal highlighted the importance and power of the host's MHC class I molecules in exerting selective pressure on pathogens to evolve [178]. By performing comparisons of HIV sequences in patients with diverse HLA types, the findings indicated that HIV adapts to the host's immune response by mutating epitopes. Significantly, the patterns of epitope alterations were associated with the individual's HLA type. HLA-driven evolution of epitope sequences is also apparent in EBV isolates. In Caucasian populations the expression of HLA-A11 is associated with an immunodominant EBV-specific CD8 T-cell response. By contrast, populations in Papua New Guinea as well as southern China are highly HLA-A11$^+$ but typically mount at best weak A11-restricted, EBV-specific responses. This can be attributed to sequence variations in anchor residues of the immunodominant EBNA 4 416 epitope in EBV isolates that circulate in these geographic regions [179, 180]. Thus, at the population level immunodominance can influence the prevalence of viral quasispecies.

Epitope sequence, immunodominance, and functional quality of the response affect the outcome of infections, influencing resolution, chronicity, and viral loads. In the case of hepatitis B virus (HBV) [181] or HCV [182] infections, narrowly focused responses are associated with the progression to a chronic rather than an acute infection. In these instances the lack of diversity drives the selection of escape mutations early after infection. Relationships between immunodominance and epitope sequences were also emphasized by profiling the responses of two brothers who were believed to have become infected with HIV-1 following treatment with the same batch of contaminated Factor VIII [183]. Despite identical HLA types, the patterns of immunodominance differed in each brother. In the first sibling, marked CTL activity was directed towards an HLA-A2-restricted p17 (SLY) epitope, but weaker responses were detectable to HLA-A3-restricted epitopes from p17 Gag, Pol, and Nef. Antiviral CTL activities in the second sibling were dominated by HLA-B8-restricted responses to a p24 Gag and a Nef epitope; however, these specificities were below limits of detection in the first sibling. So what accounts for this discordant pattern of responses? The most likely factor is amino acid alterations in the immunodominant epitopes. Although the HLA-A2-restricted, p17-specific response was prominent in the first sibling, sequence analyses of proviral DNA revealed that this epitope was significantly altered in the second sibling, likely favoring the elaboration of HLA-B8 restricted responses [183]. These findings are consistent with the notion that the selection of escape variants can reconfigure the hierarchy of the T-cell immunodominance.

Alterations in epitope sequences do not necessarily result in the expansion of novel sets of T-cell clones but instead, under certain conditions, can result in biased activation of T cells that recognize the initial epitope sequence. This concept of original antigen sin was first reported for antibody responses to influenza virus, but it has also been documented to occur with CD8 T-cell responses [184]. Asymmetric cross-reactivities were noted when mice primed with wild-type strains of LCMV were rechallenged with variant viruses that contained mutations in key epitopes. During these secondary infections, memory T cells reactive against the original epitope sequence preferentially responded, preventing the effective priming of new responses to the mutant epitope. Greater cross-reactivity was detectable if responses were first primed against the mutant sequence followed by challenge with the wild-type sequence. This suggests that the skewing of the repertoire, during the priming phase, determines the plasticity of the memory pool and the ability of this assembly of cells to respond to and combat viral variants.

6.10 Concluding Remarks

Immunodominance is an underlying aspect of CD8 T-cell responses and arises when multiple epitopes are available to initiate memory T-cell differentiation.

Striking quantitative and qualitative differences have been observed in the panoply of responses to an individual pathogen. These discordances can have profound biological consequences because they can augment or diminish the host's ability to clear primary infections, combat secondary infections, and contain persistent infections. Progress has been made in describing immunodominance and elucidating the interrelationships between some of the factors that shape this complex phenomenon. Ideally, future studies that further our understanding of this feature of CD8 T-cell biology will facilitate the development of prophylactic and therapeutic strategies for improving protective immunity.

References

1 A.R. Townsend, F.M. Gotch, J. Davey, Cytotoxic T cells recognize fragments of the influenza nucleoprotein, *Cell* **1985**, *42*, 457–467.

2 A.R. Townsend, J. Rothbard, F.M. Gotch, G. Bahadur, D. Wraith, A.J. McMichael, The epitopes of influenza nucleoprotein recognized by cytotoxic T lymphocytes can be defined with short synthetic peptides, *Cell* **1986**, *44*, 959–968.

3 P.J. Bjorkman, M.A. Saper, B. Samraoui, W.S. Bennett, J.L. Strominger, D.C. Wiley, Structure of the human class I histocompatibility antigen, HLA-A2, *Nature* **1987**, *329*, 506–512.

4 D.H. Fremont, M. Matsumura, E.A. Stura, P.A. Peterson, I.A. Wilson, Crystal structures of two viral peptides in complex with murine MHC class I H-2Kb, *Science* **1992**, *257*, 919–927.

5 M.M. Davis, P.J. Bjorkman, T-cell antigen receptor genes and T-cell recognition, *Nature* **1988**, *334*, 395–402.

6 J.L. Jorgensen, P.A. Reay, E.W. Ehrich, M.M. Davis, Molecular components of T-cell recognition, *Annu Rev Immunol* **1992**, *10*, 835–873.

7 A.W. Goldrath, M.J. Bevan, Selecting and maintaining a diverse T-cell repertoire, *Nature* **1999**, *402*, 255–262.

8 G. Glusman, L. Rowen, I. Lee, C. Boysen, J.C. Roach, A.F. Smit, K. Wang, B.F. Koop, L. Hood, Comparative genomics of the human and mouse T cell receptor loci, *Immunity* **2001**, *15*, 337–349.

9 J. Nikolich-Zugich, M.K. Slifka, I. Messaoudi, The many important facets of T-cell repertoire diversity, *Nat Rev Immunol* **2004**, *4*, 123–132.

10 D.N. Garboczi, P. Ghosh, U. Utz, Q.R. Fan, W.E. Biddison, D.C. Wiley, Structure of the complex between human T-cell receptor, viral peptide and HLA-A2, *Nature* **1996**, *384*, 134–141.

11 M.M. Davis, J.J. Boniface, Z. Reich, D. Lyons, J. Hampl, B. Arden, Y. Chien, Ligand recognition by alpha beta T cell receptors, *Annu Rev Immunol* **1998**, *16*, 523–544.

12 K.C. Garcia, M. Degano, L.R. Pease, M. Huang, P.A. Peterson, L. Teyton, I.A. Wilson, Structural basis of plasticity in T cell receptor recognition of a self peptide-MHC antigen, *Science* **1998**, *279*, 1166–1172.

13 K.C. Garcia, L. Teyton, I.A. Wilson, Structural basis of T cell recognition, *Annu Rev Immunol* **1999**, *17*, 369–397.

14 J.B. Reiser, C. Gregoire, C. Darnault, T. Mosser, A. Guimezanes, A.M. Schmitt-Verhulst, J.C. Fontecilla-Camps, G. Mazza, B. Malissen, D. Housset, A T cell receptor CDR3beta loop undergoes conformational changes of unprecedented magnitude upon binding to a peptide/MHC class I complex, *Immunity* **2002**, *16*, 345–354.

15 J.W. Yewdell, J.R. Bennink, Immunodominance in major histocompatibility complex class I-restricted T lymphocyte responses, *Annu Rev Immunol* **1999**, *17*, 51–88.

16 J.W. Yewdell, M. Del Val, Immunodominance in TCD8+ responses to viruses: cell biology, cellular immunology, and mathematical models, *Immunity* **2004**, *21*, 149–153.

17 A. Gallimore, H. Hengartner, R. Zinkernagel, Hierarchies of antigen-specific cytotoxic T-cell responses, *Immunol Rev* **1998**, *164*, 29–36.

18 R.W. Dutton, L.M. Bradley, S.L. Swain, T cell memory, *Annu Rev Immunol* **1998**, *16*, 201–223.

19 P.C. Doherty, J.P. Christensen, Accessing complexity: the dynamics of virus-specific T cell responses, *Annu Rev Immunol* **2000**, *18*, 561–592.

20 S.M. Kaech, E.J. Wherry, R. Ahmed, Effector and memory T-cell differentiation: implications for vaccine development, *Nat Rev Immunol* **2002**, *2*, 251–262.

21 R.A. Seder, R. Ahmed, Similarities and differences in CD4+ and CD8+ effector and memory T cell generation, *Nat Immunol* **2003**, *4*, 835–842.

22 D. Masopust, S.M. Kaech, E.J. Wherry, R. Ahmed, The role of programming in memory T-cell development, *Curr Opin Immunol* **2004**, *16*, 217–225.

23 L.K. Selin, R.M. Welsh, Plasticity of T cell memory responses to viruses, *Immunity* **2004**, *20*, 5–16.

24 E.J. Wherry, R. Ahmed, Memory CD8 T-cell differentiation during viral infection, *J Virol* **2004**, *78*, 5535–5545.

25 R.M. Welsh, L.K. Selin, E. Szomolanyi-Tsuda, Immunological memory to viral infections, *Annu Rev Immunol* **2004**, *22*, 711–743.

26 S. Hou, L. Hyland, K.W. Ryan, A. Portner, P.C. Doherty, Virus-specific CD8+ T-cell memory determined by clonal burst size, *Nature* **1994**, *369*, 652–654.

27 L.L. Lau, B.D. Jamieson, T. Somasundaram, R. Ahmed, Cytotoxic T-cell memory without antigen, *Nature* **1994**, *369*, 648–652.

28 D.H. Busch, I.M. Pilip, S. Vijh, E.G. Pamer, Coordinate regulation of complex T cell populations responding to bacterial infection, *Immunity* **1998**, *8*, 353–362.

29 E.A. Butz, M.J. Bevan, Massive expansion of antigen-specific CD8+ T cells during an acute virus infection, *Immunity* **1998**, *8*, 167–175.

30 K.J. Flynn, G.T. Belz, J.D. Altman, R. Ahmed, D.L. Woodland, P.C. Doherty, Virus-specific CD8+ T cells in primary and secondary influenza pneumonia, *Immunity* **1998**, *8*, 683–691.

31 K. Murali-Krishna, J.D. Altman, M. Suresh, D.J. Sourdive, A.J. Zajac, J.D. Miller, J. Slansky, R. Ahmed, Counting antigen-specific CD8 T cells: a reevaluation of bystander activation during viral infection, *Immunity* **1998**, *8*, 177–187.

32 J. Sprent, Circulating T and B lymphocytes of the mouse. I. Migratory properties, *Cell Immunol* **1973**, *7*, 10–39.

33 E.C. Butcher, L.J. Picker, Lymphocyte homing and homeostasis, *Science* **1996**, *272*, 60–66.

34 J. Banchereau, F. Briere, C. Caux, J. Davoust, S. Lebecque, Y.J. Liu, B. Pulendran, K. Palucka, Immunobiology of dendritic cells, *Annu Rev Immunol* **2000**, *18*, 767–811.

35 S.M. Kaech, R. Ahmed, Memory CD8+ T cell differentiation: initial antigen encounter triggers a developmental program in naive cells, *Nat Immunol* **2001**, *2*, 415–422.

36 M.J. van Stipdonk, E.E. Lemmens, S.P. Schoenberger, Naive CTLs require a single brief period of antigenic stimulation for clonal expansion and differentiation, *Nat Immunol* **2001**, *2*, 423–429.

37 P. Marrack, T. Mitchell, D. Hildeman, R. Kedl, T.K. Teague, J. Bender, W. Rees, B.C. Schaefer, J. Kappler, Genomic-scale analysis of gene expression in resting and activated T cells, *Curr Opin Immunol* **2000**, *12*, 206–209.

38 J.M. Grayson, K. Murali-Krishna, J.D. Altman, R. Ahmed, Gene expression in antigen-specific CD8+ T cells during viral infection, *J Immunol* **2001**, *166*, 795–799.

39 S.M. Kaech, S. Hemby, E. Kersh, R. Ahmed, Molecular and functional profiling of memory CD8 T cell differentiation, *Cell* **2002**, *111*, 837–851.

40 B.J. Johnson, E.O. Costelloe, D.R. Fitzpatrick, J.B. Haanen, T.N. Schumacher, L.E. Brown, A. Kelso, Single-cell perforin and granzyme expression reveals the anatomical localization of effector CD8+ T cells in influenza virus-infected mice, *Proc Natl Acad Sci U S A* **2003**, *100*, 2657–2662.

41 S. Holmes, M. He, T. Xu, P.P. Lee, Memory T cells have gene expression patterns intermediate between naive and effector, *Proc Natl Acad Sci U S A* **2005**, *102*, 5519–5523.

42 G. Pantaleo, J.F. Demarest, H. Soudeyns, C. Graziosi, F. Denis, J.W. Adelsberger, P. Borrow, M.S. Saag, G.M. Shaw, R.P. Sekaly, et al., Major expansion of CD8+ T cells with a predominant V beta usage during the primary immune response to HIV, *Nature* **1994**, *370*, 463–467.

43 M.F. Callan, N. Steven, P. Krausa, J.D. Wilson, P.A. Moss, G.M. Gillespie, J.I. Bell, A.B. Rickinson, A.J. McMichael, Large clonal expansions of CD8+ T cells in acute infectious mononucleosis, *Nat Med* **1996**, *2*, 906–911.

44 M.F. Callan, L. Tan, N. Annels, G.S. Ogg, J.D. Wilson, C.A. O'Callaghan, N. Steven, A.J. McMichael, A.B. Rickinson, Direct visualization of antigen-specific CD8+ T cells during the primary immune response to Epstein-Barr virus In vivo, *J Exp Med* **1998**, *187*, 1395–1402.

45 S.J. Turner, G. Diaz, R. Cross, P.C. Doherty, Analysis of clonotype distribution and persistence for an influenza virus-specific CD8+ T cell response, *Immunity* **2003**, *18*, 549–559.

46 D.R. Marshall, S.J. Turner, G.T. Belz, S. Wingo, S. Andreansky, M.Y. Sangster, J.M. Riberdy, T. Liu, M. Tan, P.C. Doherty, Measuring the diaspora for virus-specific CD8+ T cells, *Proc Natl Acad Sci U S A* **2001**, *98*, 6313–6318.

47 D. Masopust, V. Vezys, A.L. Marzo, L. Lefrancois, Preferential localization of effector memory cells in nonlymphoid tissue, *Science* **2001**, *291*, 2413–2417.

48 K.D. Klonowski, K.J. Williams, A.L. Marzo, D.A. Blair, E.G. Lingenheld, L. Lefrancois, Dynamics of blood-borne CD8 memory T cell migration in vivo, *Immunity* **2004**, *20*, 551–562.

49 V.P. Badovinac, B.B. Porter, J.T. Harty, Programmed contraction of CD8(+) T cells after infection, *Nat Immunol* **2002**, *3*, 619–626.

50 F. Sallusto, D. Lenig, R. Forster, M. Lipp, A. Lanzavecchia, Two subsets of memory T lymphocytes with distinct homing potentials and effector functions, *Nature* **1999**, *401*, 708–712.

51 E.J. Wherry, V. Teichgraber, T.C. Becker, D. Masopust, S.M. Kaech, R. Antia, U.H. von Andrian, R. Ahmed, Lineage relationship and protective immunity of memory CD8 T cell subsets, *Nat Immunol* **2003**, *4*, 225–234.

52 F. Sallusto, J. Geginat, A. Lanzavecchia, Central memory and effector memory T cell subsets: function, generation, and maintenance, *Annu Rev Immunol* **2004**, *22*, 745–763.

53 A. Khanolkar, M.J. Fuller, A.J. Zajac, CD4 T cell-dependent CD8 T cell maturation, *J Immunol* **2004**, *172*, 2834–2844.

54 C. Bouneaud, Z. Garcia, P. Kourilsky, C. Pannetier, Lineage relationships, homeostasis, and recall capacities of central- and effector-memory CD8 T cells in vivo, *J Exp Med* **2005**, *201*, 579–590.

55 H. Unsoeld, H. Pircher, Complex memory T-cell phenotypes revealed by coexpression of CD62L and CCR7, *J Virol* **2005**, *79*, 4510–4513.

56 D. Homann, L. Teyton, M.B. Oldstone, Differential regulation of antiviral T-cell immunity results in stable CD8+ but declining CD4+ T-cell memory, *Nat Med* **2001**, *7*, 913–919.

57 M.J. Fuller, A. Khanolkar, A.E. Tebo, A.J. Zajac, Maintenance, loss, and resurgence of T cell responses during acute, protracted, and chronic viral infections, *J Immunol* **2004**, *172*, 4204–4214.

58 W.E. Demkowicz, Jr., R.A. Littaua, J. Wang, F.A. Ennis, Human cytotoxic T-cell memory: long-lived responses to vaccinia virus, *J Virol* **1996**, *70*, 2627–2631.

59 E. Hammarlund, M.W. Lewis, S.G. Hansen, L.I. Strelow, J.A. Nelson, G.J. Sexton, J.M. Hanifin, M.K. Slifka, Duration of antiviral immunity after smallpox vaccination, *Nat Med* **2003**, *9*, 1131–1137.

60 S. Oehen, H. Hengartner, R.M. Zinkernagel, Vaccination for disease, *Science* **1991**, *251*, 195–198.

61 A. Bertoletti, A. Sette, F.V. Chisari, A. Penna, M. Levrero, M. De Carli, F. Fiaccadori, C. Ferrari, Natural var-

iants of cytotoxic epitopes are T-cell receptor antagonists for antiviral cytotoxic T cells, *Nature* **1994**, *369*, 407–410.
62 P. Klenerman, S. Rowland-Jones, S. McAdam, J. Edwards, S. Daenke, D. Lalloo, B. Koppe, W. Rosenberg, D. Boyd, A. Edwards, et al., Cytotoxic T-cell activity antagonized by naturally occurring HIV-1 Gag variants, *Nature* **1994**, *369*, 403–407.
63 L.A. Herzenberg, S.C. De Rosa, Monoclonal antibodies and the FACS: complementary tools for immunobiology and medicine, *Immunol Today* **2000**, *21*, 383–390.
64 S.C. De Rosa, J.M. Brenchley, M. Roederer, Beyond six colors: a new era in flow cytometry, *Nat Med* **2003**, *9*, 112–117.
65 D.H. Busch, I. Pilip, E.G. Pamer, Evolution of a complex T cell receptor repertoire during primary and recall bacterial infection, *J Exp Med* **1998**, *188*, 61–70.
66 D.J. Sourdive, K. Murali-Krishna, J.D. Altman, A.J. Zajac, J.K. Whitmire, C. Pannetier, P. Kourilsky, B. Evavold, A. Sette, R. Ahmed, Conserved T cell receptor repertoire in primary and memory CD8 T cell responses to an acute viral infection, *J Exp Med* **1998**, *188*, 71–82.
67 J.N. Blattman, D.J. Sourdive, K. Murali-Krishna, R. Ahmed, J.D. Altman, Evolution of the T cell repertoire during primary, memory, and recall responses to viral infection, *J Immunol* **2000**, *165*, 6081–6090.
68 M.J. Fuller, A.J. Zajac, Ablation of CD8 and CD4 T cell responses by high viral loads, *J Immunol* **2003**, *170*, 477–486.
69 J.L. Maryanski, C.V. Jongeneel, P. Bucher, J.L. Casanova, P.R. Walker, Single-cell PCR analysis of TCR repertoires selected by antigen in vivo: a high magnitude CD8 response is comprised of very few clones, *Immunity* **1996**, *4*, 47–55.
70 J.L. Maryanski, V. Attuil, P. Bucher, P.R. Walker, A quantitative, single-cell PCR analysis of an antigen-specific TCR repertoire selected during an in vivo CD8 response: direct evidence for a wide range of clone sizes with uniform tissue distribution, *Mol Immunol* **1999**, *36*, 745–753.
71 C. Pannetier, M. Cochet, S. Darche, A. Casrouge, M. Zoller, P. Kourilsky, The sizes of the CDR3 hypervariable regions of the murine T-cell receptor beta chains vary as a function of the recombined germ-line segments, *Proc Natl Acad Sci U S A* **1993**, *90*, 4319–4323.
72 C. Pannetier, J. Even, P. Kourilsky, T-cell repertoire diversity and clonal expansions in normal and clinical samples, *Immunol Today* **1995**, *16*, 176–181.
73 J.D. Altman, P.A. Moss, P.J. Goulder, D.H. Barouch, M.G. McHeyzer-Williams, J.I. Bell, A.J. McMichael, M.M. Davis, Phenotypic analysis of antigen-specific T lymphocytes, *Science* **1996**, *274*, 94–96.
74 A.J. McMichael, C.A. O'Callaghan, A new look at T cells, *J Exp Med* **1998**, *187*, 1367–1371.
75 P. Klenerman, V. Cerundolo, P.R. Dunbar, Tracking T cells with tetramers: new tales from new tools, *Nat Rev Immunol* **2002**, *2*, 263–272.
76 T.F. Greten, J.E. Slansky, R. Kubota, S.S. Soldan, E.M. Jaffee, T.P. Leist, D.M. Pardoll, S. Jacobson, J.P. Schneck, Direct visualization of antigen-specific T cells: HTLV-1 Tax11–19- specific CD8(+) T cells are activated in peripheral blood and accumulate in cerebrospinal fluid from HAM/TSP patients, *Proc Natl Acad Sci U S A* **1998**, *95*, 7568–7573.
77 J.R. Cochran, L.J. Stern, A diverse set of oligomeric class II MHC-peptide complexes for probing T-cell receptor interactions, *Chem Biol* **2000**, *7*, 683–696.
78 R.G. van der Most, K. Murali-Krishna, J.G. Lanier, E.J. Wherry, M.T. Puglielli, J.N. Blattman, A. Sette, R. Ahmed, Changing immunodominance patterns in antiviral CD8 T-cell responses after loss of epitope presentation or chronic antigenic stimulation, *Virology* **2003**, *315*, 93–102.
79 R.G. van der Most, A. Sette, C. Oseroff, J. Alexander, K. Murali-Krishna, L.L. Lau, S. Southwood, J. Sidney, R.W. Chesnut, M. Matloubian, R. Ahmed, Analysis of cytotoxic T cell responses to dominant and subdominant epitopes during acute and chronic lymphocytic

choriomeningitis virus infection, *J Immunol* **1996**, *157*, 5543–5554.
80 R.G. van der Most, R.J. Concepcion, C. Oseroff, J. Alexander, S. Southwood, J. Sidney, R.W. Chesnut, R. Ahmed, A. Sette, Uncovering subdominant cytotoxic T-lymphocyte responses in lymphocytic choriomeningitis virus-infected BALB/c mice, *J Virol* **1997**, *71*, 5110–5114.
81 A. Sette, J. Sidney, M.F. del Guercio, S. Southwood, J. Ruppert, C. Dahlberg, H.M. Grey, R.T. Kubo, Peptide binding to the most frequent HLA-A class I alleles measured by quantitative molecular binding assays, *Mol Immunol* **1994**, *31*, 813–822.
82 A. Sette, A. Vitiello, B. Reherman, P. Fowler, R. Nayersina, W.M. Kast, C.J. Melief, C. Oseroff, L. Yuan, J. Ruppert, et al., The relationship between class I binding affinity and immunogenicity of potential cytotoxic T cell epitopes, *J Immunol* **1994**, *153*, 5586–5592.
83 R.G. van der Most, K. Murali-Krishna, J.L. Whitton, C. Oseroff, J. Alexander, S. Southwood, J. Sidney, R.W. Chesnut, A. Sette, R. Ahmed, Identification of Db- and Kb-restricted subdominant cytotoxic T-cell responses in lymphocytic choriomeningitis virus-infected mice, *Virology* **1998**, *240*, 158–167.
84 S. Vijh, E.G. Pamer, Immunodominant and subdominant CTL responses to Listeria monocytogenes infection, *J Immunol* **1997**, *158*, 3366–3371.
85 E.G. Pamer, Direct sequence identification and kinetic analysis of an MHC class I-restricted Listeria monocytogenes CTL epitope, *J Immunol* **1994**, *152*, 686–694.
86 A.J. Sijts, A. Neisig, J. Neefjes, E.G. Pamer, Two Listeria monocytogenes CTL epitopes are processed from the same antigen with different efficiencies, *J Immunol* **1996**, *156*, 683–692.
87 W. Chen, L.C. Anton, J.R. Bennink, J.W. Yewdell, Dissecting the multifactorial causes of immunodominance in class I-restricted T cell responses to viruses, *Immunity* **2000**, *12*, 83–93.
88 A. Gallimore, T. Dumrese, H. Hengartner, R.M. Zinkernagel, H.G. Rammensee, Protective immunity does not correlate with the hierarchy of virus-specific cytotoxic T cell responses to naturally processed peptides, *J Exp Med* **1998**, *187*, 1647–1657.
89 H. Inagaki, T. Suzuki, K. Nomoto, Y. Yoshikai, Increased susceptibility to primary infection with Listeria monocytogenes in germfree mice may be due to lack of accumulation of L-selectin+ CD44+ T cells in sites of inflammation, *Infect Immun* **1996**, *64*, 3280–3287.
90 A.W. Goldrath, L.Y. Bogatzki, M.J. Bevan, Naive T cells transiently acquire a memory-like phenotype during homeostasis-driven proliferation, *J Exp Med* **2000**, *192*, 557–564.
91 K. Murali-Krishna, R. Ahmed, Cutting edge: naive T cells masquerading as memory cells, *J Immunol* **2000**, *165*, 1733–1737.
92 R.M. Welsh, L.K. Selin, No one is naive: the significance of heterologous T-cell immunity, *Nat Rev Immunol* **2002**, *2*, 417–426.
93 P. Wong, E.G. Pamer, Feedback regulation of pathogen-specific T cell priming, *Immunity* **2003**, *18*, 499–511.
94 V.P. Badovinac, A.R. Tvinnereim, J.T. Harty, Regulation of antigen-specific CD8+ T cell homeostasis by perforin and interferon-gamma, *Science* **2000**, *290*, 1354–1358.
95 F. Rodriguez, S. Harkins, M.K. Slifka, J.L. Whitton, Immunodominance in virus-induced CD8(+) T-cell responses is dramatically modified by DNA immunization and is regulated by gamma interferon, *J Virol* **2002**, *76*, 4251–4259.
96 F. Liu, J.L. Whitton, M.K. Slifka, The rapidity with which virus-specific CD8+ T cells initiate IFN-gamma synthesis increases markedly over the course of infection and correlates with immunodominance, *J Immunol* **2004**, *173*, 456–462.
97 J.K. Whitmire, J.T. Tan, J.L. Whitton, Interferon-gamma acts directly on CD8+ T cells to increase their abundance during virus infection, *J Exp Med* **2005**, *201*, 1053–1059.
98 M.Y. Lin, R.M. Welsh, Stability and diversity of T cell receptor repertoire

usage during lymphocytic choriomeningitis virus infection of mice, *J Exp Med* **1998**, *188*, 1993–2005.

99 K. Kedzierska, S.J. Turner, P.C. Doherty, Conserved T cell receptor usage in primary and recall responses to an immunodominant influenza virus nucleoprotein epitope, *Proc Natl Acad Sci U S A* **2004**, *101*, 4942–4947.

100 D.H. Busch, E.G. Pamer, T lymphocyte dynamics during Listeria monocytogenes infection, *Immunol Lett* **1999**, *65*, 93–98.

101 D.H. Busch, E.G. Pamer, T cell affinity maturation by selective expansion during infection, *J Exp Med* **1999**, *189*, 701–710.

102 S.J. Turner, K. Kedzierska, N.L. La Gruta, R. Webby, P.C. Doherty, Characterization of CD8+ T cell repertoire diversity and persistence in the influenza A virus model of localized, transient infection, *Semin Immunol* **2004**, *16*, 179–184.

103 W. Zhong, E.L. Reinherz, In vivo selection of a TCR Vbeta repertoire directed against an immunodominant influenza virus CTL epitope, *Int Immunol* **2004**, *16*, 1549–1559.

104 A.M. Deckhut, W. Allan, A. McMickle, M. Eichelberger, M.A. Blackman, P.C. Doherty, D.L. Woodland, Prominent usage of V beta 8.3 T cells in the H-2Db-restricted response to an influenza A virus nucleoprotein epitope, *J Immunol* **1993**, *151*, 2658–2666.

105 S.J. Turner, K. Kedzierska, H. Komodromou, N.L. La Gruta, M.A. Dunstone, A.I. Webb, R. Webby, H. Walden, W. Xie, J. McCluskey, A.W. Purcell, J. Rossjohn, P.C. Doherty, Lack of prominent peptide-major histocompatibility complex features limits repertoire diversity in virus-specific CD8+ T cell populations, *Nat Immunol* **2005**, *6*, 382–389.

106 P.O. Campos-Lima, V. Levitsky, M.P. Imreh, R. Gavioli, M.G. Masucci, Epitope-dependent selection of highly restricted or diverse T cell receptor repertoires in response to persistent infection by Epstein-Barr virus, *J Exp Med* **1997**, *186*, 83–89.

107 M.F. Bachmann, D.E. Speiser, P.S. Ohashi, Functional management of an antiviral cytotoxic T-cell response, *J Virol* **1997**, *71*, 5764–5768.

108 M.A. Alexander-Miller, Differential expansion and survival of high and low avidity cytotoxic T cell populations during the immune response to a viral infection, *Cell Immunol* **2000**, *201*, 58–62.

109 M.K. Slifka, J.L. Whitton, Functional avidity maturation of CD8(+) T cells without selection of higher affinity TCR, *Nat Immunol* **2001**, *2*, 711–717.

110 P.M. Gray, G.D. Parks, M.A. Alexander-Miller, High avidity CD8+ T cells are the initial population elicited following viral infection of the respiratory tract, *J Immunol* **2003**, *170*, 174–181.

111 M.J. Buchmeier, J.H. Elder, M.B. Oldstone, Protein structure of lymphocytic choriomeningitis virus: identification of the virus structural and cell associated polypeptides, *Virology* **1978**, *89*, 133–145.

112 D.D. Auperin, V. Romanowski, M. Galinski, D.H. Bishop, Sequencing studies of pichinde arenavirus S RNA indicate a novel coding strategy, an ambisense viral S RNA, *J Virol* **1984**, *52*, 897–904.

113 F.V. Fuller-Pace, P.J. Southern, Temporal analysis of transcription and replication during acute infection with lymphocytic choriomeningitis virus, *Virology* **1988**, *162*, 260–263.

114 H.C. Probst, K. Tschannen, A. Gallimore, M. Martinic, M. Basler, T. Dumrese, E. Jones, M.F. van den Broek, Immunodominance of an antiviral cytotoxic T cell response is shaped by the kinetics of viral protein expression, *J Immunol* **2003**, *171*, 5415–5422.

115 G.T. Belz, W. Xie, J.D. Altman, P.C. Doherty, A previously unrecognized H-2D(b)-restricted peptide prominent in the primary influenza A virus-specific CD8(+) T-cell response is much less apparent following secondary challenge, *J Virol* **2000**, *74*, 3486–3493.

116 S.R. Crowe, S.J. Turner, S.C. Miller, A.D. Roberts, R.A. Rappolo, P.C. Doherty, K.H. Ely, D.L. Woodland, Differential antigen presentation regulates the changing patterns of CD8+ T cell immunodominance in primary and secondary influenza virus infections, *J Exp Med* **2003**, *198*, 399–410.

117 W. Chen, K. Pang, K.A. Masterman, G. Kennedy, S. Basta, N. Dimopoulos, F. Hornung, M. Smyth, J.R. Bennink, J.W. Yewdell, Reversal in the immunodominance hierarchy in secondary CD8+ T cell responses to influenza A virus: roles for cross-presentation and lysis-independent immunodomination, *J Immunol* **2004**, *173*, 5021–5027.

118 N.L. La Gruta, S.J. Turner, P.C. Doherty, Hierarchies in cytokine expression profiles for acute and resolving influenza virus-specific CD8+ T cell responses: correlation of cytokine profile and TCR avidity, *J Immunol* **2004**, *172*, 5553–5560.

119 M.Y. Lin, L.K. Selin, R.M. Welsh, Evolution of the CD8 T-cell repertoire during infections, *Microbes Infect* **2000**, *2*, 1025–1039.

120 L.K. Selin, K. Vergilis, R.M. Welsh, S.R. Nahill, Reduction of otherwise remarkably stable virus-specific cytotoxic T lymphocyte memory by heterologous viral infections, *J Exp Med* **1996**, *183*, 2489–2499.

121 L.K. Selin, M.Y. Lin, K.A. Kraemer, D.M. Pardoll, J.P. Schneck, S.M. Varga, P.A. Santolucito, A.K. Pinto, R.M. Welsh, Attrition of T cell memory: selective loss of LCMV epitope-specific memory CD8 T cells following infections with heterologous viruses, *Immunity* **1999**, *11*, 733–742.

122 L.K. Selin, S.R. Nahill, R.M. Welsh, Cross-reactivities in memory cytotoxic T lymphocyte recognition of heterologous viruses, *J Exp Med* **1994**, *179*, 1933–1943.

123 M.A. Brehm, A.K. Pinto, K.A. Daniels, J.P. Schneck, R.M. Welsh, L.K. Selin, T cell immunodominance and maintenance of memory regulated by unexpectedly cross-reactive pathogens, *Nat Immunol* **2002**, *3*, 627–634.

124 H.D. Chen, A.E. Fraire, I. Joris, M.A. Brehm, R.M. Welsh, L.K. Selin, Memory CD8+ T cells in heterologous antiviral immunity and immunopathology in the lung, *Nat Immunol* **2001**, *2*, 1067–1076.

125 S.K. Kim, M. Cornberg, X.Z. Wang, H.D. Chen, L.K. Selin, R.M. Welsh, Private specificities of CD8 T cell responses control patterns of heterologous immunity, *J Exp Med* **2005**, *201*, 523–533.

126 H. Wedemeyer, E. Mizukoshi, A.R. Davis, J.R. Bennink, B. Rehermann, Cross-reactivity between hepatitis C virus and Influenza A virus determinant-specific cytotoxic T cells, *J Virol* **2001**, *75*, 11392–11400.

127 F.V. Chisari, C. Ferrari, Hepatitis B virus immunopathogenesis, *Annu Rev Immunol* **1995**, *13*, 29–60.

128 R.M. Welsh, Assessing CD8 T cell number and dysfunction in the presence of antigen, *J Exp Med* **2001**, *193*, F19–22.

129 D.C. Douek, L.J. Picker, R.A. Koup, T cell dynamics in HIV-1 infection, *Annu Rev Immunol* **2003**, *21*, 265–304.

130 N.H. Shoukry, A.G. Cawthon, C.M. Walker, Cell-mediated immunity and the outcome of hepatitis C virus infection, *Annu Rev Microbiol* **2004**, *58*, 391–424.

131 D. Moskophidis, F. Lechner, H. Pircher, R.M. Zinkernagel, Virus persistence in acutely infected immunocompetent mice by exhaustion of antiviral cytotoxic effector T cells, *Nature* **1993**, *362*, 758–761.

132 A. Gallimore, A. Glithero, A. Godkin, A.C. Tissot, A. Pluckthun, T. Elliott, H. Hengartner, R. Zinkernagel, Induction and exhaustion of lymphocytic choriomeningitis virus-specific cytotoxic T lymphocytes visualized using soluble tetrameric major histocompatibility complex class I-peptide complexes, *J Exp Med* **1998**, *187*, 1383–1393.

133 A.J. Zajac, J.N. Blattman, K. Murali-Krishna, D.J. Sourdive, M. Suresh, J.D. Altman, R. Ahmed, Viral immune evasion due to persistence of activated T cells without effector function, *J Exp Med* **1998**, *188*, 2205–2213.

134 E.J. Wherry, J.N. Blattman, K. Murali-Krishna, R. van der Most, R. Ahmed, Viral persistence alters CD8 T-cell immunodominance and tissue distribution and results in distinct stages of functional impairment, *J Virol* **2003**, *77*, 4911–4927.

135 S.A. Kalams, B.D. Walker, The critical need for CD4 help in maintaining effec-

tive cytotoxic T lymphocyte responses, *J Exp Med* **1998**, *188*, 2199–2204.

136 C.L. Day, B.D. Walker, Progress in defining CD4 helper cell responses in chronic viral infections, *J Exp Med* **2003**, *198*, 1773–1777.

137 M.J. Bevan, Helping the CD8(+) T-cell response, *Nat Rev Immunol* **2004**, *4*, 595–602.

138 P. Borrow, M. Oldstone, Lymphocytic choriomeningitis virus in *Viral Pathogenesis*, N. Nathanson, R. Ahmed, F. Gonzalez-Scarano, D. Griffin, K. Holmes, F. Murphy, H. Robinson, (Eds.) Lippincott-Raven, Philadelphia, PA, USA, 1997.

139 M.J. Buchmeier, A.J. Zajac, Lymphocytic choriomeningitis virus in *Persistent viral infections*, R. Ahmed, I.S.Y. Chen, (Eds.) John Wiley and Sons Ltd., West Sussex, 1999.

140 R.M. Welsh, Lymphocytic choriomeningitis virus as a model for the study of cellular immunology in *Effects of microbes on the immune system*, M.W. Cunningham, R.S. Fujinami, (Eds.) Lippincott Williams and Wilkins: Philadelphia, PA USA, 2000.

141 M.G. von Herrath, J. Dockter, M. Nerenberg, J.E. Gairin, M.B. Oldstone, Thymic selection and adaptability of cytotoxic T lymphocyte responses in transgenic mice expressing a viral protein in the thymus, *J Exp Med* **1994**, *180*, 1901–1910.

142 M.K. Slifka, J.N. Blattman, D.J. Sourdive, F. Liu, D.L. Huffman, T. Wolfe, A. Hughes, M.B. Oldstone, R. Ahmed, M.G. Von Herrath, Preferential escape of subdominant CD8+ T cells during negative selection results in an altered antiviral T cell hierarchy, *J Immunol* **2003**, *170*, 1231–1239.

143 R. Ou, S. Zhou, L. Huang, D. Moskophidis, Critical role for alpha/beta and gamma interferons in persistence of lymphocytic choriomeningitis virus by clonal exhaustion of cytotoxic T cells, *J Virol* **2001**, *75*, 8407–8423.

144 N.H. Gruener, F. Lechner, M.C. Jung, H. Diepolder, T. Gerlach, G. Lauer, B. Walker, J. Sullivan, R. Phillips, G.R. Pape, P. Klenerman, Sustained dysfunction of antiviral CD8+ T lymphocytes after infection with hepatitis C virus, *J Virol* **2001**, *75*, 5550–5558.

145 V.A. Pudney, A.M. Leese, A.B. Rickinson, A.D. Hislop, CD8+ immunodominance among Epstein-Barr virus lytic cycle antigens directly reflects the efficiency of antigen presentation in lytically infected cells, *J Exp Med* **2005**, *201*, 349–360.

146 A.D. Hislop, N.E. Annels, N.H. Gudgeon, A.M. Leese, A.B. Rickinson, Epitope-specific evolution of human CD8(+) T cell responses from primary to persistent phases of Epstein-Barr virus infection, *J Exp Med* **2002**, *195*, 893–905.

147 D. Kagi, B. Ledermann, K. Burki, P. Seiler, B. Odermatt, K.J. Olsen, E.R. Podack, R.M. Zinkernagel, H. Hengartner, Cytotoxicity mediated by T cells and natural killer cells is greatly impaired in perforin-deficient mice, *Nature* **1994**, *369*, 31–37.

148 C.M. Walsh, M. Matloubian, C.C. Liu, R. Ueda, C.G. Kurahara, J.L. Christensen, M.T. Huang, J.D. Young, R. Ahmed, W.R. Clark, Immune function in mice lacking the perforin gene, *Proc Natl Acad Sci U S A* **1994**, *91*, 10854–10858.

149 G. Pantaleo, J.F. Demarest, T. Schacker, M. Vaccarezza, O.J. Cohen, M. Daucher, C. Graziosi, S.S. Schnittman, T.C. Quinn, G.M. Shaw, L. Perrin, G. Tambussi, A. Lazzarin, R.P. Sekaly, H. Soudeyns, L. Corey, A.S. Fauci, The qualitative nature of the primary immune response to HIV infection is a prognosticator of disease progression independent of the initial level of plasma viremia, *Proc Natl Acad Sci U S A* **1997**, *94*, 254–258.

150 G.S. Ogg, X. Jin, S. Bonhoeffer, P.R. Dunbar, M.A. Nowak, S. Monard, J.P. Segal, Y. Cao, S.L. Rowland-Jones, V. Cerundolo, A. Hurley, M. Markowitz, D.D. Ho, D.F. Nixon, A.J. McMichael, Quantitation of HIV-1-specific cytotoxic T lymphocytes and plasma load of viral RNA, *Science* **1998**, *279*, 2103–2106.

151 D. Meyer-Olson, N.H. Shoukry, K.W. Brady, H. Kim, D.P. Olson, K. Hartman, A.K. Shintani, C.M. Walker, S.A. Kalams, Limited T cell receptor diversity of HCV-specific T cell responses is associated with CTL escape, *J Exp Med* **2004**, *200*, 307–319.

152 G. Pantaleo, H. Soudeyns, J.F. Demarest, M. Vaccarezza, C. Graziosi, S. Paolucci, M. Daucher, O.J. Cohen, F. Denis, W.E. Biddison, R.P. Sekaly, A.S. Fauci, Evidence for rapid disappearance of initially expanded HIV-specific CD8+ T cell clones during primary HIV infection, *Proc Natl Acad Sci U S A* **1997**, *94*, 9848–9853.
153 M. Matloubian, R.J. Concepcion, R. Ahmed, CD4+ T cells are required to sustain CD8+ cytotoxic T-cell responses during chronic viral infection, *J Virol* **1994**, *68*, 8056–8063.
154 P. Champagne, G.S. Ogg, A.S. King, C. Knabenhans, K. Ellefsen, M. Nobile, V. Appay, G.P. Rizzardi, S. Fleury, M. Lipp, R. Forster, S. Rowland-Jones, R.P. Sekaly, A.J. McMichael, G. Pantaleo, Skewed maturation of memory HIV-specific CD8 T lymphocytes, *Nature* **2001**, *410*, 106–111.
155 S. Kostense, K. Vandenberghe, J. Joling, D. Van Baarle, N. Nanlohy, E. Manting, F. Miedema, Persistent numbers of tetramer+ CD8(+) T cells, but loss of interferon-gamma+ HIV-specific T cells during progression to AIDS, *Blood* **2002**, *99*, 2505–2511.
156 V. Appay, P.R. Dunbar, M. Callan, P. Klenerman, G.M. Gillespie, L. Papagno, G.S. Ogg, A. King, F. Lechner, C.A. Spina, S. Little, D.V. Havlir, D.D. Richman, N. Gruener, G. Pape, A. Waters, P. Easterbrook, M. Salio, V. Cerundolo, A.J. McMichael, S.L. Rowland-Jones, Memory CD8+ T cells vary in differentiation phenotype in different persistent virus infections, *Nat Med* **2002**, *8*, 379–385.
157 L.G. Tussey, U.S. Nair, M. Bachinsky, B.H. Edwards, J. Bakari, K. Grimm, J. Joyce, R. Vessey, R. Steigbigel, M.N. Robertson, J.W. Shiver, P.A. Goepfert, Antigen burden is major determinant of human immunodeficiency virus-specific CD8+ T cell maturation state: potential implications for therapeutic immunization, *J Infect Dis* **2003**, *187*, 364–374.
158 S. Zhou, R. Ou, L. Huang, G.E. Price, D. Moskophidis, Differential tissue-specific regulation of antiviral CD8+ T-cell immune responses during chronic viral infection, *J Virol* **2004**, *78*, 3578–3600.
159 M.J. Fuller, D.A. Hildeman, S. Sabbaj, D.E. Gaddis, A.E. Tebo, L. Shang, P.A. Goepfert, A.J. Zajac, Cutting Edge: Emergence of CD127high Functionally Competent Memory T Cells Is Compromised by High Viral Loads and Inadequate T Cell Help, *J Immunol* **2005**, *174*, 5926–5930.
160 V. Appay, D.F. Nixon, S.M. Donahoe, G.M. Gillespie, T. Dong, A. King, G.S. Ogg, H.M. Spiegel, C. Conlon, C.A. Spina, D.V. Havlir, D.D. Richman, A. Waters, P. Easterbrook, A.J. McMichael, S.L. Rowland-Jones, HIV-specific CD8(+) T cells produce antiviral cytokines but are impaired in cytolytic function, *J Exp Med* **2000**, *192*, 63–75.
161 J. Lieberman, P. Shankar, N. Manjunath, J. Andersson, Dressed to kill? A review of why antiviral CD8 T lymphocytes fail to prevent progressive immunodeficiency in HIV-1 infection, *Blood* **2001**, *98*, 1667–1677.
162 F. Lechner, D.K. Wong, P.R. Dunbar, R. Chapman, R.T. Chung, P. Dohrenwend, G. Robbins, R. Phillips, P. Klenerman, B.D. Walker, Analysis of successful immune responses in persons infected with hepatitis C virus, *J Exp Med* **2000**, *191*, 1499–1512.
163 E.J. Wherry, D.L. Barber, S.M. Kaech, J.N. Blattman, R. Ahmed, Antigen-independent memory CD8 T cells do not develop during chronic viral infection, *Proc Natl Acad Sci U S A* **2004**, *101*, 16004–16009.
164 P.C. Doherty, S. Hou, C.F. Evans, J.L. Whitton, M.B. Oldstone, M.A. Blackman, Limiting the available T cell receptor repertoire modifies acute lymphocytic choriomeningitis virus-induced immunopathology, *J Neuroimmunol* **1994**, *51*, 147–152.
165 K. Daly, P. Nguyen, D.L. Woodland, M.A. Blackman, Immunodominance of major histocompatibility complex class I-restricted influenza virus epitopes can be influenced by the T-cell receptor repertoire, *J Virol* **1995**, *69*, 7416–7422.
166 C.C. Huang, M.A. Coppola, P. Nguyen, D. Carragher, C. Rohl, K.J. Flynn, J.D. Altman, M.A. Blackman, Effect of Staphylococcus enterotoxin B on the concurrent CD8(+) T cell response to

influenza virus infection, *Cell Immunol* **2000**, *204*, 1–10.
167 C.C. Huang, S. Shah, P. Nguyen, J.D. Altman, M.A. Blackman, Bacterial superantigen exposure after resolution of influenza virus infection perturbs the virus-specific memory CD8(+)-T-cell repertoire, *J Virol* **2002**, *76*, 6852–6856.
168 J.E. Callahan, J.W. Kappler, P. Marrack, Unexpected expansions of CD8-bearing cells in old mice, *J Immunol* **1993**, *151*, 6657–6669.
169 R. Schwab, P. Szabo, J.S. Manavalan, M.E. Weksler, D.N. Posnett, C. Pannetier, P. Kourilsky, J. Even, Expanded CD4+ and CD8+ T cell clones in elderly humans, *J Immunol* **1997**, *158*, 4493–4499.
170 J. LeMaoult, I. Messaoudi, J.S. Manavalan, H. Potvin, D. Nikolich-Zugich, R. Dyall, P. Szabo, M.E. Weksler, J. Nikolich-Zugich, Age-related dysregulation in CD8 T cell homeostasis: kinetics of a diversity loss, *J Immunol* **2000**, *165*, 2367–2373.
171 R.B. Effros, Z. Cai, P.J. Linton, CD8 T cells and aging, *Crit Rev Immunol* **2003**, *23*, 45–64.
172 I. Messaoudi, J. Lemaoult, J.A. Guevara-Patino, B.M. Metzner, J. Nikolich-Zugich, Age-related CD8 T cell clonal expansions constrict CD8 T cell repertoire and have the potential to impair immune defense, *J Exp Med* **2004**, *200*, 1347–1358.
173 A. Franco, C. Ferrari, A. Sette, F.V. Chisari, Viral mutations, TCR antagonism and escape from the immune response, *Curr Opin Immunol* **1995**, *7*, 524–531.
174 A. McMichael, T cell responses and viral escape, *Cell* **1998**, *93*, 673–676.
175 D.A. Price, U.C. Meier, P. Klenerman, M.A. Purbhoo, R.E. Phillips, A.K. Sewell, The influence of antigenic variation on cytotoxic T lymphocyte responses in HIV-1 infection, *J Mol Med* **1998**, *76*, 699–708.
176 H. Pircher, D. Moskophidis, U. Rohrer, K. Burki, H. Hengartner, R.M. Zinkernagel, Viral escape by selection of cytotoxic T cell-resistant virus variants in vivo, *Nature* **1990**, *346*, 629–633.
177 G.E. Price, R. Ou, H. Jiang, L. Huang, D. Moskophidis, Viral escape by selection of cytotoxic T cell-resistant variants in influenza A virus pneumonia, *J Exp Med* **2000**, *191*, 1853–1867.
178 C.B. Moore, M. John, I.R. James, F.T. Christiansen, C.S. Witt, S.A. Mallal, Evidence of HIV-1 adaptation to HLA-restricted immune responses at a population level, *Science* **2002**, *296*, 1439–1443.
179 P.O. de Campos-Lima, R. Gavioli, Q.J. Zhang, L.E. Wallace, R. Dolcetti, M. Rowe, A.B. Rickinson, M.G. Masucci, HLA-A11 epitope loss isolates of Epstein-Barr virus from a highly A11+ population, *Science* **1993**, *260*, 98–100.
180 P.O. de Campos-Lima, V. Levitsky, J. Brooks, S.P. Lee, L.F. Hu, A.B. Rickinson, M.G. Masucci, T cell responses and virus evolution: loss of HLA A11-restricted CTL epitopes in Epstein-Barr virus isolates from highly A11-positive populations by selective mutation of anchor residues, *J Exp Med* **1994**, *179*, 1297–1305.
181 A. Bertoletti, A. Costanzo, F.V. Chisari, M. Levrero, M. Artini, A. Sette, A. Penna, T. Giuberti, F. Fiaccadori, C. Ferrari, Cytotoxic T lymphocyte response to a wild type hepatitis B virus epitope in patients chronically infected by variant viruses carrying substitutions within the epitope, *J Exp Med* **1994**, *180*, 933–943.
182 A.L. Erickson, Y. Kimura, S. Igarashi, J. Eichelberger, M. Houghton, J. Sidney, D. McKinney, A. Sette, A.L. Hughes, C.M. Walker, The outcome of hepatitis C virus infection is predicted by escape mutations in epitopes targeted by cytotoxic T lymphocytes, *Immunity* **2001**, *15*, 883–895.
183 P.J. Goulder, A.K. Sewell, D.G. Lalloo, D.A. Price, J.A. Whelan, J. Evans, G.P. Taylor, G. Luzzi, P. Giangrande, R.E. Phillips, A.J. McMichael, Patterns of immunodominance in HIV-1-specific cytotoxic T lymphocyte responses in two human histocompatibility leukocyte antigens (HLA)-identical siblings with HLA-A*0201 are influenced by epitope mutation, *J Exp Med* **1997**, *185*, 1423–1433.
184 P. Klenerman, R.M. Zinkernagel, Original antigenic sin impairs cytotoxic T lymphocyte responses to viruses bearing variant epitopes, *Nature* **1998**, *394*, 482–485.

IV
Effects of Pathogens on the Immune Response

Immunodominance: The Choice of the Immune System. Edited by Jeffrey A. Frelinger

ISBN: 3-527-31274-9

7
Listeria monocytogenes Infection and the CD8$^+$ T-Cell Hierarchy

Brandon B. Porter and John T. Harty

7.1
Introduction

Listeria monocytogenes (LM) is a gram-positive, facultative intracellular bacterium that can lethally infect immunocompromised individuals and cause abortion in pregnant women. The bacterium is found ubiquitously in the environment. While infections with LM can be devastating, those that require medical intervention are infrequent [1]. Indeed, the ease with which most people are able to clear LM infections makes the bacterium a relatively safe laboratory tool. Presently, infection of mice with LM is widely used to investigate innate and adaptive immune system functions. The seminal studies with LM were carried out by G.B. Mackaness, who demonstrated that cellular responses and not humoral responses are responsible for acquired immunity to this pathogen [2]. The recognition that recall immunity to LM had little to do with antibody made the system ideal for investigating T-cell responses to intracellular pathogens.

Ninety percent of intravenously injected LM distributes to the liver and spleen of mice in an 80% to 20% ratio [2]. There, the bacteria are quickly engulfed by resident macrophages and, inside the phagosome, LM expresses the pore-forming protein listeriolysin O (LLO), phospholipase C (PLC), and metalloprotease (mpl) [3]. Each of these proteins assists the bacteria in escaping from the phagosome to the cytosol, where LM can readily survive and multiply. It is estimated that only about 20% of phagocytosed LM is able to escape the phagosome in non-activated macrophages, but this number is sufficient to establish intracellular infection. Soon after LM reaches the cytosol, it polarly expresses an actin-polymerizing molecule, ActA, that enables the bacterium to have motility within the host cell. Once motile, LM can propel itself into neighboring cells where it is enveloped by a double membrane that is perforated by the actions of LLO, PLC, and mpl. The bacteria then enter the cytoplasm and the process of division, propulsion, and invasion is continued in the newly infected cell. Thus, LM can spread from cell to cell without exposure to extracellular immune defense, such as antibody. LM induces an acute infection that is resolved within nine days of infection. Mice infected with 0.1 of the LM dose that is lethal to 50% of the mice (LD_{50}) have the

Immunodominance: The Choice of the Immune System. Edited by Jeffrey A. Frelinger

ISBN: 3-527-31274-9

highest numbers of bacteria in their spleens and livers three days after infection; the bacteria numbers gradually decrease down to undetectable levels by days 6–9 after infection [2]. Mice that have recovered from sublethal infection are resistant, for life, to high-dose LM challenge.

7.2
Innate Immune Response to LM

Studies by Bancroft and Unanue first demonstrated "innate" resistance to LM infection in SCID mice that lack B and T cells [4]. Since that time, the ability of LM to elicit a robust innate immune response has been well utilized as an investigative probe into the innate immune system. Indeed, peritoneal infection of mice with LM is an excellent way to elicit a large number of highly activated neutrophils and macrophages from mice for subsequent study. LM activates the innate immune response through the engagement of multiple Toll-like receptors (TLR) (likely TLRs 2, 5, 6, and 9) and scavenger receptors [5]. These receptors are found on multiple cell types, including macrophages and dendritic cells (DCs), and can cause cells to make a whole host of inflammatory cytokines (including IL-12, IL-18, TNF, IL-1, and IL-6) and chemokines (including IL-8, MIP-1α, MIP-1β, and IP-10) [5, 6]. The production of these autocrine, paracrine, and endocrine proteins serves to activate and recruit inflammatory cells and DCs into the site of infection. These inflammatory signals are also important in preparing DCs for antigen (Ag) presentation to T cells by decreasing phagocytosis, modulating Ag processing, and upregulating the production of costimulatory molecules and MHC class I and class II. The importance of signaling through TLRs, IL-1 receptor, and IL-18 receptor in LM infection is highlighted in MyD88 knockout (KO) mice, which lack a major signal transduction pathway for these receptors [5]. After LM infection, MyD88 KO mice are severely impaired in making TNF, IL-12, and IFN-γ, as measured by serum concentrations; have greatly decreased recruitment of neutrophils to sites of infection; and have greater difficulty in curtailing LM numbers in the liver and spleen than mice that lack the ability to produce IFN-γ (GKO), a vital inflammatory cytokine in the regulation of LM infection. These results emphasize that TLR, IL-1, and IL-18 signaling is central to initiating an effective innate response to LM infection.

Macrophages, neutrophils, natural killer (NK) cells, NK T cells, and $\gamma\delta$ T cells each play an innate role in combating LM infection [7]. The immediate ability of neutrophils to kill LM allows for the confinement of the bacterium microabscesses during the first 24 hours of infection. Conlan and North demonstrated a 10- to 100-fold increase in LM numbers in the liver within the first 24 hours of infection in mice treated with a neutrophil-depleting antibody. Curiously, neutrophil depletion had no effect upon the numbers of LM in the spleen during the first 24 hours of infection, suggesting that other mechanisms exist for LM containment in that organ.

In addition to direct cellular responses to LM, the cytokines IL-12 and IL-18 are extremely important in the innate response to LM [8]. These cytokines act together to stimulate NK cells, $\gamma\delta$ T cells, and non-LM-specific memory $CD8^+$ T cells to make IFN-γ during the first 2–3 days of infection [5, 9]. The importance of IFN-γ in resolving virulent LM infection cannot be overstated. Indeed, the LD_{50} for LM in GKO mice is 10 bacteria, compared to an LD_{50} of 10^4 LM for wild-type mice [10]. Likewise, TNF production is important in early protection, as demonstrated by the increased mortality of TNF KO mice in response to LM infection [5]. Each of these cytokines cooperates in activating macrophages, thereby increasing the ability of macrophages to kill LM while inhibiting the escape of the bacteria from the phagosome into the cytosol. Therefore, IFN-γ and TNF serve to severely decrease the potential reservoir where LM can reside.

7.3 Adaptive Immune Response to LM and Ag Presentation

While the innate response to LM infection is robust and necessary for survival, it is not sufficient to clear the infection [11]. It is evident that antibodies play only a small role in the adaptive response to LM because while low amounts of LM-specific antibodies are made in response to LM infection, these antibodies are not made in sufficient quantities or in the correct specificities to confer protective immunity [7]. In contrast, T cells are absolutely necessary for clearance of LM. As opposed to having receptors that recognize specific patterns that are generally found in pathogenic microorganisms, T cells rearrange their T-cell receptor (TCR) genes and thereby create a diverse array of receptors that can respond with specificity to a wide range of pathogens [12]. A very small proportion of naïve peripheral T cells are able to respond to LM infection. Upon infection, these naïve Ag-specific T cells recognize cognate Ag in the context of MHC molecules on activated DCs and expand greatly in number. During this process, the activated T cells differentiate into effector cells that secrete cytokines and/or kill cells in response to TCR ligation. Following expansion, a large percentage (~90–95%) of Ag-specific cells die, leaving a relatively high number of LM-specific memory T cells, which can persist for life and respond to subsequent TCR engagement by more rapidly dividing and executing effector functions to combat secondary infection.

Multiple experiments have been performed to determine whether $CD4^+$ T cells or $CD8^+$ T cells are more important in clearing LM infection in mice. T-cell transfer experiments into SCID mice demonstrated that either $CD4^+$ or $CD8^+$ T cells were sufficient for clearance of LM [11]. However, adoptive transfer of activated LM-specific $CD8^+$ T cells into LM-infected wild-type mice resulted in better protection than did transfer of activated LM-specific $CD4^+$ T cells. Also, LM-infected β_2 microglobulin (β_2m) KO mice, which were therefore $CD8^+$ T cell–deficient, had greater bacterial numbers and an increased duration of infection compared to MHC class II KO mice, which were devoid of $CD4^+$ T cells. These results suggested that CD8+ T cells have greater efficacy in clearing LM infection. The increased ability of Ag-specific CD8+ T cells to efficiently eradicate LM infection is likely due to the capacity of CD8+ T cells to recognize Ag in the context of MHC class I. This is advantageous because (1) the intracytoplasmic location of LM exposes the organism to the MHC class I Ag-presentation pathway and (2) many infected liver and spleen parenchymal cells express MHC class I but not MHC class II molecules and, consequently, are able to stimulate only CD8+ T cells.

The presentation of cognate Ag in the context of MHC molecules is essential to initiate a T-cell response and to elicit T-cell effector functions (cytokine secretion and/or killing) with specificity. In this review we will focus on immunodominance of the CD8+ T-cell response to LM and therefore will confine our discussion to MHC class Ia Ag display. When LM resides in the cytosol, the proteins that are secreted from viable organisms are readily available for entry into the endogenous pathway for MHC class I Ag presentation [13]. The endogenous pathway utilizes the proteasome complex (or other cytosolic proteases) for degradation of cytosolic proteins to make peptide fragments that are pumped into the endoplasmic reticulum (ER), via the transporter associated with antigen processing (TAP) complex [14]. Within the ER, high-affinity peptides bind to newly synthesized MHC class I molecules, allowing the egress of the peptide–MHC complex from the ER and presentation on the surface of the cell. In addition, there are exogenous pathways of MHC class I Ag presentation, which are restricted to phagocytic antigen-presenting cells (APCs) and allow MHC class I presentation of antigens from extracellular sources. This exogenous pathway, also called cross-presentation, can transport phagolysosome-derived Ag to MHC class I molecules through either a TAP-dependent mechanism or endosome trafficking. Potential substrates for cross-presentation are phagocytosed killed LM, apoptotic bodies from infected cells, and apoptotic bodies from neutrophils that phagocytosed and killed LM. Tvinnereim et al. demonstrated that (1) non-secreted LM Ag was not presented well by the endogenous pathway and (2) the CD8+ T-cell response to this non-secreted Ag was decreased, relative to the CD8+ T-cell response to a secreted Ag, by approximately three- to fourfold when neutrophils were depleted in mice infected with LM [15]. These results suggested that neutrophils are a major substrate of Ag for cross-presentation of non-secreted Ags in LM infection. Neutrophils likely make for a good substrate for cross-presentation because they are able to kill LM early in the infection, which exposes non-secreted Ag, and then die shortly after activation, which exposes them (and the non-secreted Ag) to scavenging by APCs.

7.4
Secreted Versus Non-secreted Ag

It is estimated that more than 70% of the BALB/c mouse Ag-specific CD8+ T-cell response to LM is against secreted protein epitopes, which make up only about 3% of total LM protein-coding genes [5]. This incongruity suggests that the location of antigen can affect its ability to elicit a $CD8^+$ T-cell response. This hypothesis was tested in experiments performed by Shen and colleagues [16, 17]. Strains of LM were engineered to express epitopes from lymphocytic choriomeningitis virus (LCMV) as part of either a secreted protein or non-secreted protein. The same promoter drove each of the introduced genes; thus, hypothetically, similar amounts of each protein were produced by the recombinant LM (rLM) strains. This was confirmed biochemically. Both rLM strains were able to activate a $CD8^+$ T-cell response against the introduced Ag; however, the $CD8^+$ T-cell response to the secreted epitope was two to five times greater in number than the response to the non-secreted epitope. The same results were found with both H-$2L^d$– and H-$2D^b$-restricted LCMV epitopes. These findings suggested that MHC class I presentation of Ag from secreted proteins was slightly more efficient than MHC class I presentation of Ag from non-secreted proteins. This notion was supported by the measurement of *in vivo* H-$2D^b$-restricted Ag presentation on DCs via CFSE dilution of naïve Ag-specific TCR-transgenic $CD8^+$ T cells [17]. Naïve $CD8^+$ T cells require Ag presentation by activated DCs to divide extensively, and thus higher numbers of transgenic T cells that divided (CFSE low) indicated higher levels of Ag presentation. These experiments revealed that the timing of Ag presentation was the same after infection with rLM that expressed the secreted or non-secreted form of the epitope. However, infection with rLM that secreted Ag resulted in greater amounts of APC-mediated Ag presentation. Therefore, secreted LM proteins are processed into MHC class I–restricted Ag with greater efficiency than are non-secreted LM proteins.

Importantly, the Ag-specific $CD8^+$ T-cell responses that are generated against both secreted and non-secreted LM proteins result in functional memory $CD8^+$ T cells [16, 17]. However, memory $CD8^+$ T cells specific for the recombinant Ag are not able to protect mice from infection with rLM strains that express the non-secreted Ag. Therefore, while non-secreted Ag is cross-presented on professional APCs, infected parenchymal cells are unable to present the non-secreted Ag in large enough quantities for the Ag-specific effector $CD8^+$ T cells to provide protection. These experiments demonstrate that the location of a bacterial Ag can determine whether or not it can be a target of protective $CD8^+$ T cells. The results also strongly suggest that vaccines should contain secreted proteins if they are designed to provide $CD8^+$ T cell–mediated protection to intracellular prokaryotes or eukaryotes.

7.5 The Hierarchy of the $CD8^+$ T-cell Responses to LM Epitopes

The LM genome contains 2853 protein-coding genes [18]. Given this information, it is astonishing that >50% of the $CD8^+$ T-cell response to LM in BALB/c mice (H-2^d) is against four H-$2K^d$-restricted LM epitopes that exist within three secreted virulence factors: LLO, a hemolysin; p60, a peptidoglycan hydrolase; and *Listeria* metalloprotease (mpl) [5]. The small number of Ags that induce a majority of the

$CD8^+$ T-cell response to LM highlights the fact that not all proteins are antigenic. The following factors are among those that determine the antigenicity of a protein and its peptide fragments [14].

1. The protein needs to be available for pruning by protease.
2. Most peptides need to bind the TAP complex in order to enter the ER (hydrophobic COOH terminal residue).
3. The peptide must bind MHC class I molecules with high affinity and exhibit a low off-rate.
4. The peptide must be presented to T cells in high enough quantities.
5. The peptide must be different enough from self that there are functional naïve T cells that can recognize the Ag in the context of MHC class I.

The hierarchy of the BALB/c $CD8^+$ T-cell response to known LM-derived Ags is as follows: LLO_{91-99}>$p60_{217-225}$>>$p60_{449-457}$>mpl_{84-92} [19]. Based upon MHC class I tetramer analysis to identify LM-specific T cells, the relative ratio of the CD62L low (denoting activated) T-cell response to these epitopes is 48:8:2:1 at 7 days after LM infection. The LLO_{91-99} and $p60_{217-225}$ T-cell responses are described to be immunodominant, while $p60_{449-457}$ and mpl_{84-92} responses are subdominant responses. However, unlike many viral infections, ablating the "immunodominant" epitopes by genetic manipulation of LM does not result in increases in the magnitude of the subdominant $CD8^+$ T-cell responses to $p60_{449-457}$ and mpl_{84-92} [14]. This suggests that the low magnitudes of the $CD8^+$ T-cell responses to $p60_{449-457}$ and mpl_{84-92} are not simply the result of competition between other $CD8^+$ T-cell responses, but are likely due to these epitopes' physical properties or low numbers of functional naïve Ag-specific $CD8^+$ T cells in the T-cell repertoire. These conclusions are supported by an observation in our lab that introduction of a dominant $H\text{-}2L^d$-restricted Ag from lymphocytic choriomeningitis virus (LCMV), $NP_{118-126}$, as a secreted fusion protein in LM results in no difference in the magnitudes of the T-cell responses to the immunodominant epitopes LLO_{91-99} and $p60_{217-225}$ or to the subdominant epitope $p60_{449-457}$ (Badovinac, unpublished). Infection with this $NP_{118-126}$-secreting strain of LM results in an $NP_{118-126}$ response that is similar in magnitude to the LLO_{91-99} response, which represents 3–5% of the $CD8^+$ T cells at the peak of the response [20]. This number pales in comparison to the >50% of total $CD8^+$ T cells that are $NP_{118-126}$ specific 8 days after LCMV infection [21, 22]. These data demonstrate that the precise sequence of the epitope is not the only characteristic that influences the size of the T-cell response. Indeed, the pathogen carrying the specific epitope also has a large impact upon the magnitude of the $CD8^+$ T-cell response, perhaps as a consequence of differential production of antigen or stimulation of innate immunity. The results also emphasize that the hierarchy of the $CD8^+$ T-cell response to LM Ags is not due to T-cell responses to dominant epitopes causing diminished magnitudes in the $CD8^+$ T-cell responses to subdominant epitopes. Instead, the magnitude of the $CD8^+$ T-cell response to each Ag is likely the result of a combination between the number of Ag-specific precursors that are able to respond to each epitope and the biological properties of each epitope.

The ability of a peptide to elicit a $CD8^+$ T-cell response depends heavily upon its dissociation rate from the MHC class I molecule [14]. This relationship holds true in LM infection of $H\text{-}2^d$ mice. Here, the hierarchy of the $CD8^+$ T-cell response to LM correlates well with the reciprocal of the rates of peptide–MHC class I dissociation; LLO_{91-99} and $p60_{217-225}$ are the best at stabilizing MHC class I, with a $t_{1/2}$ of ~6 h, while mpl_{84-92} and $p60_{449-457}$ are the worst, with an estimated $t_{1/2}$ of ~1–2 h and ~1 h, respectively [23]. It is important to note that this correlation is imperfect: the $CD8^+$ T-cell response to $p60_{449-457}$ is of greater magnitude than the response to mpl_{84-92}, and thus the other factors probably contribute to the $CD8^+$ T-cell hierarchy.

As mentioned previously, one factor that can influence immunodominance is the number of Ag-specific precursors to specific epitopes. Busch and Pamer tried to address this issue through *in vitro* expansion, for 3 weeks, of the Ag-specific $CD8^+$ T-cell populations to each LM epitope and determination of the number of TCR-Vβ chains that were utilized by these cytotoxic lymphocyte (CTL) lines [24]. The expectation was that there would be a lower variability in TCR-Vβ usage in T-cell responses that were the result of a low number of recruited Ag-specific precursors. Consistent with the $CD8^+$ T-cell response hierarchy, they found that the LLO_{91-99}– and $p60_{217-225}$-specific $CD8^+$ T-cell lines had greater utilization of multiple TCR-Vβ chains than did the T-cell lines against the $p60_{449-457}$ and mpl_{84-92} epitopes. Also, more TCR-Vβ variety was found in $p60_{449-457}$ CTL than in mpl_{84-92} CTL, implying that the hierarchy of the $CD8^+$ T-cell response to LM was due to the number of available precursors to each LM epitope. However, it was also possible that these results were due to the higher probability of the selection of single CTL clones from *in vitro* expansion of Ag-specific T cells that existed in very low frequencies.

Another possible explanation for the hierarchy of the $CD8^+$ T-cell response to LM is that the T-cell epitopes are presented to naïve T cells in different quantities. Pamer's lab took great pains to measure the amounts of MHC class Ia–restricted Ags that were produced by *in vitro* infection of a macrophage-like cell line, J774 [14]. They used an Ag detection method developed by Rammensee and colleagues that involves extraction of MHC class I–restricted Ag from infected cells with acid followed by HPLC separation of peptides and subsequent measurement of Ag by sensitization of MHC class I–expressing cells to *in vitro* cytotoxicity assays with Ag-specific CTL lines. Their results were interesting because the estimated quantities of MHC class I–restricted peptides were inversely related to the magnitude of the $CD8^+$ T-cell response to each epitope; 700 LLO_{91-99} peptides, 2700 $p60_{217-225}$ peptides, and 9000 $p60_{449-457}$ peptides were recovered per cell after 6 hours of infection. These results suggested that, in LM infection, the quantities of epitope generated by infected cells had no relationship to the magnitude of the $CD8^+$ T-cell responses.

One potential limitation of these *in vitro* studies is that Ag presentation by infected cell lines may not appropriately mimic the dynamics of Ag presentation of *in vivo* infections. Also, when using the acid extraction method of determining MHC class I–restricted epitope quantities, it is unknown whether the epitope is actually presented on the surface of the cell. For these reasons, Skoberne et al. developed a direct *ex vivo* assay to measure total Ag levels in splenocytes after LM infection [25]. In these studies, mice were infected with LM and the spleens were harvested 2 days after infection. The splenocytes were then used as APCs for CTL lines of different Ag specificities, and ELISPOT assays for IFN-γ were performed. In this assay, the number of cells that made IFN-γ was proportional to the amount of Ag presented. Importantly, the CTL lines had sensitivities similar to those of their cognate Ag, which allowed the authors to make conclusions about relative amounts of Ag presentation. IFN-γ production by LLO_{91-99}-, $p60_{217-225}$-, $p60_{476-484}$-, and $p60_{449-457}$-specific CTL lines in response to splenocytes from LM-infected mice correlated well with the previously demonstrated $CD8^+$ T-cell response hierarchy (LLO_{91-99}>$p60_{217-225}$> $p60_{476-484}$>$p60_{449-457}$), suggesting that epitope levels *in vivo* do correlate with the $CD8^+$ T-cell hierarchy [25]. While these results were opposite to earlier findings in infected cell lines, they were not due to differences in methodologies. The researchers went on to demonstrate that LM-infected J774 cells activated their $CD8^+$ T-cell lines in a hierarchy similar to that observed by Pamer's group: $p60_{449-457}$>$p60_{217-225}$>LLO_{91-99}. Another LM-infected macrophage-like cell line, P388, activated its $CD8^+$ T-cell lines in the following relative ratios: $p60_{449-457}$=$p60_{217-225}$>LLO_{91-99}. Acid extraction of Ag resulted in the same hierarchy of $p60_{449-457}$=$p60_{217-225}$>LLO_{91-99}. These results indicate that the hierarchy of MHC class I–restricted Ag production after LM infection can be cell type dependent. Importantly, Skoberne et al. also showed that acid extraction of Ag from the spleens of infected mice resulted in Ag measurements that were similar to the direct *ex vivo* ELISPOT results: LLO_{91-99}>$p60_{217-225}$> undetectable levels of $p60_{476-484}$ and $p60_{449-457}$. These results demonstrated that the hierarchy of MHC class I–restricted epitopes created by cell lines *in vitro* does not approximate the hierarchy of MHC class I–restricted epitopes created *in vivo*. Possible explanations for these differences in Ag presentation between direct *ex vivo* splenocytes and *in vitro* cell lines include (1) cell type–specific differences in bacterial expression of LLO and p60 and (2) cell type–specific differential processing of Ag, perhaps because of variations in the proteasome subunit composition in each cell type. Also, inflammatory cytokines (especially IFN-γ) that would be present *in vivo*, but not after *in vitro* infection, influence the preference of proteasome-mediated endopeptidase cleavage of proteins [26]. All together, these data demonstrate that the hierarchy of *in vivo* epitope presentation correlates with the hierarchy of the $CD8^+$ T-cell response to LM.

7.6
IFN-γ and the $CD8^+$ T-cell Hierarchy

IFN-γ is a pluripotent cytokine that is reported to affect the transcription of hundreds of genes [27, 28]. In LM infection, IFN-γ is best known for its important role in activating macrophages to kill the bacteria before they escape the phagosome [7]. IFN-γ also plays an important role in regulating multiple aspects of Ag presentation [28]. In MHC class I Ag presentation, IFN-γ serves to increase the expression of MHC class I, β_2-microgobulin; TAP-1 and TAP-2; proteasome β-subunits (LMP-2, LMP-7, MECL-1), which alter the preference of proteasome endopeptidase cleavage to enhance MHC class I binding; proteasome activator (PA28), which optimizes the size of proteasome-cleaved peptides for TAP transport and MHC class I binding; tapasin (GP48), an integral protein for MHC class I assembly; and ER lumenal stress protein (GP96), which might protect TAP-transported peptides from peptidase. The end result of IFN-γ-mediated increases in these genes is higher expression of MHC class I molecules on cells, and, in some cases, the quantities of MHC class I presentation of specific peptides are changed.

As noted above, IFN-γ-deficient (GKO) mice are highly susceptible to LM infection [10]. For this reason, an attenuated strain (*actA*-) of LM lacking a virulence factor that assists in cell–cell transmission is used to examine the adaptive immune response in GKO mice. Notably, the LD_{50} of *actA*- LM in GKO mice ($10^{6.6}$) is similar to that in wild-type mice ($10^{6.9}$). This suggests that IFN-γ has little influence upon the immune response to *actA*- LM. However, infection of GKO mice with 0.1 LD_{50} of *actA*- LM does result in a modest delay in bacterial clearance compared to infection of wild-type mice with the same dose of *actA*- LM. Therefore, IFN-γ assists in the clearance of *actA*- LM, while factors other than IFN-γ are able to protect mice from succumbing to *actA*- LM infection.

Despite the absence of IFN-γ, GKO mice are able to manifest a robust CD8$^+$ T-cell response [29] and create memory to infection with *actA*- LM [10]. This result was demonstrated in our lab by measuring the LLO_{91-99}– and $p60_{217-225}$-specific CD8$^+$ T-cell responses to *actA*- LM by using peptide-stimulated intracellular cytokine staining (ICS) to detect Ag-specific TNF production at the single-cell level [29]. Curiously, the ratio of LLO_{91-99}:$p60_{217-225}$-specific CD8$^+$ T cells was different in wild-type mice (4.7:1) and GKO mice (1.9:1). It was thought that the altered ratio of LLO_{91-99}:$p60_{217-225}$-specific CD8$^+$ T cells was due to increased numbers of $p60_{217-225}$-specific CD8$^+$ T cells in the absence of IFN-γ, because the magnitudes of the LLO_{91-99} CD8$^+$ T-cell responses to *actA*- LM were the same in GKO and wild-type mice. It was further hypothesized that the higher numbers of $p60_{217-225}$-specific CD8$^+$ T cells were due to increased presentation of $p60_{217-225}$ in the absence of IFN-γ. However, it is also possible that the presentation of LLO was decreased relative to $p60_{217-225}$ presentation in GKO mice and that the increased magnitude of the CD8$^+$ T-cell response to $p60_{217-225}$ was due to prolonged infection.

Skoberne and Geninat investigated the effect of IFN-γ on Ag presentation in LM infection of BALB/c and GKO mice by using both peptide extraction and their direct *ex vivo* ELISPOT method of Ag detection [30]. Each method revealed that, while LLO_{91-99} was presented in similar amounts on day 2, the amount of $p60_{217-225}$ presentation was increased in LM-infected GKO mice compared to wild-type mice. Unfortunately, these methods could not determine whether the increase in $p60_{217-225}$ presentation in GKO mice was due to increased efficiency in creating $p60_{217-225}$ or the increased numbers of bacteria found in GKO spleen after LM infection.

Attempts were made to address this question *in vitro*. In these experiments, the P388 macrophage-like cell line was pretreated with IFN-γ for 24 h and infected with LM for 6 h. Bacteria numbers were then counted, and $p60_{217-225}$ and LLO_{91-99} Ags were measured after acid extraction. The recovered bacterial numbers were similar in the control and IFN-γ-treated P388 cells, and acid-extracted p60 Ag amounts per cell were slightly decreased in the IFN-γ-treated group compared to the control group, while LLO_{91-99} Ag amounts per cell were increased two- to threefold in the IFN-γ-pretreated group. These results suggested that IFN-γ served to increase the production of LLO_{91-99} and, therefore, implied that the decreased LLO_{91-99}:$p60_{217-225}$ ratio of Ag-specific $CD8^+$ T cells in LM-infected GKO mice was the result of diminished presentation of LLO_{91-99} rather than substantially increased $p60_{217-225}$ presentation. However, IFN-γ treatment did not change the Ag presentation patterns of LM-infected P388 cells to the hierarchy found in the spleens of LM-infected mice. The LLO_{91-99}:$p60_{217-225}$ peptide ratio was >1:2 in spleens 2 days after infection, while a ratio of <1:300 was found in IFN-γ-treated P388 cells. In our opinion, the discrepancies in the relative Ag amounts created by infected spleens and cell lines suggest that *in vitro* infection and Ag processing in cell lines do not approximate the dynamics of *in vivo* infection and Ag processing. Therefore, it would be useful to determine whether another method of inquiry will yield the same results, i.e., IFN-γ increases the production of LLO_{91-99} by two- to threefold.

As mentioned earlier, IFN-γ induces the exchange of proteasome β-subunits that can change the proteolytic preferences of the proteasome. Non-cellular systems exist to examine the efficiency of epitope creation by different mixtures of proteasomes. One possible experiment would utilize this non-cellular system to determine whether there is a difference in the efficiency of LLO_{91-99} or $p60_{217-225}$ epitope production by isolated 20S proteasomes that contain constitutive subunits or IFN-γ-inducible subunits (LMP2, LMP7, MECL-1). If the efficiency of LLO_{91-99} epitope creation is increased by proteasomes with the IFN-γ-inducible subunits, these experiments would further support the *in vitro* results with infected P388 cells and suggest that the decreased LLO_{91-99}:$p60_{217-225}$ ratio of Ag-specific $CD8^+$ T cells in LM-infected GKO mice is the result of diminished presentation of LLO_{91-99}.

7.7
Timing of Ag Presentation and the $CD8^+$ T-cell response

It is possible that the timing of Ag display might affect the hierarchy of $CD8^+$ T-cell responses to infectious pathogens. This possibility is highlighted by the observation that the magnitude of the $CD8^+$ T-cell response is diminished in conditions where the length of Ag display is attenuated. Examples exist in antibiotic treatment of LM-infected BALB/c mice [20] and surgical excision of the herpes simplex virus-1 inoculation site [31], both 24 hours post infection. In each case, the length of Ag display is decreased and the magnitude of the $CD8^+$ T-cell response is lessened compared to the Ag-specific T-cell response detected in untreated mice. Prolonged Ag display may increase the magnitude of a $CD8^+$ T-cell response by increasing Ag-specific precursor recruitment, by increasing the number of divisions that activated Ag-specific $CD8^+$ T cells carry out, or by increasing the survival of activated T cells. It is presently unknown which of these possibilities contributes most to increased magnitudes of T-cell responses in conditions of prolonged Ag display.

There is some evidence that the hierarchy of the $CD8^+$ T-cell response to LM is correlated with the timing of Ag display. Skoberne et al. examined the time course of LLO_{91-99}, $p60_{217-225}$, $p60_{476-484}$, and $p60_{449-457}$ display after LM infection using their ELISPOT method of Ag detection [25]. They found that the relative levels of Ag presentation changed over time. At 6 and 24 hours after infection, the number of IFN-γ-producing CTL line cells was similar for each Ag, demonstrating that Ag presentation was similar for each epitope. By 48 hours after infection, presentation of the subdominant epitopes $p60_{476-484}$ and $p60_{449-457}$ dropped to almost undetectable levels, while Ag presentation of the dominant epitopes LLO_{91-99} and $p60_{217-225}$ increased slightly from the 24-hour time point. This change in Ag display correlated well with the ability of each epitope to stabilize H2-K^d molecules. However, there is a possibility that the splenocytes from 6 and 24 hours after infection induced similar numbers of LLO_{91-99}-, $p60_{217-225}$-, $p60_{476-484}$-, and $p60_{449-457}$-specific CTL line cells to make IFN-γ as a result of non-Ag-specific activation, through IL-12 and IL-18 stimulation. Given this possibility, it is important to verify these data by other methods of Ag detection. Also, in these studies, Ag measurements were performed on total splenocytes after LM infection. DCs are the most important cells in stimulating a $CD8^+$ T-cell response and represent only about 5–10% of the cells in the spleen [5]. Therefore, examination of the timing of dominant and subdominant Ag presentation on DCs after LM infection is probably the most relevant investigation to address.

It is possible to measure Ag presentation on DCs in direct *ex vivo* splenocytes from infected mice by isolating the DCs with magnetic beads and then using the ELISPOT method of Ag detection, or simply by measuring IFN-γ production in indicator T-cell lines by ICS. These methods of Ag detection are useful because they provide a "snapshot" of Ag presentation during specific time points after infection. In contrast, there is also an *in vivo* method of measuring Ag presentation on DCs that is a more qualitative, but it cuts to the crux of DC Ag display by evaluating whether specific conditions support the activation of naïve T cells. In this assay, naïve CFSE-labeled, Ag-specific, TCR-transgenic T cells are transferred into infected mice at different times after infection, are harvested 2–4 days later, and then are evaluated for mitotic activity by monitoring CFSE dilution. The more TCR-transgenic T cells that dilute the CFSE labeling, the better the quality of DC-mediated Ag presentation. Importantly, no CFSE dilution is observed in LM-specific TCR-transgenic T cells when they are transferred into LM-infected mice that are $CD11c^+$ cell depleted. These data demonstrate that Ag presentation on $CD11c^+$ cells, which are primarily DCs, is required to initiate naïve $CD8^+$ T-cell division.

Wong and Pamer utilized two lines of CD8$^+$ TCR-transgenic mice to evaluate the time course of *in vivo* DC-mediated presentation of $p60_{217-225}$ and $p60_{449-457}$ Ag after LM infection in BALB/c mice [32]. They found that naïve CD8$^+$ T cells specific for each LM Ag, both dominant and subdominant, were induced to divide after transfer into mice infected for 1–3 days. The pattern of CFSE dilution was similar for each specificity of naïve CD8$^+$ TCR-transgenic T cell on each day of T-cell transfer. These results are in contrast to the serial direct *ex vivo* measurements of Ag display and suggest that the timing of DC-mediated presentation of $p60_{217-225}$ and $p60_{449-457}$ is the same after LM infection.

The authors went on to measure *in vivo* Ag presentation on both professional and nonprofessional cells using two different types of Ag-specific T cells as probes. First, they used TCR-transgenic CD8$^+$ T cells that were specific for $p60_{449-457}$ and cultured, *in vitro*, with IL-2 to generate "effector" memory–like cells, as defined by their expression of low amounts of CD62L. Second, they used a polyclonal CD8$^+$ T-cell line specific for LLO_{91-99}. Each of these cell types did not require DCs to divide in response to encountering cognate Ag in the context of MHC class I molecules. When the $p60_{449-457}$-specific effector memory cells and the polyclonal CD8$^+$ T-cell line specific for LLO_{91-99} were CFSE labeled and transferred, individually, into LM-infected mice, these cells demonstrated similar CFSE profiles on each day of T-cell transfer. In each case, Ag display that could induce T-cell division continued until day 3 after infection and was undetectable when Ag-specific T cells were transferred into mice that were infected for 4 days. This combination of the results with naïve and effector Ag-specific CD8$^+$ T-cell probes demonstrated that DC $p60_{449-457}$ display and total $p60_{449-457}$ display follow similar kinetics and that the presentation of LLO_{91-99}, $p60_{217-225}$, and $p60_{449-457}$ continues until at least 3 days after LM infection. It is important to recognize that the avidity for the TCR-transgenic T cells and polyclonal T-cell lines utilized in these studies may not be similar. For this reason, it is impossible to make conclusions about relative amounts of Ag presentation using these transfer assays. However, it is quite evident that $p60_{449-457}$-specific CD8$^+$ T cells (naïve and effector memory–like cells) were stimulated to divide in mice 1, 2, and 3 days after LM infection. Therefore, this subdominant epitope is presented for at least 72 hours *in vivo*, as are the dominant epitopes LLO_{91-99} and $p60_{217-225}$. These results suggest that the hierarchy of the CD8$^+$ T-cell response to these Ags is not due to differences in the timing of their presentation to CD8$^+$ T cells. This is consistent with the finding that the timing of the CD8$^+$ T-cell response to dominant LM epitopes is coordinated with the T-cell response to subdominant LM epitopes [19].

7.8 Conclusions

LM infection of mice has been an invaluable tool in investigating the relationships between infection, Ag presentation, and the magnitude of the CD8$^+$ T-cell response. Unlike what is found in many viral infections, CD8$^+$ T-cell responses to "immunodominant" LM Ags do not overwhelm the $CD8^+$ T-cell responses to subdominant LM Ags [14]. This relationship allows our conclusions about $CD8^+$ T-cell response hierarchies in response to LM infection to focus on the biological properties of individual epitopes without concern about interrelationships between different $CD8^+$ T-cell epitopes. The results from just over 10 years of investigations into the hierarchy of the $CD8^+$ T-cell response to LM Ags demonstrate that the T-cell hierarchy is primarily based upon (1) whether the protein is secreted by LM; (2) intrinsic properties of the Ag (e.g., dissociation rates from the MHC molecule and endopeptidase processing efficiency); (3) the quantity of Ag presented *in vivo*; and (4), perhaps, the number of naïve Ag-specific precursors that are available in the $CD8^+$ T-cell repertoire. Investigations into LM and the $CD8^+$ T-cell response hierarchies assist in our understanding of what influences the magnitude of a $CD8^+$ T-cell response. This knowledge is important not only for its aesthetic value, in that it is a fascinating and elegant process of life, but also for its potential practical value. Protection from subsequent infection is directly proportional to the number of memory Ag-specific $CD8^+$ T cells. Understanding what influences the magnitude of a $CD8^+$ T-cell response may help us to design more effective methods of promoting large $CD8^+$ T-cell responses to vaccination. In the case of protection from LM infection, the answer is very predictable: vaccination with more Ag will elicit greater numbers of Ag-specific $CD8^+$ T cells. What is not yet easily predictable is which molecules from a complex pathogen such as LM will be antigenic. The answer to this question requires a thorough understanding of the complex interrelationships between the pathogen, the innate immune response, and the $CD8^+$ T-cell repertoire. These relationships will continue to be the subject of future study.

Acknowledgments

The authors would like to thank past and present members of the Harty laboratory for helpful discussions, K.A.N. Messingham and J.S. Haring for helpful comments, and the NIH for support. We would also like to apologize for our inability to cite all of the primary papers that contributed to this field.

References

1 Southwick, F. S. and Purich, D. L. New England Journal of Med. **1996**, 334, 770–776.
2 Mackaness, G. B. Journal of Experimental Medicine. **1962**, 116, 381–406.
3 Portnoy, D. A., Auerbuch, V. and Glomski, I. J. Journal of Cell Biology. **2002**, 158, 409–414.
4 Bancroft, G. J., Schreiber, R. D. and Unanue, E. R. Immunological Reviews. **1991**, 124, 5–24.
5 Pamer, E. G. Nature Reviews Immunology. **2004**, 4, 812–823.
6 Luster, A. D. Current Opinion in Immunology. **2002**, 14, 129–135.
7 Unanue, E. R. Immunological Reviews. **1997**, 158, 11–25.
8 Swain, S. L. Journal of Experimental Medicine. **2001**, 194, F11-F14.
9 Berg, R. E., Crossley, E., Murray, S. and Forman, J. Journal of Experimental Medicine. **2003**, 198, 1583–1593.

10 Harty, J. T. and Bevan, M. J. Immunity. **1995**, 3, 109–117.

11 North, R. J., Dunn, P. L. and Conlan, J. W. Immunological Reviews. **1997**, 158, 27–36.

12 Ahmed, R. and Gray, D. Science. **1996**, 272, 54–60.

13 Harty, J. T. and Bevan, M. J. Current Opinion in Immunology. **1999**, 11, 89–93.

14 Yewdell, J. W. and Bennink, J. R. Annual Review of Immunology. **1999**, 17, 51–88.

15 Tvinnereim, A. R., Hamilton, S. E. and Harty, J. T. Journal of Immunology. **2004**, 173, 1994–2002.

16 Shen, H., Miller, J. F., Fan, X., Kolwyck, D., Ahmed, R. and Harty, J. T. Cell. **1998**, 92, 535–545.

17 Zenewicz, L. A., Foulds, K. E., Jiang, J., Fan, X. and Shen, H. Journal of Immunology. **2002**, 169, 5805–5812.

18 Glaser, P., Frangeul, L., Buchrieser, C., Rusniok, C., Amend, A., Baquero, F., Berche, P., Bloecker, H., Brandt, P., Chakraborty, T., Charbit, A., Chetouani, F., Couve, E., de Daruvar, A., Dehoux, P., Domann, E., Dominguez-Bernal, G., Duchaud, E., Durant, L., Dussurget, O., Entian, K. D., Fsihi, H., Garcia-del Portillo, F., Garrido, P., Gautier, L., Goebel, W., Gomez-Lopez, N., Hain, T., Hauf, J., Jackson, D., Jones, L. M., Kaerst, U., Kreft, J., Kuhn, M., Kunst, F., Kurapkat, G., Madueno, E., Maitournam, A., Vicente, J. M., Ng, E., Nedjari, H., Nordsiek, G., Novella, S., de Pablos, B., Perez-Diaz, J. C., Purcell, R., Remmel, B., Rose, M., Schlueter, T., Simoes, N., Tierrez, A., Vazquez-Boland, J. A., Voss, H., Wehland, J. and Cossart, P. Science. **2001**, 294, 849–852.

19 Busch, D. H., Pilip, I. M., Vijh, S. and Pamer, E. G. Immunity. **1998**, 8, 353–362.

20 Badovinac, V. P., Porter, B. B. and Harty, J. T. Nature Immunology. **2002**, 3, 619–626.

21 Butz, E. A. and Bevan, M. J. Immunity. **1998**, 8, 167–175.

22 Murali-Krishna, K., Altman, J. D., Suresh, M., Sourdive, D. J., Zajac, A. J., Miller, J. D., Slansky, J. and Ahmed, R. Immunity. **1998**, 8, 177–187.

23 Pamer, E. G., Sijts, A. J., Villanueva, M. S., Busch, D. H. and Vijh, S. Immunological Reviews. **1997**, 158, 129–136.

24 Busch, D. H. and Pamer, E. G. Journal of Immunology. **1998**, 160, 4441–4448.

25 Skoberne, M., Holtappels, R., Hof, H. and Geginat, G. Journal of Immunology. **2001**, 167, 2209–2218.

26 Kloetzel, P. M. and Ossendorp, F. Current Opinion in Immunology. **2004**, 16, 76–81.

27 Schroder, K., Hertzog, P. J., Ravasi, T. and Hume, D. A. Journal of Leukocyte Biology. **2004**, 75, 163–189.

28 Boehm, U., Klamp, T., Groot, M. and Howard, J. C. Annual Review of Immunology. **1997**, 15, 749–795.

29 Badovinac, V. P., Tvinnereim, A. R. and Harty, J. T. Science. **2000**, 290, 1354–1358.

30 Skoberne, M. and Geginat, G. Journal of Immunology. **2002**, 168, 1854–1860.

31 Stock, A. T., Mueller, S. N., van Lint, A. L., Heath, W. R. and Carbone, F. R. Journal of Immunology. **2004**, 173, 2241–2244.

32 Wong, P. and Pamer, E. G. Immunity. **2003**, 18, 499–511.

8
Immunodominance in Tuberculosis

David M. Lewinsohn and JoAnne L. Flynn

8.1
Immune Responses to *Mycobacterium tuberculosis*

Tuberculosis (TB) kills approximately 2 million people per year. There are 8 million new cases of tuberculosis each year, and although there are drugs that can cure this disease, the regimen is complex and lengthy. Infection with the causative agent of TB, *Mycobacterium tuberculosis* (Mtb), is reflected in a positive tuberculin skin test (TST), which indicates a cell-mediated immune response to the pathogen. These individuals are at risk for progression to active TB, particularly during the first two years following skin test conversion. However, most individuals with a positive skin test remain disease-free throughout their lifetime and hence are thought to be latently infected. Reactivation of latently infected persons can occur, suggesting that latency reflects chronic persistent infection with Mtb. Although HIV infection and aging are major risk factors for reactivation of latent tuberculosis, in some cases the immunosuppression responsible for disease is not obvious. Furthermore, for those living in Mtb-endemic areas, re-infection may account for a significant proportion of tuberculosis in those with TST reactivity.

Cell-mediated immunity is necessary for control of Mtb infection. Macrophages are an important reservoir of Mtb in the host, and activation of these cells through production of cytokines by T cells can lead to control, although not elimination, of the infection. Several macrophage antibacterial responses affect outcome of infection. Macrophages produce reactive oxygen and nitrogen intermediates that can limit mycobacterial replication. Indeed, loss of inducible nitric oxide synthase dramatically increases susceptibility to Mtb in a murine model [1, 2]. There is evidence for reactive nitrogen intermediates produced during human infection as well, although the circumstances under which this occurs remain unclear [3]. In addition, phagolysosome fusion can impair survival of *M. tuberculosis*. An immune evasion strategy of Mtb is inhibition of phagolysosome fusion [4]; macrophage activation by T cells can overcome this inhibition. A new antimicrobial mechanism of macrophages involving the Lrg47 gene is also important in controlling intracellular replication, in an IFN-γ-dependent but NOS2-independent manner [5]. It is clear that effective activation of macrophages is crucial in control of

Immunodominance: The Choice of the Immune System. Edited by Jeffrey A. Frelinger

ISBN: 3-527-31274-9

the infection, but it is equally clear that Mtb has various strategies to counteract the antimicrobial mechanisms of these cells.

8.2 B Cells

Although an antibody response against various Mtb antigens is mounted, there is little evidence that this response participates in the containment of infection. However, there are often many B cells present at the site of infection in the lungs, and they may play a role in pathology, immune regulation, or protection that has not yet been elucidated. In humans, serological reactivity to Mtb is associated with clinical tuberculosis and likely reflects the high bacterial burden associated with the disease. However, a Th2 isotype profile has not been described. The antibody response to infection with mycobacteria is reasonably broad, such that the challenge of identifying useful markers has been finding serologic markers capable of distinguishing infection with Mtb from exposure to other mycobacteria. However, in this regard, a panel of antibodies of potential diagnostic utility have recently emerged (reviewed in Ref. [6]).

8.3 CD4 T Cells

CD4 T cells are accepted as a major mediator of control of Mtb. HIV infection substantially increases the risk of developing tuberculosis, both primary and reactivation disease. It is believed that one contributing factor in this enhanced susceptibility is the loss of CD4 T cells. In the murine model, CD4 T cell–deficient mice succumb to Mtb infection within ~8 weeks [7]. In a murine model of chronic infection, depletion of CD4 T cells results in increased bacterial growth in the lungs and death of the mice [8]. CD4 T cells seem to be necessary for the cytotoxic function of CD8 T cells in tuberculosis [9]. In terms of effector function, CD4 T cells in tuberculosis are primarily Th1, secreting IFN-γ and TNF. These cytokines are necessary for activation of macrophages, and TNF plays additional roles, such as directing granuloma formation and maintenance (reviewed in Ref. [10]). Mice deficient in IFN-γ or TNF are the most susceptible and succumb to the infection after 3–4 weeks with high bacterial numbers [11–14].

Many of the important observations derived from genetically disrupted mice have proven to have strong clinical correlates with human TB. Clinical conditions that impair $CD4^+$ T-cell immunity, such as HIV infection, dramatically increase the likelihood of developing active TB [15]. Patients in whom the gene for the IFN-γ receptor is mutated are susceptible to infection with atypical mycobacteria [16]. Strong TH1-type, antigen-specific IFN-γ-secreting T cells are found in peripheral blood mononuclear cells (PBMCs) from healthy individuals with latent Mtb infection but are diminished in individuals with pulmonary TB [17–19]. Furthermore, those with the most severe disease have the most impaired responses [20, 21]. Mtb-specific IFN-γ-secreting cells are detected in bronchoalveolar lavage (BAL) cells from individuals with LTBI [22] and in individuals with pulmonary TB [23]. Patients with more severe disease may have decreased local Mtb-specific TH1 immunity, as indicated by decreased IFN-γ-secreting cells in BAL cells compared to patients with less severe disease [24]. Data from experimental animals and in those with $CD4^+$ T-cell depletion due to infection with HIV suggest that CD4 T cells, in addition to production of cytokines, have other roles, including direct interactions with macrophages and facilitating induction or maintenance of CD8 T-cell responses.

In recent years, it has become clear that neutralization of TNF, as a treatment for various inflammatory diseases such as rheumatoid arthritis and Crohn's disease, can result in reactivation of latent tuberculosis [25]. It is likely, although it has not been directly demonstrated in humans, that use of these TNF-neutralizing therapies will also increase susceptibility to primary tuberculosis. The role of TNF in control of Mtb infection was first demonstrated in mouse models, and the various effector mechanisms of this cytokine are still being studied in animal models.

8.4 CD8 T Cells

Because Mtb resides within a vacuole in the macrophage, it was believed that antigens would not be presented efficiently by MHC class I molecules. However, there is substantial evidence to the contrary, and it now appears the MHC class I molecules can be loaded with mycobacterial antigens via both cytosolic and non-cytosolic pathways. In humans, Mtb-specific CD8 T cells have been identified in Mtb-infected individuals and include $CD8^+$ T cells that are classically Ia-restricted [26–32] and non-classically restricted by HLA-E [33, 34] and by CD1 [35–37]. In mice, CD8 T cells specific for Mtb antigens have also been documented [38, 39].

Adoptive transfer or *in vivo* depletion of CD8 T cells showed that this subset could confer protection against subsequent challenge, although the effects were much less pronounced than that seen with CD4 T cells [40–42]. More definitive evidence was provided by the study of $\beta 2$ microglobulin–deficient mice. $\beta 2$ microglobulin is an essential component of the MHC class I molecular complex, and mice deficient for $\beta 2$ microglobulin ($\beta 2m^{-/-}$) are more susceptible to Mtb [43] and to large doses of BCG [44] than their wild-type littermates. This finding has been confirmed in other mice deficient in class I processing and presentation, including TAP1-deficient ($TAP1^{-/-}$) mice [45, 46], CD8-deficient ($CD8^{-/-}$) mice [46, 47], and class Ia–deficient ($K^{b-/-}/D^{b-/-}$) mice [48, 49], as these animals are all more susceptible to infection than wild-type animals.

Interestingly, there are significant differences in the degree of susceptibility to Mtb infection observed in these studies, depending on which "CD8-deficient" model is examined. When survival of $\beta 2m^{-/-}$, $TAP^{-/-}$, and $CD8^{-/-}$ mice were compared, β2m-deficient mice were clearly the most susceptible to Mtb infection, followed by TAP-deficient mice and then CD8α-deficient mice [46]. While $\beta 2m^{-/-}$ mice are deficient in expression of both MHC-Ia and MHC-Ib molecules, $TAP^{-/-}$ mice can still express these molecules bound to antigen processed in a TAP-independent manner. In fact, it has been demonstrated that several MHC-Ib molecules can present antigen in a TAP-independent manner, including H2-M3 [50], Qa-1 [51], and CD1 [52]. Therefore, protection by MHC-Ib-restricted CD8 T cells could possibly account for the difference in susceptibility observed in $TAP^{-/-}$ compared to $\beta 2m^{-/-}$ mice. This explanation is further supported by the observation that class Ia– and class Ib–deficient ($\beta 2m^{-/-}$) mice are more susceptible to infection than class Ia–deficient mice ($K^{b-/-}/D^{b-/-}$) [48, 49]. Taken together, these studies suggest a protective role for both classically and non-classically restricted CD8 T cells in Mtb infection.

Mycobacteria-specific CD8 T cells produce IFN-γ and TNF in response to antigen presented by infected macrophages or dendritic cells. These cells can also function as cytotoxic T lymphocytes (CTLs) and kill infected cells via the perforin–granzyme pathway of directed apoptosis (see review in Ref. [53]). Human cytotoxic T cells also express granulysin, and this molecule can be delivered to infected cells to kill the intracellular mycobacteria. Furthermore, CD8 T cells preferentially recognize heavily infected cells and preferentially induce target cell apoptosis [54]. It has been postulated that the apoptotic macrophage is an unfavorable environment associated with diminished intracellular growth of Mtb [55]. Thus, CD8 T cells have the potential to contribute significantly to control of Mtb infection.

It is clear that the immune response to this tenacious pathogen is complex and wide-ranging. Sorting out which responses and effector functions are important at different times during infection, as well as the antigens recognized by the T cells over the course of infection and disease, will be important steps in designing effective vaccine strategies.

8.5
Antigen Processing and Presentation of Mtb Antigens

One of the unusual features of Mtb is that it resides in an organelle with features of an arrested early endosome [56, 57]. Following phagocytosis, the Mtb-containing phagosome neither acidifies fully nor undergoes fusion with the lysosome. While the mechanisms of possible MHC-I processing will be discussed more fully below, the intracellular location of Mtb has important implications for the evolution and maintenance of the Mtb-specific T-cell repertoire. With regard to MHC-II-restricted responses, Mtb-derived antigens can be processed and presented both by cells that are infected and by those that might pick up antigen from the extracellular milieu through a pinocytotic mechanism. Ramachandra et al. have demonstrated that the Mtb-containing phagosome contains MHC-II–peptide complexes and have suggested that this phagosome retains some of its MHC-II processing capability [58]. Given the altered pH of this environment, and the as yet unknown repertoire of available cathepsins and other lysosomally derived proteases, it would appear that antigens processed within this environment may be distinct from those derived by pinocytosis. This raises the interesting possibility that MHC-II epitopes may uniquely identify an Mtb-infected cell. However, experimental demonstration of this possibility is lacking. In addition to the altered endosomal environment, Mtb has the ability to generate post-translational modifications, such as mannosylation [59], acylation [60], and methylation, that could alter protein stability and hence the generation of antigenic peptides. It has recently been demonstrated that these modifications, at least in the case of methylation, can constitute a motif for TCR-specific binding [61].

Because Mtb is believed to reside within the phagosome, a persistent question concerning the $CD8^+$ response to Mtb infection is how Mtb-derived antigens present in this compartment can gain access to the cytosolic MHC-I processing machinery. Data obtained by examining processing and presentation of particulate-associated antigen, such as latex beads or bacterially associated proteins, have suggested two separate but non-mutually exclusive pathways. These have been broadly characterized as cytosolic and vacuolar models. In the cytosolic model, phagosomal antigens escape into the cytosol, where they are processed by the conventional MHC-I machinery [62]. Alternatively, the vacuolar model relies on the generation of peptide epitopes and MHC-I loading in a distinct endosomal compartment.

In support of the cytosolic model, mycobacterial infection can facilitate the presentation of soluble ovalbumin to MHC-Ia-restricted CD8 T cells. Additionally, cytosolically delivered fluorescent markers up to 70 kDa localized to phagosomes containing live but not heat-killed Mtb [63, 64]. From these results, it has been argued that infection with live bacteria results in formation of a pore in the Mtb-resident phagosomes, giving antigen access to the cytosol. However, in a recent study, detection of radioactive or fluorescently labeled Mtb antigens was low within the cytoplasm, suggesting that relatively low amounts of Mtb proteins localize to the cytosol [65].

Consistent with the cytosolic pathway is recent evidence suggesting that peptides may be generated and loaded onto MHC class I within phagosomes [66–71]. Following fusion with ER-derived membrane, phagosomes acquire the protein translocation complex Sec61 in conjunction with the class I processing molecules TAP, calnexin, tapasin, calreticulin, Erp57, and MHC-I. It is hypothesized that antigens may escape the phagosome using Sec61 and be degraded by the proteasome. Then resultant peptides may be transported back into the phagosome via TAP, where loading of class I MHC occurs. Furthermore, trafficking of phagosome-derived MHC-I–peptide complexes to the cell surface is only partially inhibited by treatment with BFA, suggesting that ER-Golgi transport is not required for cell surface presentation. Together, these data characterize the role of phagosomes as autonomous organelles for cross-presentation. In this regard, our laboratory and others have reported the existence of lactacystin-sensitive [33], BFA-insensitive [33, 72] processing pathways that would be consistent with this newly described pathway. In support of a cytosolic model, we have also demonstrated that processing of the Mtb-secreted protein CFP10 follows the traditional MHC-I processing pathway, in that processing is dependent on proteasomal degradation and is inhibited with BFA (D.M.L., unpublished data).

In contrast, the vacuolar pathway of MHC-I presentation is TAP and proteasome independent as well as brefeldin A insensitive [73–75]. Mechanistically, it has been suggested that exogenous antigens can be loaded onto MHC-I molecules through a vesicular recycling pathway. Recently, Shen et al. demonstrated that cathepsin S is required for the generation of certain endosomally derived peptides [76].

Mtb lipids and proteins gain access to small exosomes [77–79], which have been demonstrated to lead to bystander activation of T cells in a TAP-independent manner [80]. Recently, Schaible et al. showed that Mtb-induced apoptosis led to the release of apoptotic bodies containing Mtb-derived lipids and proteins [65]. Cross-presentation of these vesicles after phagocytosis by bystander cells, presumably dendritic cells, was dependent on phagosomal acidification but not on proteasomal degradation.

From these data, a complex and dynamic picture of the mechanisms by which particulate-associated antigens gain access to the MHC-I processing pathway has emerged. Although the relevance of any of these processing pathways in presentation of Mtb-derived antigens *in vivo* is unknown, it is likely that a combination of these pathways is utilized for the priming of naïve $CD8^+$ T cells and presentation of antigen to memory/effector $CD8^+$ T cells. These data raise the possibility that the physical characteristics of the antigen may govern which pathway is used. For example, secreted proteins may preferentially be processed by the cytosolic pathway, whereas the non-secreted antigens may be preferentially processed in a phagosomal/late endosomal compartment. Finally, uninfected dendritic cells that have taken up secreted exosomes or apoptotic bodies may preferentially present lipids, lipoproteins, and hydrophobic proteins that could lead to cross-priming of $CD8^+$ T cells or to local production of IFN-γ by antigen-specific T cells.

8.6
How Does Infection with Mtb Differ from Other Acute or Chronic Infections?

Most infections that have been intensely studied in terms of immunology are those that cause acute infections that are cleared by the host and induce a strong memory immune response, such as influenza virus, *Listeria monocytogenes*, and some strains of lymphocytic choriomeningitis virus. However, chronic infections are likely to interact with the immune system differently, and the rules that are set in place by studying acute infections may not hold when it comes to chronic or persistent infections. While the natural history of latent infection with Mtb remains incompletely understood, it is difficult to demonstrate eradication of the bacteria in any experimental system. Similarly, while little is known about the bacterial burden in humans latently infected with Mtb, it is clear that the organism can persist for many years [81, 82]. Hence, latent infection with Mtb would be more accurately described as chronic persistent infection. Mice infected with a low dose of Mtb via the respiratory route experience growth of the organisms in the lungs, with dissemination to the mediastinal lymph node, spleen, and liver. The bacterial burden in the lungs peaks at ~10^6 bacteria by four weeks post-infection, and this bacterial burden is maintained for many months, although the pathology in the lungs continues to increase. The control of the growth of the organisms coincides with the infiltration of T cells into the lungs. The mouse is truly a model of chronic infection, with relatively high bacterial loads.

Immunologically, the persistence of Mtb within the host means that the immune system is chronically exposed over a period of many years, a possibility that has important implications for the generation and maintenance of effector memory and central memory T cells. The maintenance of very high-frequency CD4 and CD8 T-cell responses in humans whose likely Mtb exposure likely occurred in the remote past [83] suggests that chronic antigenic exposure is a central feature of the immune response to Mtb. In humans, there have been no longitudinal descriptions of the patterns of immunodominance. In the mouse model, tetramer and transgenic reagents suitable for these studies are being developed and likely will shed light on the impact of chronic antigenic exposure to the ontogeny of the immune response. In this chapter, we will discuss the available data on immunodominance in tuberculosis, including both CD4 and CD8 T-cell responses.

8.7 Immunodominance in the CD4 T-cell Response

8.7.1 Human

CD4 T cells have been unambiguously identified as central effectors in the host response to Mtb. As a result, extensive research has been performed to identify those antigens that are both immunodominant and associated with protective immunity. Here, immunodominance is defined as those antigens for which a majority of the individuals infected with Mtb have circulating memory/effector cells at frequencies suggestive of a memory response. An early observation was that live, but not heat-killed Mtb, could elicit protective immunity in the mouse model. It should be noted, however, that protective immunity in this model is defined as a reduction in lung bacterial burden compared to naïve mice, but the result remains a chronic and progressive infection. From this observation, it was hypothesized that proteins secreted from actively dividing Mtb might be protective [84]. Indeed, vaccination of guinea pigs and mice with early secreted proteins conferred protection to subsequent challenge [85–87]. These initial observations led to the identification and purification of several important CD4 T-cell antigens, including ESAT6 and Ag85, that are strongly recognized by T cells in persons with Mtb infection and disease and that provide some protection in animal models when used as a subunit vaccine [87, 88].

More recently, alternate antigen discovery approaches have demonstrated that secreted antigens are not uniquely capable of conferring protective immunity. For example, expression cloning has been used to identify immunologically important antigens. In this approach, an Mtb genomic library is cloned into *E. coli*, and proteins expressed in the bacteria are screened with sera or T cells from vaccinated or infected humans or experimental animals. For example, by using rabbits immunized with secreted protein from Mtb, Skeiky et al. identified the serine protease Mtb32 [89], which is currently a component of the Corixa/GlaxoSmithKline vaccine candidate [90]. Using human serum, Dillon et al. identified both the low-molecular-weight secreted protein CFP10 [91] and Mtb39 (a non-secreted protein that is part of the PPE family of TB proteins) [92]. Mtb39 is strongly recognized by persons infected with Mtb [92] and constitutes the remaining portion of the Corixa/GSK vaccine candidate. Finally, by using DCs pulsed with *E. coli* expressing Mtb genomic DNA, the Mtb9.9 family of proteins was identified [93]. The strength of the expression cloning approach is that it allows for the identification of antigens based upon a preexisting immune response.

While these efforts have defined a set of antigens that are strongly recognized by the human immune system, experiments designed to characterize the dominance hierarchy within an individual have demonstrated that this response is generally broad and polyspecific. For example, Pathan et al. used overlapping 15-mer peptides of ESAT-6 to define reactivity in a panel of subjects and found generally broad recognition [94]. However, $ESAT_{1-16}$ and $ESAT_{71-85}$ were the most commonly recognized peptides, and recognition was in some cases blocked by antibodies to HLA-DQ. Similarly broad patterns of reactivity have been reported for the Mtb antigen CFP10 [28], PPE protein Rv3873 [95], MBP70 [96], Rv2654 and Rv2653 [97], TB27 (Rv3878) [98], Ag85 complex proteins [99], and ESAT-6 family members TB 10.3, 10.4, and 12.9 [100].

One notable exception to this broad pattern of reactivity is the 16-kDa protein HspX (*acr*, Rv2031c). Of 55 HspX-reactive $CD4^+$ T-cell clones derived from subjects with active tuberculosis, 28 were reactive to the $HspX_{91-110}$ epitope [101]. While a common TCR usage was not identified, approximately 50% of the clones were uniquely restricted by a single parental HLA-DR molecule. Four of the 18 clones could recognize the epitope in the context of either parental HLA-DR, and five were capable of responding to either parental or allogeneic HLA-DR molecules. These data suggest substantial promiscuity in both peptide binding and TCR recognition [102]. Similarly, Shams, et al. have described $CFP10_{71-85}$ as a commonly recognized peptide that contains two epitopes, T1 (IRQAGVQYSR) and T6 (EISTNIRQA). Both these peptides could be presented by both HLA-DR and HLA-DQ alleles, and a T1-specific T-cell clone could recognize the peptides in the context of both HLA-DR and HLA-DQ alleles [103].

8.7.2 Mouse

In the mouse system, the CD4 T-cell response is quite broad, and a number of antigens are recognized by infected mice. However, ESAT-6 stands out as a mycobacterial antigen that is recognized not only by several mouse strains and humans but also by guinea pigs, non-human primates, and cattle [104, 105]. This is clearly a very immunogenic protein. Interestingly, the first 20 amino acids of ESAT-6 contain a major CD4 epitope recognized by mice of the H-2^b and H-2^d haplotypes, as well as many Mtb-infected humans [94] and guinea pigs and *M. bovis*-infected cattle.

In C57BL/6 mice, 2–3% of CD4 T cells in the lungs and ~1% of CD4 T cells in the spleen respond to epitope ESAT-6_{1-20} at the peak of the response (3–4 weeks post-infection) [106]. There was a modest contraction of the response, but the frequency of ESAT-6 responsive cells remained at 1.5–2% during the chronic phase of infection. When Mtb-infected or ESAT-6 peptide-pulsed dendritic cells were used as stimulators in an ELISPOT assay, similar numbers of IFN-γ-producing cells were detected early in infection, suggesting that ESAT-6 was a major antigen detected early by CD4 T cells [107]. However, as the infection progressed, Mtb-infected dendritic cells stimulated a higher frequency of IFN-γ-producing cells than did ESAT-6 peptide-pulsed dendritic cells, suggesting a broadening of the response to include additional antigens. The identities of those other antigens were not determined in that study.

Additional antigens recognized by CD4 T cells have been identified, but no studies comparing the frequency of responses over the course of infection in mice to a large panel of potential CD4 T-cell epitopes and antigens are available. Antigens that appear to generate relatively high-frequency responses in at least some strains of mice include CFP10 (in C3H/SnJ mice) [108], PPE68 (in C57BL/6 mice) [109], antigen 85A (in C57Bl/6 and BALB/c mice) [110], and PstS-3 [111]. Interestingly, Mtb infection results in a much more restricted response to the Ag85 complex of proteins than does DNA vaccination [112].

8.8 Immunodominance in the CD8 T-cell Response

8.8.1 Human

At present, the relationship between CD8 T-cell response and protective immunity in humans remains unclear. For example, it is possible that a robust, high-frequency $CD8^+$ T-cell response is necessary for preventing the replication of Mtb and hence may prevent the development of clinical tuberculosis. Alternately, it is possible that increasing bacterial burden will be associated with increasing $CD8^+$ T-cell frequencies. By defining immunodominant $CD8^+$ T-cell antigens, it will become possible to employ direct *ex vivo* analysis to define the abundance of Mtb-specific T-cell responses in subject groups of particular interest. Surprisingly little is known about common Mtb antigens recognized by human $CD8^+$ T cells (Table 8.1). With two exceptions [113, 114], definition of $CD8^+$ antigens has relied upon using antigens known to be involved in $CD4^+$ T-cell recognition. Moreover, many of the epitopes that have been identified were discovered through the synthesis of peptides predicted to bind a specific HLA-I allele (HLA-A2 in most cases) [29, 115, 116].

Table 8.1 Dominant human CD8 T-cell epitopes*

Protein	Epitope amino acids	Restricting allele	Frequency	Identified by	Reference
16-kDa protein (HspX)	21–29 120–128	HLA-A2 HLA-A2	1/1000 CD8+[a] 1/800 CD8+[a]	Synthesis of predicted HLA-A2-binding peptides	101
Ag85A	48–56 242–250	HLA-A2 HLA-A2	1/20,693 PBMC by ELISPOT 1/3300 by tetramer[c] 1/23,779 PBMC by ELISPOT, 1/3750 by tetramer[c]	Screening of overlapping peptides spanning the entire protein	156
Ag85 Complex	264–272 204–212	HLA-B35 HLA-B35	Not determined 1/13,700–1/22,200 PBMC[c]	Synthesis of predicted HLA-B35-binding peptides from Ag85A, -B, and -C	157
ESAT-6	69–76	HLA-B53	1/23,000 PBMC[a]	Synthesis of predicted HLA-A2, -B7, -B8, -B35, -B52, and -B53-binding peptides	29
ESAT-6	21–29	HLA-A6802	1/2500 PBMC[b]	Screening of overlapping peptides spanning the entire protein	94
CFP10	2–11 85–94 49–58 2–10 3–11	HLA-B44 HLA-B14 HLA-B3514 HLA-B4501 HLA-A0101	1/700 CD8+[b] 1/2100 CD8+[b] 1/1000 CD8+[a] 1/55 CD8+[a] 1:800 CD8+[a]	T-cell clones from Mtb-infected donor	28 Unpublished (D.M.L.) Unpublished (D.M.L.) Unpublished (D.M.L.) Unpublished (D.M.L.)
Mtb8.4	32–40	HLA-B3514	1/8500 PBMC[a]	T-cell clones from Mtb-infected donor	Unpublished (D.M.L.)
Mtb39	144–153 346–355	HLA-B44 HLA-B44	1/3000 CD8+ by ELISPOT[b]	T-cell clones from Mtb-infected donor	26

* Dominance defined as a response where *ex vivo* frequency is estimated to be over 1/10,000 CD8+ T cells

a Frequency determined in active TB patients.

b Frequency determined in PPD+ patients.

c Frequency determined in BCG-vaccinated patients.

While most studies have highlighted the natural generation of the identified class I epitopes during the course of Mtb infection *in vitro*, a weakness of these results is the low to undetectable frequencies of CD8 T-cell precursors specific for many epitopes in the blood of Mtb-exposed individuals. However, CD8 T-cell responses to a subset of antigens (CFP10, ESAT-6, Ag85, Mtb39, Mtb8.4, and the 16-kDa antigen) have been detected at a high frequency in persons infected with Mtb. These data would suggest that these responses are acquired in response to infection with Mtb (Table 8.1). Indeed, we have used overlapping peptide pools to assess the frequency of antigen-specific responses *ex vivo* in persons with latent infection or active tuberculosis and in those without evidence of Mtb infection. In these experiments, Mtb-associated responses are not detected in uninfected persons, while high-frequency responses are often detected in those with Mtb infection (D.M.L., unpublished data). While no single antigen appears dominant, it is clear that HLA-Ia-restricted T-cell responses are associated with infection with Mtb, and these observations are consistent with the hypothesis that these responses reflect an adaptively acquired response to infection with Mtb. More recently, Tully et al. have synthesized HLA-I and HLA-II tetramers based on HLA-A2 predicted binding motifs [117]. These investigators were able to demonstrate the presence of Mtb-specific CD8 T cells in the peripheral blood and to identify these cells within lung granulomas. The clear advantage of this approach is the ability of tetramers to define the frequency and phenotype of these cells. With regard to immunodominance, it is not known whether or not the responses that were evaluated reflected dominant or subdominant responses.

In contrast to the relatively broad and likely promiscuous responses described for CD4 T-cell antigens, the pattern of immunodominance for CD8 T-cell responses is narrowly focused. For example, in one subject with active tuberculosis, two CFP10-specific T-cell clones were used to define two distinct epitopes within this protein [28]: $CFP10_{2-11}$, restricted by HLA-B44, and $CFP10_{85-94}$, restricted by HLA-B14. Both were found to be present at a high frequency, and nearly the entire response to this protein was to these two epitopes. By using a similar T cell–driven approach, high-frequency epitopes for the Mtb antigens CFP10, Mtb8.4, Mtb39, and Ag85 have been defined (Table 8.1) (Ref. [26] and unpublished data). In each case, these epitopes have proven to be the dominant epitopes for the antigen of interest (D.M.L., unpublished data).

While the complete repertoire of CD8 T-cell responses remains uncharacterized, the following conclusions can be drawn.

CD8 T-cell responses are present in persons infected with Mtb at a frequency comparable to that seen following viral infection. This conclusion is based on both the pooled peptide experiments described above and the observation that when they are defined, dominant epitopes are present at a high frequency (Table 8.1). Conversely, we have not observed high-frequency responses in those without evidence of infection with Mtb. This observation strongly supports the hypothesis that these responses reflect an adaptively acquired response to infection with Mtb rather than an innate response.

Most of the epitopes that have been mapped to date are restricted by HLA-B molecules. The reasons for this possible skewing are not yet clear.

While the immune response within an individual to a given antigen is tightly focused, dominant epitopes that would be useful for population-based analysis have yet to be defined. This conclusion is based on the fact that none of the HLA-A2 epitopes described to date have proved to be widely recognized and, most importantly, on the observation that a wide variety of HLA alleles appear to be used in the recognition of Mtb antigens. The antigen CFP10 is an excellent case in point. As is demonstrated in Table 8.1, T-cell clones have been used to define high-frequency epitopes restricted by a variety of HLA alleles. However, in each case the epitope defined has been unique to the individual from which the T cells were derived.

When computer algorithms are used to predict HLA binding [118, 119], many of the epitopes (HLA-B3514::$CFP10_{2-10}$; HLA-B3514::$CFP10_{32-40}$; HLA-A0101::$CFP10_{3-11}$) have no predicted binding affinity. While these data have not been experimentally confirmed, it supports the ongoing efforts of the NIH-sponsored Large-scale Antibody and T Cell Epitope Discovery Program to refine these prediction tools.

8.8.2 Mouse

Mtb infection induces a CD8 T-cell response in mice, as evidenced by cytotoxic activity of these cells against Mtb-infected macrophages *in vitro* or *in vivo* [108, 111, 146, 147] and by the induction of IFN-γ from CD8 T cells from infected mice by Mtb-infected dendritic cells [38, 147]. In a recent study from our group, the early CD8 T-cell response in the lungs appeared to be cytotoxic, with evolution of the IFN-γ-producing CD8 T-cell response as the infection progressed to the chronic stage [147]. The mechanisms behind this apparent switch in CD8 T-cell function are unknown, but they may be related to antigen load in the lungs during chronic infection.

There are data to support that the majority of CD8 T cells in the lungs during Mtb infection capable of producing IFN-γ are specific for Mtb [147]. However, it has been difficult to identify CD8 T-cell epitopes, and this has made it very challenging to perform state-of-the-art immunologic investigations into the priming and evolution of the immune response. Recently, there have been several studies where CD8 T-cell epitopes have been identified (Table 8.2), but there are obviously more antigens that are recognized by CD8 T cells during Mtb infection in the mouse than those currently known. One study examined peptides from culture filtrate proteins of Mtb (those that are found in the culture supernatant) and reported that these antigens were recognized primarily in the context of H-$2D^b$ and that there appeared to be two immunodominant peptides recognized [148]; the peptides were not identified in this study.

Although ESAT-6 has been shown to be recognized by CD8 T cells in humans, there are no data suggesting that CD8 T-cell epitopes from ESAT-6 are recognized

Table 8.2 Major CD8 T-cell antigens in the murine model of TB.

Antigen	Epitope amino acids	Restriction	Identified by	Reference
CFP10	32–39	H-2^k	Screening overlapping peptides	108
TB10.3/10.4	20–28	H-2K^d	Predicted H-2K^d binding of peptide sequences from 400 ORFs	149
PstS-3	285–293	H-2D^b	DNA vaccination and screening consensus peptides from PstS-3	111

in mice. However, CFP-10, which is associated with ESAT-6 in Mtb, and the closely related proteins Tb10.3 (Rv3019c) and Tb10.4 (Rv0288), are recognized by CD8 T cells from infected mice [108, 149]. The Tb10.3/10.4 proteins share the same epitope recognized by H-2^d-restricted CD8 T cells (aa 20–28). This epitope was identified following screening of 400 open reading frames of Mtb genes for peptides that would bind to H-2K^d [149]. Of 11 selected from this program, only one was identified as an immunogenic CTL epitope.

In a separate study, peptides spanning CFP-10 were tested in H-2^b, H-2^d, and H-2^k mice; only CD8 T cells from H-2^k mice (C3H/SnJ) responded to any peptide as measured by IFN-γ production [108]. The epitope recognized (CFP10$_{32-39}$) turned out to be a major CD8 T-cell epitope in this mouse strain. By four weeks post-infection, ~1% of the CD8 T cells in the lungs produced IFN-γ in response to this peptide, and tetramer staining indicated that 8–10% of CD8 T cells in the lungs at 4–5 weeks post-infection recognized this epitope. Tetramer staining indicated that 3.2% of CD8 T cells in the spleens of infected BALB/c mice recognized the Tb10.3/10.4 epitope, while the frequency of such cells in C3H mice recognizing CFP10$_{32-39}$ was ~1% [108]. Clearly, there is an enrichment of CD8 T cells recognizing the CFP10 epitope in the lungs, compared to the lymphoid tissues (including lymph nodes). *In vivo* cytotoxic activity in Mtb-infected mice was demonstrated using the CFP10 epitope, confirming that CD8 T cells generated during infection can recognize this epitope *in vivo*.

Another recently identified CD8 T-cell epitope is from a putative phosphate transport receptor (PstS-3) [111]. This antigen was identified as potentially protective through DNA vaccination studies [150], and the CD4 and CD8 T-cell epitopes were mapped using DNA vaccination and peptide screening and were confirmed in mice infected with Mtb [111]. The PstS-3 antigen is immunogenic only in mice of the H-2^b, H-2^p, and H-2^r haplotypes. The frequency of the CD8 T-cell response to PstS-3$_{285-293}$ was not precisely determined, but these CD8 T cells could also recognize and lyse peptide-pulsed targets *in vivo*. This is the only CD8 T-cell epitope identified to date for C57Bl/6 mice.

Although the Ag85 complex is immunogenic, infection with Mtb did not induce CD8 T-cell responses against any of the three Ag85 proteins in BALB/c or C57BL/

6 mice [112]. In contrast, DNA immunization with Ag85A resulted in CD8 T-cell responses to this antigen [112]. This may reflect the amount of antigen present during natural infection or the presentation of DNA-delivered antigens versus native antigens. Nonetheless, induction of stronger CD8 T-cell responses may enhance protection against Mtb infection and may be a major factor in vaccine strategies.

There is much more work to be done on identifying CD8 T-cell epitopes for study in the versatile mouse model. Identification of epitopes, including those recently described, will enhance our understanding of the immune response to Mtb, including initiation, expansion, contraction, effector functions, clonal exhaustion, and changes in antigens seen over the course of infection. In addition, these studies will provide important and novel information about the immune response to a persistent bacterial pathogen and offer insight into how this differs from responses to acute infection.

8.9
Non-classically Restricted T Cells in TB

In addition to HLA-Ia-restricted responses, infection with Mtb has been associated with robust, HLA-Ib-restricted response or, in the case of $\gamma\delta$ T cells, an absent or unknown restricting molecule. In general, these systems utilize molecules of limited polymorphism to present antigens uniquely characteristic of an infectious pathogen. *Listeria*-derived peptides can be presented to CD8 T cells through the murine molecule H2-M3 [120]. CD8 T cells restricted to H2-M3 recognize short (approximately five amino acids) peptides containing an initial formylmethionine, which is found only in bacterial proteins. Several predicted H2-M3 Mtb epitopes were tested in the mouse model of TB. A subset of those peptides were capable of stimulating T cells from infected mice, but these cells were present at low levels [121]. At present, no known human homologue of H2-M3 has been described.

T cells expressing the $\gamma\delta$ T-cell receptor have been implicated in the control of TB disease. Mice lacking $\gamma\delta$ T cells exhibited altered granuloma formation [122] but were no more susceptible to low dose inocula of Mtb than wild-type mice were [123]. In macaques, Vγ9Vδ2 T-cell expansion following infection with BCG was coincident with the clearance of BCG bacteremia [124]. In humans, Vγ9Vδ2 T cells represent the majority of the Mtb-specific $\gamma\delta$ T-cell response [125]. These cells require presentation by antigen-presenting cells but are not restricted by a known MHC class I [126] or MHC class II molecule [127]. A majority of those cells are reactive to non-peptidic mycobacterial antigens [128]. Mtb-specific $\gamma\delta$ T cells are expanded in humans infected with Mtb [129, 130], suggesting the possibility that $\gamma\delta$ T cells reflect a recall response to mycobacterial exposure. Furthermore, decreased frequencies of Mtb-reactive $\gamma\delta$T cells in patients with active disease [131] and in those co-infected with HIV [132, 133] suggest that these cells could play a role in the control of TB in humans.

The group I CD1 antigens (CD1a, -b, and -c) of humans have been extensively characterized. By virtue of an unusually deep binding groove, the CD1 molecule has been shown to both bind and present lipid, glycolipid, and lipopeptide mycobacterially derived antigens to various subsets of T cells [37, 134–139]. CD1-dependent antigen processing can occur via exogenous or endogenous routes, but it requires processing in an acidic, late endosomal compartment and is proteasome independent [140]. Moody and colleagues have demonstrated that CD1c responses may represent a pathogen-specific recall response [36]. Mice do not have CD1 group I homologues, making it difficult to test the importance of these cells in protection against infection. However, guinea pigs do have CD1 group 1 molecules, and immunization of guinea pigs with mycobacterial lipids provided improvement of pulmonary pathology after Mtb challenge [141]. At present, a dominance hierarchy for the CD1 antigens in persons infected with Mtb has not been described.

Expressed on a wide variety of cell types, HLA-E is capable of binding leader peptides from HLA-A, -B, -C, and -G alleles [142, 143]. HLA-E is recognized by the NK receptor CD94/NKG2A, and thus serves to inhibit NK cell function [142, 143]. A role for HLA-E in host response to infectious agents has been described in tuberculosis [34] and in *Salmonella* [144]. In the case of *Salmonella*, the antigen is a heat shock protein, GroEL. While evolutionarily distinct, the murine ortholog Qa-1 binds hydrophobic nonameric peptides that are nearly identical to those bound by HLA-E. Lo and colleagues have described $CD8^+$ T cells restricted by Qa-1 generated in response to *Salmonella* infection [145]. When panels of human Mtb-specific clones are characterized, it becomes clear that a sizable fraction of these clones are neither HLA-Ia- nor CD1-restricted [27], suggesting that these cells could comprise a portion of the dominant response to infection with Mtb. However, the antigenic specificity for the HLA-E-restricted cells has not been defined.

8.10 Conclusions and Implications for Future Research

By using a T cell–driven approach to epitope identification, it is possible to define dominant epitopes in humans infected with Mtb. While the current observations are limited to a small panel of known CD4 antigens, work is underway to perform a genome-wide survey of dominant antigens in Mtb. Definition of this panel will likely prove useful in the further study of the natural history of infection with Mtb as well as in the design of novel vaccines and diagnostics. Current observations, however, would suggest that approaches based strictly on HLA-binding motifs to design either peptide or tetramer reagents are likely to define subdominant epitopes and should therefore be regarded with caution. Even in the mouse system, this has met with only limited success.

One limitation to our current knowledge is that the responses in humans have been made at a single time point. In this regard, a feature of Mtb is the chronic exposure to antigen that may persist for many years. How this chronic infection influences the shaping of the immune response and dominance repertoire is an important question that remains unresolved. For example, chronic antigenic exposure seems likely to alter the affinity of the T-cell response over time. Furthermore, it is possible that such long-term infection might lead to clonal exhaustion, as has been described for chronic viral infection. Finally, given the very high-frequency responses that we and others have observed to Mtb-infected DCs and to single antigens, it appears that the immune response to Mtb occupies a sizable fraction of the host's immunological activity. If so, this may have important implications for the aging immune system. As a result, this static picture leaves open important questions as to the evolution of the Mtb-specific response and its relationship with chronic Mtb infection. With the advent of new reagents for immunologic studies of Mtb infection in the mouse model, these questions should be addressed.

Finally, it seems likely that Mtb has evolved potent mechanisms to modulate the immune response. At present, specific mechanisms for MHC-I immune modulation have not been described. However, it appears that TLR2 stimulation via the Mtb-derived 19-kDa lipoprotein can modulate both MHC-I and MHC-II antigen processing and can interfere with IFN-γ signaling [151–155]. Further work on the T-cell subsets important in Mtb, including the immunodominant epitopes, will extend our understanding of the immunology of tuberculosis and potentially contribute to the development of a vaccine against this major killer.

Acknowledgments

This work was supported by NIH grants AI37859–08 (JLF) and 50732–04 (JLF); the American Lung Association Research and Career Investigator Award (DML and JLF); NIH grants K08AI01644, R01AI48090, and N01AI40081 (DML); the VA Merit Review (DML); the Portland VA Medical Center (JLF); and the Ellison Medical Foundation (JLF).

References

1 MacMicking, J., North, RJ, LaCourse, R, Mudgett, JS, Shah, SK, Nathan, CF. 1997. Identification of nitric oxide synthase as a protective locus against tuberculosis. *Proc. Natl. Acad. Sci. USA 94:5243.*

2 Flynn, J. L., C. A. Scanga, K. E. Tanaka, and J. Chan. 1998. Effects of aminoguanidine on latent murine tuberculosis. *J. Immunol. 160:1796.*

3 Choi, H. S., P. R. Rai, H. W. Chu, C. Cool, and E. D. Chan. 2002. Analysis of nitric oxide synthase and nitrotyrosine expression in human pulmonary tuberculosis.[comment]. *American Journal of Respiratory & Critical Care Medicine. 166:178.*

4 Russell, D., Sturgill-Koszycki, S, Vanheyningen, T, Collins, H, Schaible, UE. 1997. Why intracellular parasitism need not be a degrading experience for Mycobacterium. *Philos. Trans. R. Soc. Lond 352:1303.*

5 MacMicking, J. D., G. A. Taylor, and J. D. McKinney. 2003. Immune control of tuberculosis by IFN-gamma-inducible LRG-47. *Science 302:654.*

6 Gennaro, M. L. 2000. Immunologic diagnosis of tuberculosis. *Clin Infect Dis 30 Suppl 3:S243.*

7 Caruso, A. M., N. Serbina, E. Klein, K. Triebold, B. R. Bloom, J. L. Flynn. 1999. Mice deficient in CD4 T cells have only transiently diminished levels of

IFN-γ, yet succumb to tuberculosis. *J. Immunol. 162:5407.*

8 Scanga, C. A., V. P. Mohan, K. Yu, H. Joseph, K. Tanaka, J. Chan, J. L. Flynn. 2000. Depletion of $CD4^+$ T cells causes reactivation of murine persistent tuberculosis despite continued expression of IFN-γ and NOS2. *J. Exp Med. 192:347.*

9 Serbina, N. V., V. Lazarevic, J.L. Flynn. 2001. $CD4^+$ T cells are required for the development of cytotoxic $CD8^+$ T cells during *Mycobacterium tuberculosis* infection. *J.Immunol 167:6991.*

10 Flynn, J. L., J. Chan. 2001. Immunology of tuberculosis. *Ann Rev Immunol 19:93.*

11 Cooper, A. M., D.K. Dalton, T.A. Stewart, J.P. Griffen, D.G. Russell, I.M. Orme. 1993. Disseminated tuberculosis in IFN-γ gene-disrupted mice. *J. Exp. Med. 178:2243.*

12 Flynn, J. L., J. Chan, K.J. Triebold, D.K. Dalton, T.A. Stewart and B.R. Bloom. 1993. An essential role for Interferon-γ in resistance to *Mycobacterium tuberculosis* infection. *J. Exp. Med. 178:2249.*

13 Flynn, J. L., M. M. Goldstein, J. Chan, K.J. Triebold, K. Pfeffer, C.J. Lowenstein, R. Schreiber, T.W. Mak, and B.R. Bloom. 1995. Tumor necrosis factor-α is required in the protective immune response against *M. tuberculosis* in mice. *Immunity 2:561.*

14 Bean, A. G. D., D.R. Roach, H. Briscoe, M.P. France, H. Korner, J.D. Sedgwick, W.J. Britton. 1999. Structural deficiencies in granuloma formation in TNF gene-targeted mice underlie the heightened susceptibility to aerosol *Mycobacterium tuberculosis* infection, which is not compensated for by lymphotoxin. *J. Immunol. 162:3504.*

15 Barnes, P. F., A. B. Bloch, P.T. Davidson, D. E. Snider, Jr. 1991. Tuberculosis in patients with human immunodeficiency virus infection. *N. Engl. J. Med. 234:1644.*

16 Newport, M. J., C. M. Huxley, S. Huston, C. M. Hawrylowicz, B. A. Oostra, R. Williamson, and M. Levin. 1996. A mutation in the interferon-gamma-receptor gene and susceptibility to mycobacterial infection. *New England Journal of Medicine. 335:1941.*

17 Garcia, M., J. A. Vargas, R. Castejon, E. Navas, and A. Durantez. 2002. Flow-cytometric assessment of lymphocyte cytokine production in tuberculosis. *Tuberculosis (Edinb) 82:37.*

18 Zhang, M., Y. Lin, D. V. Iyer, J. gong, J. S. Abrams, P. F. Barnes. 1995. T cell cytokine responses in human infection with *Mycobacterium tuberculosis. Infect. Immun. 63:3231.*

19 Hirsch, C. S., Z. Toossi, C. Othieno, J. L. Johnson, S. K. Schwander, S. Robertson, R. S. Wallis, K. Edmonds, A. Okwera, R. Mugerwa, P. Peters, and J. J. Ellner. 1999. Depressed T-cell interferon-gamma responses in pulmonary tuberculosis: analysis of underlying mechanisms and modulation with therapy. *J Infect Dis 180:2069.*

20 Sodhi, A., J. Gong, C. Silva, D. Qian, and P. F. Barnes. 1997. Clinical correlates of interferon gamma production in patients with tuberculosis. *Clin Infect Dis 25:617.*

21 Ellner, J. J., C. S. Hirsch, and C. C. Whalen. 2000. Correlates of protective immunity to Mycobacterium tuberculosis in humans. *Clin Infect Dis 30 Suppl 3:S279.*

22 Schwander, S. K., M. Torres, C. C. Carranza, D. Escobedo, M. Tary-Lehmann, P. Anderson, Z. Toossi, J. J. Ellner, E. A. Rich, and E. Sada. 2000. Pulmonary mononuclear cell responses to antigens of Mycobacterium tuberculosis in healthy household contacts of patients with active tuberculosis and healthy controls from the community. *J Immunol 165:1479.*

23 Schwander, S. K., M. Torres, E. Sada, C. Carranza, E. Ramos, M. Tary-Lehmann, R. S. Wallis, J. Sierra, and E. A. Rich. 1998. Enhanced responses to Mycobacterium tuberculosis antigens by human alveolar lymphocytes during active pulmonary tuberculosis. *J Infect Dis 178:1434.*

24 Condos, R., W. N. Rom, Y. M. Liu, and N. W. Schluger. 1998. Local immune responses correlate with presentation and outcome in tuberculosis. *Am J Respir Crit Care Med 157:729.*

25 Keane, J. 2004. Tumor necrosis factor blockers and reactivation of latent tuberculosis. *Clin Infect Dis 39:300.*

26 Lewinsohn, D. A., R. A. Lines, and D. M. Lewinsohn. 2002. Human dendritic cells presenting adenovirally expressed antigen elicit Mycobacterium tuberculosis–specific CD8$^+$ T cells. *Am J Respir Crit Care Med 166:843.*

27 Lewinsohn, D. M., A. L. Briden, S. G. Reed, K. H. Grabstein, M. R. Alderson. 2000. Mycobacterium tuberculosis-reactive CD8$^+$ T lymphocytes: the relative contribution of classical versus nonclassical HLA restriction. *J. Immunol. 165:925.*

28 Lewinsohn, D. M., L. Zhu, V. J. Madison, D. C. Dillon, S. P. Fling, S. G. Reed, K. H. Grabstein, and M. R. Alderson. 2001. Classically restricted human CD8$^+$ T lymphocytes derived from Mycobacterium tuberculosis-infected cells: definition of antigenic specificity. *J Immunol 166:439.*

29 Lalvani, A., R. Brookes, R. Wilkinson, A. Malin, A. Pathan, P. Andersen, H. Dockrell, G. Pasvol, A. Hill. 1998. Human cytolytic and interferon gamma-secreting CD8$^+$ T lymphocytes specific for Mycobacterium tuberculosis. *Proc. Natl. Acad. Sci. USA`95:270.*

30 Mohagheghpour, N., D. Gammon, L. M. Kawamura, A. van Vollenhoven, C. J. Benike, E. G. Engleman. 1998. CTL response to *Mycobacterium tuberculosis*: identification of an immunogenic epitope in the 19kDa lipoprotein. *J. Immunol. 161:2400.*

31 Tan, J. S., D. H. Canady, W. H. Boom, K. N. Balaji, S. K. Schwander, and E. A. Rich. 1997. Human alveolar T lymphocyte responses to *Mycobacterium tuberculosis* infection. *J. Immunol. 159:290.*

32 Turner, J., and H. M. Dockrell. 1996. Stimulation of human peripheral blood mononuclear cells with live Mycobacterium bovis BCG activates cytolytic CD8$^+$ T cells in vitro. *Immunology 87:339.*

33 Lewinsohn, D., M. Alderson, A. Briden, S. Riddell, S. Reed, K. Grabstein. 1998. Characterization of human CD8$^+$ T cells reactive with Mycobacterium tuberculosis-infected antigen presenting cells. *J. Exp. Med. 187:1633.*

34 Heinzel, A. S., J. E. Grotzke, R. A. Lines, D. A. Lewinsohn, A. L. McNabb, D. N. Streblow, V. M. Braud, H. J. Grieser, J. T. Belisle, and D. M. Lewinsohn. 2002. HLA-E-dependent presentation of Mtb-derived antigen to human CD8$^+$ T cells. *J Exp Med 196:1473.*

35 Beckman, J. S., and W. H. Koppenol. 1996. Nitric oxide, superoxide, and peroxynitrite: the good, the bad, and ugly. *American Journal of Physiology. 271:C1424.*

36 Moody, D. B., T. Ulrichs, W. Muhlecker, D. C. Young, S. S. Gurcha, E. Grant, J. P. Rosat, M. B. Brenner, C. E. Costello, G. S. Besra, and S. A. Porcelli. 2000. CD1c-mediated T-cell recognition of isoprenoid glycolipids in Mycobacterium tuberculosis infection. *Nature 404:884.*

37 Rosat, J. P., E. P. Grant, E. M. Beckman, C. C. Dascher, P. A. Sieling, D. Frederique, R. L. Modlin, S. A. Porcelli, S. T. Furlong, and M. B. Brenner. 1999. CD1-restricted microbial lipid antigen-specific recognition found in the CD8$^+$ alpha beta T cell pool. *J Immunol 162:366.*

38 Serbina, N. V., J. L. Flynn. 1999. Early emergence of CD8$^+$ T cells primed for production of Type 1 cytokines in the lungs of *Mycobacterium tuberculosis*-infected mice. *Infect. Immun. 67:3980.*

39 De Libero, G., I. Flesch, S.H. E. and Kaufmann. 1988. Mycobacteria-reactive Lyt-2$^+$ T cell lines. *Eur. J. Immunol. 18:59.*

40 Orme, I., and F. Collins. 1984. Adoptive protection of the *Mycobacteria tuberculosis*-infected lung. *Cell. Immun. 84:113.*

41 Muller, I., S. Cobbold, H. Waldmann, S.H.E. Kaufmann. 1987. Impaired resistance to Mycobacterium tuberculosis infection after selective in vivo depletion of L3T4$^+$ and Lyt2$^+$ T cells. *Infect. and Immunol. 55:2037.*

42 Silva, C. L., M.F. Silva, R. Pietro, D. B. Lowrie. 1994. Protection against tuberculosis by passive transfer with T-cell clones recognizing mycobacterial heat shock protein 65. *Immunology 83:341.*

43 Flynn, J. L., M.M. Goldstein, K.J. Triebold, B. Koller, and B.R. Bloom. 1992. Major histocompatibility complex class I-restricted T cells are required for resistance to *Mycobacterium tuberculosis* infection. *Proc. Natl. Acad. Sci. USA 89:12013.*

44 Ladel, C. H., S. Daugelat, and S. H. E. Kaufmann. 1995. Immune response to *Mycobacterium bovis* bacille Calmette Guerin infection in major histocompatibility complex class I- and II-deficient knock-out mice: contribution of CD4 and CD8 T cells to acquired resistance. *Eur. J. Immunol. 25:377.*

45 Behar, S. M., C. C. Dascher, M. J. Grusby, C. R. Wang, and M. B. Brenner. 1999. Susceptibility of mice deficient in CD1D or TAP1 to infection with *Mycobacterium tuberculosis. J. Exp. Med. 189:1973.*

46 Sousa, A. O., R. J. Mazzaccaro, R. G. Russell, F. K. Lee, O. C. Turner, S. Hong, L. Van Kaer, and B. R. Bloom. 2000. Relative contributions of distinct MHC class I-dependent cell populations in protection to tuberculosis infection in mice. *Proc Natl Acad Sci U S A 97:4204.*

47 Mogues, T., M. E. Goodrich, L. Ryan, R. LaCourse, and R. J. North. 2001. The relative importance of T cell subsets in immunity and immunopathology of airborne Mycobacterium tuberculosis infection in mice. *Journal of Experimental Medicine. 193:271.*

48 Rolph, M. S., B. Raupach, H. H. Kobernick, H. L. Collins, B. Perarnau, F. A. Lemonnier, and S. H. Kaufmann. 2001. MHC class Ia-restricted T cells partially account for beta2-microglobulin-dependent resistance to Mycobacterium tuberculosis. *Eur J Immunol 31:1944.*

49 Urdahl, K. B., D. Liggitt, and M. J. Bevan. 2003. $CD8^+$ T cells accumulate in the lungs of Mycobacterium tuberculosis-infected $K^{b-/-}D^{b-/-}$ mice, but provide minimal protection. *J Immunol 170:1987.*

50 Lenz, L. L., and M. J. Bevan. 1996. H2-M3 restricted presentation of Listeria monocytogenes antigens. *Immunol Rev 151:107.*

51 Tompkins, S. M., J. R. Kraft, C. T. Dao, M. J. Soloski, and P. E. Jensen. 1998. Transporters associated with antigen processing (TAP)-independent presentation of soluble insulin to alpha/beta T cells by the class Ib gene product, Qa-1(b). *J Exp Med 188:961.*

52 Brutkiewicz, R. R., J. R. Bennink, J. W. Yewdell, and A. Bendelac. 1995. TAP-independent, beta 2-microglobulin-dependent surface expression of functional mouse CD1.1. *J Exp Med 182:1913.*

53 Lazarevic, V., and J. Flynn. 2002. $CD8(^+)$ T cells in tuberculosis. *Am J Respir Crit Care Med 166:1116.*

54 Lewinsohn, D. A., A. S. Heinzel, J. M. Gardner, L. Zhu, M. R. Alderson, and D. M. Lewinsohn. 2003. Mycobacterium tuberculosis-specific $CD8^+$ T cells preferentially recognize heavily infected cells. *Am J Respir Crit Care Med 168:1346.*

55 Keane, J., H. G. Remold, and H. Kornfeld. 2000. Virulent Mycobacterium tuberculosis strains evade apoptosis of infected alveolar macrophages. *J Immunol 164:2016.*

56 Sturgill-Koszycki, S., Schlesinger, PH, Chakraborty, P, Haddix, PL, Collins, HL, Fok, AK, Allen, RD, Gluck, SL, Heuser, J, Russell, DG. 1994. Lack of acidification in Mycobacterium phagosomes produced by exclusion of the vesicular proton-ATPase. *Science 263:678.*

57 Clemens, D. L., and M. A. Horowitz. 1995. Characterization of the Mycobacterium tuberculosis phagosome and evidence that phagosomal maturation is inhibited. *J. Exp. Med. 181:257.*

58 Ramachandra, L., E. Noss, W. H. Boom, and C. V. Harding. 2001. Processing of Mycobacterium tuberculosis antigen 85B involves intraphagosomal formation of peptide-major histocompatibility complex II complexes and is inhibited by live bacilli that decrease phagosome maturation. *J Exp Med 194:1421.*

59 Dobos, K. M., K. Swiderek, K. H. Khoo, P. J. Brennan, and J. T. Belisle. 1995. Evidence for glycosylation sites on the 45-kilodalton glycoprotein of Mycobacterium tuberculosis. *Infect Immun 63:2846.*

60 Belisle, J. T., M. R. McNeil, D. Chatterjee, J. M. Inamine, and P. J. Brennan. 1993. Expression of the core lipopeptide of the glycopeptidolipid surface antigens in rough mutants of Mycobacterium avium. *J Biol Chem 268:10510.*

61 Temmerman, S., K. Pethe, M. Parra, S. Alonso, C. Rouanet, T. Pickett,

A. Drowart, A. S. Debrie, G. Delogu, F. D. Menozzi, C. Sergheraert, M. J. Brennan, F. Mascart, and C. Locht. 2004. Methylation-dependent T cell immunity to Mycobacterium tuberculosis heparin-binding hemagglutinin. *Nat Med 10:935.*

62 Kovacsovics-Bankowski, M., and K. L. Rock. 1995. A phagosome-to-cytosol pathway for exogenous antigens presented on MHC Class I molecules. *Science 267:243.*

63 Mazzaccaro, R. J., M. Gedde, E. R. Jensen, H. M. van Santem, H. L. Ploegh, K. L. Rock, and B. R. Bloom. 1996. Major histocompatibility class I presentation of soluble antigen facilitated by *Mycobacterium tuberculosis* infection. *Proc. Natl. Acad. Sci. USA 93:11786.*

64 Teitelbaum, R., M. Cammer, M. L. Maitland, N. E. Freitag, J. Condeelis, and B. R. Bloom. 1999. Mycobacterial infection of macrophages results in membrane-permeable phagosomes. *Proc Natl Acad Sci U S A 96:15190.*

65 Schaible, U. E., F. Winau, P. A. Sieling, K. Fischer, H. L. Collins, K. Hagens, R. L. Modlin, V. Brinkmann, and S. H. Kaufmann. 2003. Apoptosis facilitates antigen presentation to T lymphocytes through MHC-I and CD1 in tuberculosis. *Nat Med 9:1039.*

66 Guermonprez, P., L. Saveanu, M. Kleijmeer, J. Davoust, P. Van Endert, and S. Amigorena. 2003. ER-phagosome fusion defines an MHC class I cross-presentation compartment in dendritic cells. *Nature 425:397.*

67 Desjardins, M. 2003. ER-mediated phagocytosis: a new membrane for new functions. *Nat Rev Immunol 3:280.*

68 Gagnon, E., J. J. Bergeron, and M. Desjardins. 2005. ER-mediated phagocytosis: myth or reality? *J Leukoc Biol 77:843.*

69 Gagnon, E., S. Duclos, C. Rondeau, E. Chevet, P. H. Cameron, O. Steele-Mortimer, J. Paiement, J. J. Bergeron, and M. Desjardins. 2002. Endoplasmic reticulum-mediated phagocytosis is a mechanism of entry into macrophages. *Cell 110:119.*

70 Houde, M., S. Bertholet, E. Gagnon, S. Brunet, G. Goyette, A. Laplante, M. F. Princiotta, P. Thibault, D. Sacks, and M. Desjardins. 2003. Phagosomes are competent organelles for antigen cross-presentation. *Nature 425:402.*

71 Ackerman, A. L., and P. Cresswell. 2004. Cellular mechanisms governing cross-presentation of exogenous antigens. *Nat Immunol 5:678.*

72 Canaday, D. h., C. Ziebold, E. H. Noss, K. A. Chervenak, C. V. Harding, W. H. Boom. 1999. Activation of human $CD8^+$ $\alpha\beta$ TCR^+ cells by *Mycobacterium tuberculosis* via an alternate Class I MHC antigen processing pathway. *J. Immunol. 162:372.*

73 Song, R., and C. V. Harding. 1996. Roles of proteasomes, transporter for antigen presentation (TAP), and beta 2-microglobulin in the processing of bacterial or particulate antigens via an alternate class I MHC processing pathway. *J Immunol 156:4182.*

74 Pfeifer, J. D., M. J. Wick, R. L. Roberts, K. Findlay, S. J. Normark, and C. V. Harding. 1993. Phagocytic processing of bacterial antigens for class I MHC presentation to T cells. *Nature 361:359.*

75 Chefalo, P. J., and C. V. Harding. 2001. Processing of exogenous antigens for presentation by class I MHC molecules involves post-Golgi peptide exchange influenced by peptide-MHC complex stability and acidic pH. *J Immunol 167:1274.*

76 Shen, L., L. J. Sigal, M. Boes, and K. L. Rock. 2004. Important role of cathepsin S in generating peptides for TAP-independent MHC class I crosspresentation in vivo. *Immunity 21:155.*

77 Beatty, W. L., E. R. Rhoades, H. J. Ullrich, D. Chatterjee, J. E. Heuser, and D. G. Russell. 2000. Trafficking and release of mycobacterial lipids from infected macrophages. *Traffic 1:235.*

78 Beatty, W. L., and D. G. Russell. 2000. Identification of mycobacterial surface proteins released into subcellular compartments of infected macrophages. *Infect Immun 68:6997.*

79 Beatty, W. L., H. J. Ullrich, and D. G. Russell. 2001. Mycobacterial surface moieties are released from infected macrophages by a constitutive exocytic event. *Eur J Cell Biol 80:31.*

80 Neyrolles, O., K. Gould, M. P. Gares, S. Brett, R. Janssen, P. O'Gaora, J. L. Herrmann, M. C. Prevost, E. Perret, J. E. Thole, and D. Young. 2001. Lipoprotein access to MHC class I presentation during infection of murine macrophages with live mycobacteria. *J Immunol 166:447.*

81 Hernandez-Pando, R., M. Jeyanathan, G. Mengistu, D. Aguilar, H. Orozco, M. Harboe, G. A. Rook, and G. Bjune. 2000. Persistence of DNA from Mycobacterium tuberculosis in superficially normal lung tissue during latent infection. *Lancet 356:2133.*

82 Lillebaek, T., A. Dirksen, I. Baess, B. Strunge, V. O. Thomsen, and A. B. Andersen. 2002. Molecular evidence of endogenous reactivation of Mycobacterium tuberculosis after 33 years of latent infection. *J Infect Dis 185:401.*

83 Lalvani, A., L. Richeldi, and H. Kunst. 2005. Interferon gamma assays for tuberculosis. *Lancet Infect Dis 5:322.*

84 Orme, I., E. Miller, A. Roberts, S. Furney, J. Griffen, K. Dobos, D. Chi, B. Rivoire, P. Brennan. 1992. T Lymphocytes mediating protection and cellular cytolysis during the course of *Mycobacterium tuberculosis* infection. *J. Immunol. 148:189.*

85 Pal, P. G., and M. A. Horwitz. 1992. Immunization with extracellular proteins of Mycobacterium tuberculosis induces cell-mediated immune responses and substantial protective immunity in a guinea pig model of pulmonary tuberculosis. *Infect Immun 60:4781.*

86 Roberts, A. D., M. G. Sonnenberg, D. J. Ordway, S. K. Furney, and P. J. Brennan. 1995. Characteristics of protective immunity engendered by vaccination of mice with purified culture filtrate protein antigens of Mycobacterium tuberculosis. *Immunology 85:502.*

87 Andersen, P., D. Askgaard, A. Gottschau, J. Bennedsen, S. Nagai, and I. Heron. 1992. Identification of immunodominant antigens during infection with Mycobacterium tuberculosis. *Scand J Immunol 36:823.*

88 Andersen, P., A. B. Andersen, A. L. Sorensen, S. Nagai. 1995. Recall of long-lived immunity to *Mycobacterium tuberculosis* infection in mice. *J. Immunol 154:3359.*

89 Skeiky, Y. A., P. J. Ovendale, S. Jen, M.R. Alderson, D.C. Dillon, S. Smith, C.B. Wilson, I.M. Orme, S. G. Reed, A. Campos-Neto. 2000. T cell expression cloning of a Mycobacterium tuberculosis gene encoding a protective antigen associated with early control of the infection. *J. Immunol 165:7140.*

90 Reed, S. G., M. R. Alderson, W. Dalemans, Y. Lobet, and Y. A. Skeiky. 2003. Prospects for a better vaccine against tuberculosis. *Tuberculosis (Edinb) 83:213.*

91 Dillon, D. C., M. R. Alderson, C. H. Day, T. Bement, A. Campos-Neto, Y. A. Skeiky, T. Vedvick, R. Badaro, S. G. Reed, and R. Houghton. 2000. Molecular and immunological characterization of Mycobacterium tuberculosis CFP-10, an immunodiagnostic antigen missing in Mycobacterium bovis BCG. *J Clin Microbiol 38:3285.*

92 Dillon, D. C., M. R. Alderson, C.H. Day, D.M. Lewinsohn, R. N. Coler, T. Bement, A. Campos-Neto, Y.A.W. Skeiky, I.M. Orme, A. Roberts, S. Steen, W. Dalemans, R. Badaro, S.G.Reed. 1999. Molecular characterization and human T cell responses to a member of a novel Mycobacterium tuberculosis mtb39 gene family. *Infect. Immun. 67:2941.*

93 Alderson, M. R., T. Bement, C. H. Day, L. Zhu, D. Molesh, Y. A. Skeiky, R. Coler, D. M. Lewinsohn, S. G. Reed, and D. C. Dillon. 2000. Expression cloning of an immunodominant family of Mycobacterium tuberculosis antigens using human $CD4(^{+})$ T cells. *J Exp Med 191:551.*

94 Pathan, A. A., K. A. Wilkinson, R. J. Wilkinson, M. Latif, H. McShane, G. Pasvol, A. V. S. Hill, A. Lalvani. 2000. High frequencies of circulating IFN-g secreting CD8 cytotoxic T cells specific for a novel MHC Class I restricted Mycobacterium tuberculosis epitope in M. tuberculois-infected subjects without disease. *Eur J Immunol 30:2713.*

95 Okkels, L. M., I. Brock, F. Follmann, E. M. Agger, S. M. Arend, T. H. Ottenhoff, F. Oftung, I. Rosenkrands, and P. Andersen. 2003. PPE protein (Rv3873) from DNA segment RD1 of Mycobacterium tuberculosis: strong recognition of both specific T-cell epitopes and epitopes conserved within the PPE family. *Infect Immun 71:6116.*

96 Al-Attiyah, R., F. A. Shaban, H. G. Wiker, F. Oftung, and A. S. Mustafa. 2003. Synthetic peptides identify promiscuous human Th1 cell epitopes of the secreted mycobacterial antigen MPB70. *Infect Immun 71:1953.*

97 Aagaard, C., I. Brock, A. Olsen, T. H. Ottenhoff, K. Weldingh, and P. Andersen. 2004. Mapping immune reactivity toward Rv2653 and Rv2654: two novel low-molecular-mass antigens found specifically in the Mycobacterium tuberculosis complex. *J Infect Dis 189:812.*

98 Agger, E. M., I. Brock, L. M. Okkels, S. M. Arend, C. S. Aagaard, K. N. Weldingh, and P. Andersen. 2003. Human T-cell responses to the RD1-encoded protein TB27.4 (Rv3878) from Mycobacterium tuberculosis. *Immunology 110:507.*

99 Black, G. F., R. E. Weir, S. D. Chaguluka, D. Warndorff, A. C. Crampin, L. Mwaungulu, L. Sichali, S. Floyd, L. Bliss, E. Jarman, L. Donovan, P. Andersen, W. Britton, G. Hewinson, K. Huygen, J. Paulsen, M. Singh, R. Prestidge, P. E. Fine, and H. M. Dockrell. 2003. Gamma interferon responses induced by a panel of recombinant and purified mycobacterial antigens in healthy, non-mycobacterium bovis BCG-vaccinated Malawian young adults. *Clin Diagn Lab Immunol 10:602.*

100 Skjot, R. L., I. Brock, S. M. Arend, M. E. Munk, M. Theisen, T. H. Ottenhoff, and P. Andersen. 2002. Epitope mapping of the immunodominant antigen TB10.4 and the two homologous proteins TB10.3 and TB12.9, which constitute a subfamily of the esat-6 gene family. *Infect Immun 70:5446.*

101 Caccamo, N., A. Barera, C. Di Sano, S. Meraviglia, J. Ivanyi, F. Hudecz, S. Bosze, F. Dieli, and A. Salerno. 2003. Cytokine profile, HLA restriction and TCR sequence analysis of human $CD4^+$ T clones specific for an immunodominant epitope of Mycobacterium tuberculosis 16-kDa protein. *Clin Exp Immunol 133:260.*

102 Caccamo, N., S. Meraviglia, C. La Mendola, S. Bosze, F. Hudecz, J. Ivanyi, F. Dieli, and A. Salerno. 2004. Characterization of HLA-DR- and TCR-binding residues of an immunodominant and genetically permissive peptide of the 16-kDa protein of Mycobacterium tuberculosis. *Eur J Immunol 34:2220.*

103 Shams, H., P. Klucar, S. E. Weis, A. Lalvani, P. K. Moonan, H. Safi, B. Wizel, K. Ewer, G. T. Nepom, D. M. Lewinsohn, P. Andersen, and P. F. Barnes. 2004. Characterization of a Mycobacterium tuberculosis peptide that is recognized by human $CD4^+$ and $CD8^+$ T cells in the context of multiple HLA alleles. *J Immunol 173:1966.*

104 Vordermeier, M., A. O. Whelan, and R. G. Hewinson. 2003. Recognition of mycobacterial epitopes by T cells across mammalian species and use of a program that predicts human HLA-DR binding peptides to predict bovine epitopes. *Infect Immun 1:1980.*

105 Langermans, J. A. M., P. Andersen, D. van Soolingen, R. A. W. Vervenne, P.A. Frost, T. van der Laan, L. A.H. van Pinsteren, J. van den Hombergh, S. Kroom, I. Peekel, S. Florquin, A. W. Thomas. 2001. Divergent effect of bacillus Calmette-Guerin (BCG) vaccination on Mycobacterium tuberculosis infection in highly related macaque species: Implications for primate models in tuberculosis vaccine research. *Proc. Nat. Acad. Sci. USA 98:11497.*

106 Winslow, G. M., A. D. Roberts, M. A. Blackman, and D. L. Woodland. 2003. Persistence and turnover of antigen-specific CD4 T cells during chronic tuberculosis infection in the mouse. *J Immunol 170:2046.*

107 Lazarevic, V., D. J. Yankura, S. J. DiVito, and J. L. Flynn. 2005. Induction of Mycobacterium tuberculosis-specific primary and secondary T-cell responses in interleukin-15-deficient mice. *Infect Immun 73:2910.*

108 Kamath, A. B., J. Woodworth, X. Xiong, C. Taylor, Y. Weng, and S. M. Behar. 2004. Cytolytic $CD8^+$ T cells recognizing CFP10 are recruited to the lung after Mycobacterium tuberculosis infection. *J Exp Med 200:1479.*

109 Demangel, C., P. Brodin, P. J. Cockle, R. Brosch, L. Majlessi, C. Leclerc, and S. T. Cole. 2004. Cell envelope protein PPE68 contributes to Mycobacterium tuberculosis RD1 immunogenicity independently of a 10-kilodalton culture filtrate protein and ESAT-6. *Infect Immun 72:2170.*

110 D'Souza, S., V. Rosseels, M. Romano, A. Tanghe, O. Denis, F. Jurion, N. Castiglione, A. Vanonckelen, K. Palfliet, and K. Huygen. 2003. Mapping of murine Th1 helper T-Cell epitopes of mycolyl transferases Ag85A, Ag85B, and Ag85C from Mycobacterium tuberculosis. *Infect Immun 71:483.*

111 Romano, M., O. Denis, S. D'Souza, X. M. Wang, T. H. Ottenhoff, J. M. Brulet, and K. Huygen. 2004. Induction of in vivo functional Db-restricted cytolytic T cell activity against a putative phosphate transport receptor of Mycobacterium tuberculosis. *J Immunol 172:6913.*

112 Denis, O., A. Tanghe, K. Palfliet, F. Jurion, T. P. van den Berg, A. Vanonckelen, J. Ooms, E. Saman, J. B. Ulmer, J. Content, and K. Huygen. 1998. Vaccination with plasmid DNA encoding mycobacterial antigen 85A stimulates a $CD4^+$ and $CD8^+$ T-cell epitopic repertoire broader than that stimulated by Mycobacterium tuberculosis H37Rv infection. *Infect Immun 66:1527.*

113 Flyer, D. C., V. Ramakrishna, C. Miller, H. Myers, M. McDaniel, K. Root, C. Flournoy, V. H. Engelhard, D. H. Canaday, J. A. Marto, M. M. Ross, D. F. Hunt, J. Shabanowitz, and F. M. White. 2002. Identification by mass spectrometry of CD8(+)-T-cell Mycobacterium tuberculosis epitopes within the Rv0341 gene product. *Infect Immun 70:2926.*

114 Klein, M. R., A. S. Hammond, S. M. Smith, A. Jaye, P. T. Lukey, and K. P. McAdam. 2002. HLA-B*35-restricted CD8(+)-T-cell epitope in Mycobacterium tuberculosis Rv2903c. *Infect Immun 70:981.*

115 Charo, J., A. Geluk, M. Sundback, B. Mirzai, A. D. Diehl, K. J. Malmberg, A. Achour, S. Huriguchi, K. E. van Meijgaarden, J. W. Drijfhout, N. Beekman, P. van Veelen, F. Ossendorp, T. H. Ottenhoff, and R. Kiessling. 2001. The identification of a common pathogen-specific HLA class I A*0201-restricted cytotoxic T cell epitope encoded within the heat shock protein 65. *Eur J Immunol 31:3602.*

116 Geluk, A., K. E. van Meijgaarden, K. L. Franken, J. W. Drijfhout, S. D'Souza, A. Necker, K. Huygen, and T. H. Ottenhoff. 2000. Identification of major epitopes of Mycobacterium tuberculosis AG85B that are recognized by HLA-A*0201-restricted $CD8^+$ T cells in HLA-transgenic mice and humans. *J Immunol 165:6463.*

117 Tully, G., C. Kortsik, H. Hohn, I. Zehbe, W. E. Hitzler, C. Neukirch, K. Freitag, K. Kayser, and M. J. Maeurer. 2005. Highly focused T cell responses in latent human pulmonary Mycobacterium tuberculosis infection. *J Immunol 174:2174.*

118 Nielsen, M., C. Lundegaard, P. Worning, S. L. Lauemoller, K. Lamberth, S. Buus, S. Brunak, and O. Lund. 2003. Reliable prediction of T-cell epitopes using neural networks with novel sequence representations. *Protein Sci 12:1007.*

119 Nielsen, M., C. Lundegaard, P. Worning, C. S. Hvid, K. Lamberth, S. Buus, S. Brunak, and O. Lund. 2004. Improved prediction of MHC class I and class II epitopes using a novel Gibbs sampling approach. *Bioinformatics 20:1388.*

120 Pamer, E. G., C. R. wang, L. Flaherty, K. F. Lindahl, M. J. Bevan. 1992. H2-M3 presents a Listeria monocytogenes peptide to cytotoxic T lymphocytes. *Cells 70:215.*

121 Chun, T., N. V. Serbina, D. Nolt, B. Wang, N. M Chiu, J. L. Flynn, C.-R. Wang. 2001. Induction of M3-restricted cytotoxic T lymphocyte responses by N-formylated peptides derived from *Mycobacterium tuberculosis. J. Exp Med. 193:1213.*

122 Ladel, C. H., C. Blum, A. Dreher, K. Reifenberg, and S. H. Kaufmann. 1995. Protective role of gamma/delta T cells and alpha/beta T cells in tuberculosis. *Eur J Immunol 25:2877.*

123 Mogues, T., M.E. Goodrich, L. Ryan, R. LaCourse, R. J. North. 2001. The relative importance of T cell subsets in immunity and immunopathology of airborne Mycobacterium tuberculosis infection in mice. *J. Exp Med. 193:271.*

124 Shen, Y., D. Zhou, L. Qiu, x. Lai, M. Simon, L. Shen, Z. Kou, Q. Wang, L. Jiang, J. Estep, R. Hunt, M. Clagett, P.K. Sehgal, Y. Li, X. Zeng, C. T. Morita, M. B. Brenner, N. L. Letvin, Z.W. Chen. 2002. Adaptive immune response of $V\gamma2V\delta2^{+}$T cells during mycobacterial infections. *Science 295:2255.*

125 Kabelitz, D., A. Bender, T. Prospero, S. Wesselborg, O. Janssen, and K. Pechhold. 1991. The primary response of human gamma/delta + T cells to Mycobacterium tuberculosis is restricted to V gamma 9-bearing cells. *J Exp Med 173:1331.*

126 De Libero, G., G. Casorati, C. Giachino, C. Carbonara, N. Migone, P. Matzinger, and A. Lanzavecchia. 1991. Selection by two powerful antigens may account for the presence of the major population of human peripheral gamma/delta T cells. *J Exp Med 173:1311.*

127 Kabelitz, D., A. Bender, S. Schondelmaier, B. Schoel, S.H.E. Kaufmann. 1990. A large fraction of human peripheral blood gd T cells is activated by Mycobacterium tuberculosis but not by its 65 KD heat shock protein. *J. Exp. Med. 171:667.*

128 Constant, P., Y. Poquet, M. A. Peyrat, F. Davodeau, M. Bonneville, and J. J. Fournie. 1995. The antituberculous Mycobacterium bovis BCG vaccine is an attenuated mycobacterial producer of phosphorylated nonpeptidic antigens for human gamma delta T cells. *Infect Immun 63:4628.*

129 Havlir, D. V., J. J. Ellner, K. A. Chervenak, and W. H. Boom. 1991. Selective expansion of human gamma delta T cells by monocytes infected with live Mycobacterium tuberculosis. *J Clin Invest 87:729.*

130 Ito, M., N. Kojiro, T. Ikeda, T. Ito, J. Funada, and T. Kokubu. 1992. Increased proportions of peripheral blood gamma delta T cells in patients with pulmonary tuberculosis. *Chest 102:195.*

131 Li, B., M. D. Rossman, T. Imir, A. F. Oner-Eyuboglu, C. W. Lee, R. Biancaniello, and S. R. Carding. 1996. Disease-specific changes in gammadelta T cell repertoire and function in patients with pulmonary tuberculosis. *J Immunol 157:4222.*

132 Pellegrin, J. L., J. L. Taupin, M. Dupon, J. M. Ragnaud, J. Maugein, M. Bonneville, and J. F. Moreau. 1999. Gammadelta T cells increase with Mycobacterium avium complex infection but not with tuberculosis in AIDS patients. *Int Immunol 11:1475.*

133 Carvalho, A. C., A. Matteelli, P. Airo, S. Tedoldi, C. Casalini, L. Imberti, G. P. Cadeo, A. Beltrame, and G. Carosi. 2002. gammadelta T lymphocytes in the peripheral blood of patients with tuberculosis with and without HIV co-infection. *Thorax 57:357.*

134 Beckman, E. M., A. Melian, S. M. Behar, P. A. Sieling, D. Chatterjee, S. T. Furlong, R. Matsumoto, J. P. Rosat, R. L. Modlin, and S. A. Porcelli. 1996. CD1c restricts responses of mycobacteria-specific T cells. Evidence for antigen presentation by a second member of the human CD1 family. *J Immunol 157:2795.*

135 Beckman, E. M., S. A. Porcelli, C. T. Morita, S. M. Behar, S. T. Furlong, and M. B. Brenner. 1994. Recognition of a lipid antigen by CD1-restricted alpha beta+ T cells. *Nature 372:691.*

136 Ernst, W. A., J. Maher, S. Cho, K. R. Niazi, D. Chatterjee, D. B. Moody, G. S. Besra, Y. Watanabe, P. E. Jensen, S. A. Porcelli, M. Kronenberg, and R. L. Modlin. 1998. Molecular interaction of CD1b with lipoglycan antigens. *Immunity 8:331.*

137 Sugita, M., S. A. Porcelli, and M. B. Brenner. 1997. Assembly and retention of CD1b heavy chains in the endoplasmic reticulum. *J Immunol 159:2358.*

138 Moody, D. B., D. C. Young, T. Y. Cheng, J. P. Rosat, C. Roura-Mir, P. B. O'Connor, D. M. Zajonc, A. Walz,

M. J. Miller, S. B. Levery, I. A. Wilson, C. E. Costello, and M. B. Brenner. 2004. T cell activation by lipopeptide antigens. *Science 303:527.*

139 Ulrichs, T., D. B. Moody, E. Grant, S . H. Kaufmann, and S. A. Porcelli. 2003. T-cell responses to CD1-presented lipid antigens in humans with Mycobacterium tuberculosis infection. *Infect Immun 71:3076.*

140 Porcelli, S. A., R. L. Modlin. 1999. The CD1 system: Antigen-presenting molecules for T cell recognition of lipids and glycolipids. *Ann. Rev. Immunol. 17:297.*

141 Dascher, C. C., K. Hiromatsu, X. Xiong, C. Morehouse, G. Watts, G. Liu, D. N. McMurray, K. P. LeClair, S. A. Porcelli, and M. B. Brenner. 2003. Immunization with a mycobacterial lipid vaccine improves pulmonary pathology in the guinea pig model of tuberculosis. *Int Immunol 15:915.*

142 Lee, N., M. Llano, M. Carretero, A. Ishitani, F. Navarro, M. Lopez-Botet, and D. E. Geraghty. 1998. HLA-E is a major ligand for the natural killer inhibitory receptor CD94/NKG2A. *Proc Natl Acad Sci U S A 95:5199.*

143 Braud, V. M., D. S. Allan, C. A. O'Callaghan, K. Soderstrom, A. D'Andrea, G. S. Ogg, S. Lazetic, N. T. Young, J. I. Bell, J. H. Phillips, L. L. Lanier, and A. J. McMichael. 1998. HLA-E binds to natural killer cell receptors CD94/NKG2A, B and C. *Nature 391:795.*

144 Salerno-Goncalves, R., M. Fernandez-Vina, D. M. Lewinsohn, and M. B. Sztein. 2004. Identification of a human HLA-E-restricted $CD8^+$ T cell subset in volunteers immunized with Salmonella enterica serovar Typhi strain Ty21a typhoid vaccine. *J Immunol 173:5852.*

145 Lo, W. F., H. Ong, E. S. Metcalf, and M. J. Soloski. 1999. T cell responses to Gram-negative intracellular bacterial pathogens: a role for $CD8^+$ T cells in immunity to Salmonella infection and the involvement of MHC class Ib molecules. *J Immunol 162:5398.*

146 Serbina, N. V., C.-C. Liu, C. A. Scanga, J. L. Flynn. . 2000. $CD8^+$ cytotoxic T lymphocytes from lungs of *M. tuberculosis* infected mice express perforin in vivo and lyse infected macrophages. *J. Immunol. 165:353.*

147 Lazarevic, V., D. Nolt, and J. L. Flynn. 2005. Long-Term Control of Mycobacterium tuberculosis Infection Is Mediated by Dynamic Immune Responses. *J Immunol 175:1107.*

148 Denis, O., V. Stroobant, D. Colau, S. D'Souza, and K. Huygen. 2002. Culture filtrate specific H-2(b) restricted $CD8^+$ T cells activated in vivo by Mycobacterium tuberculosis or bovis BCG recognize a restricted number of immunodominant peptides. *Immunol Lett 81:115.*

149 Majlessi, L., M. J. Rojas, P. Brodin, and C. Leclerc. 2003. $CD8^+$-T-cell responses of Mycobacterium-infected mice to a newly identified major histocompatibility complex class I-restricted epitope shared by proteins of the ESAT-6 family. *Infect Immun 71:7173.*

150 Tanghe, A., P. Lefevre, O. Denis, S. D'Souza, M. Braibant, E. Lozes, M. Singh, D. Montgomery, J. Content, and K. Huygen. 1999. Immunogenicity and protective efficacy of tuberculosis DNA vaccines encoding putative phosphate transport receptors. *J Immunol 162:1113.*

151 Noss, E. H., C. V. Harding, and W. H. Boom. 2000. Mycobacterium tuberculosis inhibits MHC class II antigen processing in murine bone marrow macrophages. *Cell Immunol 201:63.*

152 Noss, E. H., R. K. Pai, T. J. Sellati, J. D. Radolf, J. Belisle, D. T. Golenbock, W. H. Boom, and C. V. Harding. 2001. Toll-like receptor 2-dependent inhibition of macrophage class II MHC expression and antigen processing by 19-kDa lipoprotein of Mycobacterium tuberculosis. *J Immunol 167:910.*

153 Tobian, A. A., N. S. Potter, L. Ramachandra, R. K. Pai, M. Convery, W. H. Boom, and C. V. Harding. 2003. Alternate class I MHC antigen processing is inhibited by Toll-like receptor signaling pathogen-associated molecular patterns: Mycobacterium tuberculosis 19-kDa lipoprotein, CpG DNA, and lipopolysaccharide. *J Immunol 171:1413.*

154 Gehring, A. J., R. E. Rojas, D. H. Canaday, D. L. Lakey, C. V. Harding, and W. H. Boom. 2003. The Mycobacterium tuberculosis 19-kilodalton lipoprotein inhibits gamma interferon-regulated HLA-DR and Fc gamma R1 on human macrophages through Toll-like receptor 2. *Infect Immun 71:4487.*

155 Fulton, S. A., S. M. Reba, R. K. Pai, M. Pennini, M. Torres, C. V. Harding, and W. H. Boom. 2004. Inhibition of major histocompatibility complex II expression and antigen processing in murine alveolar macrophages by Mycobacterium bovis BCG and the 19-kilodalton mycobacterial lipoprotein. *Infect Immun 72:2101.*

156 Smith, S. M., M. R. Klein, A. S. Malin, J. Sillah, K. Huygen, P. Andersen, K. P. McAdam, and H. M. Dockrell. 2000. Human CD8(+) T cells specific for Mycobacterium tuberculosis secreted antigens in tuberculosis patients and healthy BCG-vaccinated controls in The Gambia. *Infect Immun 68:7144.*

157 Klein, M. R., S. M. Smith, A. S. Hammond, G. S. Ogg, A. S. King, J. Vekemans, A. Jaye, P. T. Lukey, and K. P. McAdam. 2001. HLA-B*35-restricted CD8 T cell epitopes in the antigen 85 complex of Mycobacterium tuberculosis. *J Infect Dis 183:928.*

9
T-Cell Specificity and Respiratory Virus Infections

Sherry R. Crowe and David L. Woodland

9.1
Introduction

The complex and delicate architecture of the lung poses a substantial problem for an immune system that must eliminate invading pathogens without inducing excessive damage. Yet for the most part, the immune system is able to successfully control pulmonary infections and maintain a sterile environment in the lower lung. Central to this immune control is the T-cell response, which is able to rapidly eliminate invading pathogens and establish long-lived memory that can mediate accelerated control of secondary infections. It has been established that T cells recognize pathogen antigens that have been degraded into short peptides and presented on the surface of antigen-presenting cells in the context of major histocompatibility antigens. Typically, relatively few antigens from the pathogen are targeted by the T cells, a phenomenon referred to as immunodominance. However, the factors that regulate immunodominance hierarchies are incompletely understood, particularly with respect to peripheral sites such as the lung. In the current chapter, the specificity of T-cell responses will be explored in the context of mouse models of respiratory virus infection. In particular, the factors that affect the specificity of primary T-cell responses, memory T-cell pools, and the recall T-cell response will be addressed.

Over recent years, substantial progress has been made in understanding how cellular immune responses are able to control and eliminate pathogens. We now realize that secondary lymphoid organs, such as the lymph nodes and spleen, play a key role in the initiation and regulation of T-cell responses. These organs control the response by recruiting antigen-charged dendritic cells from the site of infection and promoting their interaction with naïve T cells. In addition, they provide highly specialized anatomical niches for the proliferation of activated T cells and their maturation into effector cells capable of mediating anti-pathogen responses at peripheral sites. A key aspect of the T-cell response is that it is highly specific for the invading pathogen [1]. In addition, the response is typically dominated by T cells specific for a relatively limited number of the available antigenic epitopes, a phenomenon referred to as immunodominance [2, 3]. Epitopes that drive major

Immunodominance: The Choice of the Immune System. Edited by Jeffrey A. Frelinger

ISBN: 3-527-31274-9

or minor fractions of the T-cell response are considered to be immunodominant and subdominant respectively, resulting in a hierarchy of T-cell responses. Several factors such as antigen availability, antigen processing, epitope stability, and T-cell repertoire all appear to play critical roles in determining the immunodominance patterns during T-cell responses, and these are discussed in detail in other chapters of this book. Here, we will specifically focus on immunodominance as it relates to viral pathogens that infect a key peripheral site, the lung. Using mouse models of influenza and parainfluenza virus infection, we will discuss the specificity of T-cell responses in the lung and the impact of immunodominance on the specificity and function of the primary T-cell response, the pool of memory T cells established, and the recall of memory T cells to a secondary viral challenge. Understanding how these immunodominance hierarchies are established and regulated at these different stages of the immune response is important for the development of effective cellular-based vaccines.

9.2
Primary Immune Responses to Respiratory Virus Infections

Respiratory virus infections are a major cause of morbidity and mortality. For example, yearly epidemics of influenza A are responsible for an average of 30,000 deaths each year in the United States, and there is a constant threat that newly emerging variants, such as highly virulent avian influenza viruses, might trigger a major pandemic [4, 5]. Immune control of respiratory virus infections is mediated by a variety of immune mechanisms, central among which is the T-cell response [6, 7]. Thus, there has been substantial interest in determining how T cells recognize antigen, become activated, mediate effector functions, and establish memory. However, we still have a relatively poor understanding of how these responses are organized and how they orchestrate control of a pathogen whose replication is restricted exclusively to the lung. Mouse-adapted influenza viruses and murine parainfluenza (Sendai) virus represent useful models for the study of the immunology of respiratory virus infections [8–10]. Intranasal infection of mice results in the infection and inflammation of the respiratory tract, with little evidence of productive infection in the lymphoid tissues, and is primarily controlled by major histocompatibility complex (MHC) class I–restricted $CD8^+$ cytotoxic T lymphocytes [8, 11]. This response is initiated by dendritic cells present in the lung epithelium that acquire viral antigens, become activated by different viral components, and subsequently traffic into the local draining lymph nodes (cervical and mediastinal lymph nodes, which drain the upper and lower respiratory tracts, respectively) [12]. Once in the lymph nodes, these primed dendritic cells present viral antigens to naïve T cells [13]. Those T cells that are able to recognize viral antigens and that receive appropriate costimulatory signals are activated and initiate a program of proliferation and maturation that culminates in the acquisition of various antiviral effector functions and the capacity to migrate to inflammatory sites. These cells initially appear in the lung by day 7 post-infection, and their numbers usually peak around day 9 or 10 post-infection [8]. Viral clearance is normally achieved by day 10 post-infection and generally depends on either perforin- or Fas-mediated mechanisms, although antiviral cytokines, such as IFN-γ and TNF-α, can also play a critical role [14]. $CD4^+$ T cells are also involved in the inflammatory response in the lung, but their major role is thought to be promotion of effective $CD8^+$ memory T-cell generation and regulation of antibody responses [15, 16]. Antibody is generated late in the primary response and plays a significant role in clearing the infection if the virus is highly virulent [17].

9.3 Specificity of the Primary Immune Response

T-cell recognition of viral antigens is mediated by the clonally distributed T-cell antigen receptor, which recognizes short viral peptides associated with self-MHC molecules on the surface of antigen-presenting cells. The epitopes that drive the T-cell response are selected on the basis of a number of parameters, including the abundance of the source protein, efficient processing of the target peptide, the affinity of the peptide for the MHC molecule, and the frequency and avidity of T cells that recognize the peptide MHC complex [2, 18]. The specificity of the primary T-cell response to influenza virus has been extensively studied in mouse models. For example, it has been established that the MHC class I–restricted $CD8^+$ T-cell response to influenza virus infection of C57BL/6 mice is driven by at least 16 H-2K^b– and H-2D^b-restricted epitopes [19]. T cells specific for three of these epitopes ($NP_{366-374}/D^b$, $PA_{224-233}/D^b$, $PB1_{703-711}/K^b$) dominate the response, accounting for over 50% of the $CD8^+$ T cells that accumulate in the lung during the infection (Figure 9.1A). These cells, combined with the T cells specific for the remaining 13 subdominant epitopes that have been identified, account for over 70% of the total $CD8^+$ T cells in the lung airways during the peak of the T-cell response. Flow cytometric data showing IFN-γ production by T cells specific for MHC class I–restricted epitopes are presented in Figure 9.2A. The biasing of the $CD8^+$ T-cell response to a limited number of antigens is a common feature of respiratory virus infections. An extreme example is the $CD8^+$ T-cell response to Sendai virus in C57BL/6 mice. In this case, 70% of the $CD8^+$ T-cell response in the lung airways is directed against a single nucleoprotein epitope, $NP_{324-332}/K^b$, and there is evidence that the remaining 30% of the response represents a bystander response that is not specific for the virus [9]. One possible explanation for the limited number of epitopes that dominate the response in these cases is that the viruses involved are relatively small. However, large DNA viruses that mediate respiratory infections have also been shown to drive highly immunodominant T-cell responses. For example, γHV-68, a DNA virus with a genome of 135 kb, nevertheless drives a $CD8^+$ T-cell response similar to that described for influenza virus in terms of the numbers of epitopes recognized and the immunodominance patterns [20]. Thus, it appears that immunodominance is not merely a result of limited numbers of proteins in the pathogen.

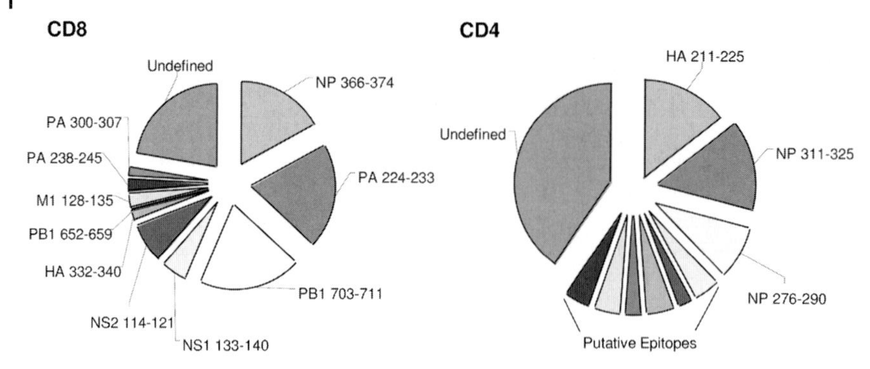

Figure 9.1 Antigen specificity in the lung airways following influenza virus infection. Three dominant epitopes and seven subdominant epitopes account for approximately 78% of the total $CD8^+$ T-cell response to influenza virus infection in the lung airways. The remaining 22% of the $CD8^+$ T-cell response is undetermined at this time. Two dominant and at least seven subdominant epitopes account for approximately 60% of the total $CD4^+$ T-cell response to influenza virus infection in the lung airways. The remaining 40% of the $CD4^+$ T-cell response is undetermined at this time. Some of the $CD4^+$ T-cell epitopes are listed as putative because detailed characterization of the epitopes has not yet been performed. (This figure also appears with the color plates.)

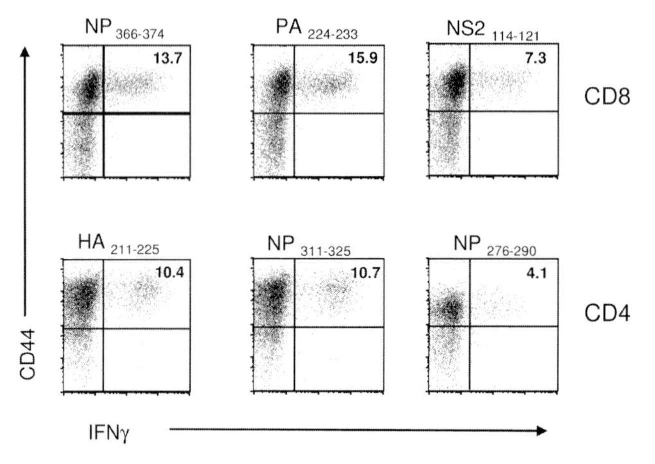

Figure 9.2 IFN-γ production by T cells present in the lung airways following influenza virus infection. T cells from the lung airways were isolated and stimulated with peptides that define known MHC class I and class II epitopes in a standard intracellular cytokine assay. The cells were analyzed flow cytometrically, and the data show dual staining with anti-CD44 and anti-IFN-γ in cells stimulated with the dominant ($NP_{366-374}$ and $PA_{224-233}$) and subdominant ($NS2_{114-121}$) MHC class I–restricted peptides (top row) and the dominant ($HA_{211-225}$ and $NP_{311-325}$) and subdominant ($NP_{276-290}$) MHC class II–restricted peptides (bottom row). The data are gated on $CD8^+$ T cells in the top row and on $CD4^+$ T cells in the bottom row. The numbers in the top right corner of the panels are the percentages of $CD44^+$ T cells that produce IFN-γ among the $CD8^+/CD44^+$ T cells (top row) and $CD4^+/CD44^+$ T cells (bottom row).

It has generally been believed that the specificity of $CD4^+$ T-cell responses is more diverse than that of $CD8^+$ T cells. This idea was based, in part, on the fact that T-cell recognition of MHC class II–peptide complexes was somewhat degenerate. However, our laboratory has recently examined the specificity of the $CD4^+$ T-cell response to influenza virus in C57BL/6 mice [21]. These studies have identified approximately 20 I-A^b-restricted epitopes, two of which, $HA_{211-225}/A^b$ and $NP_{311-325}/A^b$, clearly dominate the response. These two epitopes, along with the previously defined $NP_{276-290}/A^b$ epitope, account for approximately 34% of the total CD4 T-cell response in the lung airways at the peak of infection (Figures 9.1B and 9.2). Comparison of the specificity patterns of $CD4^+$ and $CD8^+$ T-cell responses to influenza suggests that the responses are similar in terms of the numbers of epitopes recognized and the immunodominance hierarchy (Figure 9.1). Similar patterns of immunodominance have also been described for the $CD4^+$ T-cell response to Sendai virus in C57BL/6 mice [10]. In this case, the response is largely dominated by T cells specific for a hemagglutinin-neuraminidase epitope restricted by I-A^b. However, it should be stressed that there is only one MHC class II molecule available in C57BL/6 mice, which may limit the repertoire of epitopes available. It remains to be seen whether the breadth of the response differs in strains of mice that express two MHC class II molecules or in outbred human populations. Overall, the conclusion is that patterns of immunodominance in the $CD4^+$ and $CD8^+$ T-cell subsets may be quite similar in terms of the numbers of epitopes recognized and the relative fractions of the T-cell response that those epitopes stimulate.

The fact that relatively few epitopes drive a major fraction of the $CD8^+$ and $CD4^+$ T-cell responses to respiratory virus infections raises the question as to which proteins are normally targeted. It is reasonable to assume that the key factors in determining the likelihood that a protein carries a dominant epitope are the length of the protein (increasing the chance that a strong MHC-binding peptide is present) and its relative abundance in infected cells. In the case of influenza virus, the nucleoprotein is abundant in both the virion and infected cells and is correspondingly found to frequently contribute dominant $CD8^+$ T-cell epitopes. Similarly, class II epitopes tend to be derived from the abundant nucleoprotein and hemagglutinin epitopes (Figure 9.1). However, it is clear that immunodominant epitopes can also be derived from other proteins, some of which are comparatively poorly expressed in infected mice (Figure 9.1), suggesting the contribution of factors other than protein size and abundance.

Although the factors that determine the immunodominance hierarchy of T-cell epitopes are extremely complex, they can be broken down into two key aspects. The first aspect is the level at which the immunodominant epitope is expressed on the surface of appropriate antigen-presenting cells. This depends on protein abundance during infection, access of the protein into different processing pathways (MHC class I versus class II, immunoproteasomes, cross-presentation, etc.), the capacity of the peptide to be cleaved from the protein (which may depend on flanking regions), and the affinity of the peptide for the relevant MHC molecule. The mechanisms underlying these different features of antigen processing are discussed in detail in other chapters. The second aspect is the capacity of the epitope to compete at the level of T-cell recognition. This depends on the numbers of precursor T cells that recognize the epitope and the affinity of T-cell antigen receptors that bind the peptide–MHC complex. Several factors have been shown to influence the repertoire of T cells available to a given epitope. The structure of the epitope itself dictates the breadth of T-cell receptors that it recognizes. For example, it is clear that MHC class I–restricted peptides that have prominent central residues (where the peptide bulges out of the peptide-binding groove) drive strong T-cell responses with high T-cell receptor diversity [22–26]. A good example of this is the $PA_{224-233}/D^b$ epitope, which drives a T-cell response characterized by a prominent bulge of the P6 and P7 residues out of the MHC groove [23–25]. Another example is the highly immunodominant Sendai virus $NP_{324-332}/K^b$ epitope, which features a very prominent central peptide bulge and stimulates a diverse repertoire of T cells in C57BL/6 mice [26]. Another factor that may affect the availability of T cells is self-tolerance. Epitopes that are similar to self-antigens may elicit comparatively poor T-cell responses because of holes in the repertoire. Such epitopes may drive relatively weak T-cell responses despite strong expression on the antigen-presenting cell. There is substantial evidence for this phenomenon in peptide vaccination models [27–29]. Yet another factor that may affect the specificity of T-cell responses is the presence of memory T cells to other pathogens. Studies by Selin and colleagues have shown that T-cell specificity is relatively degenerate and that frequent T-cell cross-reactivities can be identified between otherwise unrelated pathogens [30, 31]. These cross-reactive T cells mount stronger responses than naïve T cells and can radically change the expected pattern of immunodominance, at both the level of T-cell receptor usage and the specific epitopes recognized. For example, Pichinde virus (PV) and lymphocytic choriomeningitis (LCMV) virus share a subdominant epitope. LCMV infection of mice that had recovered from a prior PV infection elicited a significantly altered pattern of immunodominance compared to LCMV infection of naïve mice. T cells specific for the cross-reactive epitope dominated the response, whereas T cells specific for a normally dominant (but non-cross-reactive) LCMV epitope were suppressed [31]. Thus, immunodominance is significantly affected by the host's prior exposure to unrelated antigens. Moreover, because the ratio of memory to naïve T cells in an individual substantially increases with age, the influence of preexisting memory T cells on immunodominance patterns is likely to increase over time. Finally, another factor that may influence immunodominance hierarchies is the activity of regulatory T cells (Treg) [32]. These cells are able to modulate the strength of T-cell responses and appear to act primarily to dampen autoimmune responses. However, they also reduce the strength of T-cell responses to infection, and in the influenza virus system it has been reported that Treg selectively dampen the T-cell responses to immunodominant epitopes. However, there is currently no evidence that this mechanism fundamentally alters the immunodominance hierarchy [32–34].

A particularly interesting aspect of immunodominance hierarchies is the competitive nature of the T-cell response to different epitopes. Many years ago, it was noticed that an epitope that was dominant in a particular strain of mouse may become subdominant in the context of an F_1 mouse that expressed a second MHC haplotype [35]. The overall strength of the T-cell response to the epitope (in terms of the numbers of T cells that responded) did not appear to be a fixed characteristic of the epitope, but rather depended on whether other competing T cells were present [36]. For example, as discussed above, T cells specific for the $NP_{324-332}/K^b$ epitope dominate the $CD8^+$ T-cell response to Sendai virus in C57BL/6 (H-2^b) mice. However, this response is relegated to subdominant status in (B6 × CBA) F_1 (H-$2^{b/k}$) mice [37]. This is not simply due to lower levels of epitope expression in heterozygous mice. Rather, it seems to reflect a competition between T cells for resources. The basis of this competition is not understood, but the possibilities include limited availability of antigen-presenting cells, limited availability of necessary cytokines, or perhaps simply limited "space" for the cells to proliferate. If T-cell competition is a major factor in immunodominance, it is possible that there may also be progressive changes in the avidity of the responding T cells. Indeed, there is substantial evidence for this. Studies in mice that have been infected with simian virus 5 have shown that there is a gradual change in the avidity of the T-cell response throughout the course of the infection [38]. However, it is not known whether these changes directly impact immunodominance patterns.

9.4 T-Cell Memory to Respiratory Virus Infections

Mice that have recovered from a prior respiratory virus infection are better able to resist secondary viral challenge because of the persistence of neutralizing antibody and memory T cells [39]. In situations where neutralizing antibody is ineffective, such as following challenge with serologically distinct virus strains, memory T cells play the key role in controlling secondary viral challenge [17]. These cells are able to respond more rapidly and in greater numbers than naïve T cells, resulting in accelerated viral clearance. Memory T cells can be distinguished from naïve T cells based on several characteristics (Figure 9.3A) [39]. First, in contrast to naïve T cells, memory T cells persist at a high frequency, enabling them to mount very rapid responses to secondary antigen exposure. Second, memory T cells require less costimulation than naïve T cells, effectively lowering their threshold of activation. Third, subsets of memory T cells have the capacity to rapidly mediate effector functions, such as cytolytic activity and cytokine production. Finally, subpopulations of memory T cells are distributed in non-lymphoid peripheral tissues, where they can mediate immediate responses at the site of initial pathogen entry [40–42].

It was originally thought that memory cells reside exclusively in secondary lymphoid tissues such as the spleen and lymph nodes. However, it has now emerged that large numbers of memory T cells also persist in non-lymphoid organs and other peripheral sites [42–44]. For example, as many as half of the long-lived memory $CD8^+$ T cells in animals that have recovered from an influenza virus infection are located at peripheral sites, such as the lungs [44]. In the case of influ-

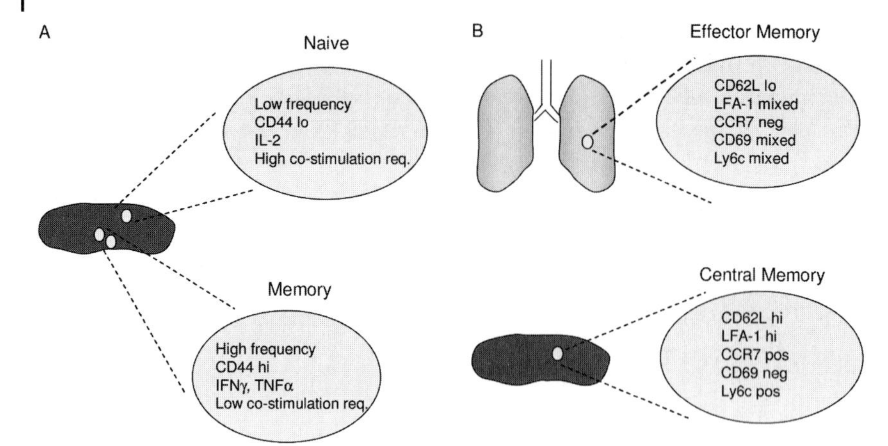

Figure 9.3 Characteristics of naïve and memory T cells. (A) Distinguishing characteristics between naïve and memory T cells. Naïve T cells are found at low frequency, have low expression of the activation marker CD44, are limited in their cytokine production, and have very high costimulation requirements. Memory T cells, on the other hand, are found at high frequencies, have high expression of CD44, are capable of producing multiple cytokines including IFN-γ and TNF-α, and require little costimulation. (B) Distinguishing characteristics between effector and central memory T cells. Effector memory T cells are found in peripheral organs and tissues such as the lung airways. These cells typically express low levels of CD62L and CCR7 and have mixed expression of LFA-1, CD69, and Ly6c (depending on location). Central memory cells typically express high levels of CD62L and CCR7, which allows them to enter and localize within the T-cell zones of lymphoid tissue such as the spleen. Additionally, these cells usually express high levels of LFA-1 and Ly6c, but are CD69 negative.

enza and Sendai virus infections of mice, as many as 40,000 antigen-specific $CD8^+$ T cells can be recovered from the lung airways of mice a month following resolution of the infections [42]. These numbers tend to rapidly decline over the first few months post-infection then stabilize at much lower numbers (typically only a few thousand cells per animal). Interestingly, this decline and stabilization in the number of memory T cells correlates with a progressive loss of protection against a secondary viral challenge [45]. These observations have led to the idea that memory T cells can be divided into two major subsets, central memory or effector memory, based on their ability to traffic to either secondary lymphoid organs or peripheral tissues (Figure 9.3B) [46]. This difference in trafficking appears to be associated with the expression of specific adhesion molecules and chemokine receptors that target the cells to different tissues [46]. Central memory T cells typically express CD62L and CCR7 and are found primarily in lymphoid organs such as the draining lymph nodes and spleen[1]. The expression of these

1) It should be noted that the expression of CD62L and CCR7 on central and effector memory T cells is not perfectly coordinated [47]. In addition, there is currently some question as to whether the division of memory T cells into effector and central subtypes is really a relevant paradigm.

molecules promotes migration of the cells through high endothelial venules and localization within the T zone of lymphoid tissue. Effector memory T cells typically lack CD62L and CCR7 but express alternative adhesion molecules and chemokine receptors (which vary based on the site of initial infection) that promote a wider distribution of these cells into the peripheral organs and tissues. Similar rules appear to apply for both memory $CD4^+$ and $CD8^+$ T cells. However, while the general phenotypic and functional characteristics of these cells are similar, the absolute numbers of memory $CD4^+$ T cells tend to be much lower than those for memory $CD8^+$ T cells [48, 49].

Once established, memory T-cell pools persist for years, or even the lifetime of the individual. Several factors have been identified as important for maintaining these memory T-cell pools. Both IL-7 and IL-15 have been shown to play an important role in driving the slow, but continual, homeostatic proliferation of memory $CD8^+$ T cells [50–52]. The cells appear to retain their functional characteristics for life, although there is a progressive switch on a population basis from a predominantly effector memory phenotype to a central memory phenotype. It has recently been shown that the development of appropriate effector functions in memory T-cell pools depends on $CD4^+$ T helper cells present during the initial acute response. For example, depletion of $CD4^+$ T cells during primary infection leads to the development of functionally impaired memory T cells [53, 54]. There is also evidence that memory $CD4^+$ responses require B cells, because B cell–deficient mice have decreased numbers of memory $CD4^+$ T cells. Thus, there is evidence for complex interplay between different immune cell subsets in the establishment of effective pools of memory T cells.

9.5 The Specificity of Memory T Cells

Many studies have established that the specificity of memory T cells that persist following resolution of a primary infection usually reflects the specificity of the acute response (sometimes referred to as the clonal burst hypothesis) [55]. This is true in both peripheral sites such as the lung airways or pleural cavity and secondary lymphoid organs such as the lymph nodes and spleen. However, there are some instructive exceptions to this rule. One example comes from the Sendai virus system. As discussed earlier, the $NP_{324-332}/K^b$ epitope drives the dominant $CD8^+$ T-cell response in Sendai virus-infected C57BL/6 mice. Interestingly, this same $NP_{324-332}$ peptide also binds with high affinity to the D^b molecule in C57BL/6 mice to generate an $NP_{324-332}/D^b$ epitope [37]. T cells specific for this $NP_{324-332}/D^b$ epitope cannot be detected in the primary acute response to infection, indicating that it is strongly subdominant to the $NP_{324-332}/K^b$ epitope. However, memory T cells specific for the $NP_{324-332}/D^b$ epitope can be detected in the spleen and lymph nodes, but not in peripheral sites such as the lung, following resolution of the infection [37, 42]. The mechanism by which this occurs is unclear but may be related to differences in the distribution of specificity patterns in central and effector memory T-cell pools. In support of this general idea, it has recently been shown that at least some central and effector memory T cells may represent distinct lineages and that they may be differentially generated during the acute response [56, 57]. Thus, one possibility is that T cells specific for the $NP_{324-332}/D^b$ epitope are generated preferentially in the central memory pool and therefore are found exclusively in the secondary lymphoid organs.

Most of the discussion thus far has focused on viruses that are cleared from the host, leaving behind a stable pool of memory T cells. However, some viruses establish persistent infections or latency and interact with the host immune system over protracted periods of time. Continual antigen exposure can have significant effects on the properties of memory T cells and patterns of immunodominance. For example, the specificity of the T-cell response to γHV-68 has been shown to change over time, with the dominant response switching from the $ORF6_{487-495}/D^b$ to the $ORF61_{524-531}/K^b$ epitope [20]. This pattern appears to be due to different antigens being presented during the early phase of infection (characterized by a high level of viral reactivation from latency) and the late phase of infection (characterized by stable latent infection) [58]. In other situations of persistent infection (such as persistent LCMV infection), changes in the hierarchy may be related to clonal exhaustion and loss of T cells specific for dominant epitopes, followed by the emergence of T cells specific for subdominant epitopes [59, 60].

Another situation in which the specificity of the memory T-cell pool can change over time is following the outgrowth of age-associated $CD8^+$ T-cell clones [61, 62]. For example, in C57BL/6 mice that have recovered from a prior Sendai virus infection, $CD8^+$ T-cell outgrowths begin to appear after 10 months (i.e., the mice are about 12 months of age), and many of these outgrowths are specific for the immunodominant $NP_{324-332}/K^b$ epitope (D.L. Woodland, unpublished data). The mechanism underlying this phenomenon is unknown, although it appears to be due to the loss of control of homeostatic proliferation of the memory T cells [62, 63]. In mice that have recovered from a prior infection, these outgrowths result in grossly distorted T-cell receptor usage among T cells specific for a particular antigen. The same phenomenon occurs in human, and it is generally believed that these outgrowths are detrimental in terms of the individual's capacity to mount recall responses. However, it is unclear whether this is fundamentally due to the specificity change or some other defect in these cells.

9.6 Recall Responses to Secondary Infections

The rules governing the recall of memory T cells during secondary pathogen challenge are poorly understood. While it is clear that many distinct pools of memory T cells persist following resolution of the primary infection (in terms of phenotype, function, and anatomical distribution), the relative contributions of these cells to the recall response have not been determined. A major component of the recall response is the reactivation of the effector and central memory T cells in secondary lymphoid organs. These cells are stimulated by antigen and proliferate for several days prior to being recruited into the lungs [64]. In addition to this, antigen-specific T cells present in the lung airways also appear to contribute to the control of secondary respiratory virus infections. For example, in the Sendai virus model, it has been shown that intratracheal transfer of these cells into naïve mice results in substantial control of a subsequent virus infection, suggesting that they represent the first line of defense against secondary infection [41]. We have also demonstrated the recruitment of non-proliferating memory T cells to the lungs during the first week of infection [65]. These cells appear to be circulating effector memory T cells that are recruited to the lungs without having been activated by viral antigen. Based on these observations, we have speculated that the memory recall response can be divided into three major phases [66]. First, there is the immediate response of effector memory T cells already present at the site of infection that are able to respond when viral loads are very low. These cells are unable to proliferate in response to infection, most likely because of the constraints of the airway environment, but they can produce cytokines that may limit viral replication and spread. Second, there is an intermediate phase in which non-proliferating memory cells are recruited to the lung airways during the first week of the infection. These cells presumably act to further constrain viral replication until proliferating effector cells from the lymph nodes begin to arrive in the lungs. Third, memory cells divide in response to antigen in the local draining lymph nodes to generate a prolonged supply of new effector cells. One of the key features of this model is that these three phases are integrated to produce a sustained response in the lung airways that limits viral replication and mediates viral clearance. It is important to note that the first two phases of the response do not involve T-cell expansion, but the T cells do engage the pathogen at the site of infection when viral loads are relatively low. Presumably, this reduces the amount of virus encountered by T cells comprising the third phase of the response. Becuase this latter phase involves proliferating cells, it is able to mediate a sustained response to the infection.

9.7 Immunodominance Patterns in Recall Responses

As discussed above, the bulk of the primary $CD8^+$ T-cell response to respiratory virus infections is driven by relatively few antigenic epitopes, and the immunodominance pattern established during the acute response is usually retained in the memory T-cell pool. Analysis of the recall response suggests that the same pattern of specificity is also retained when these cells are reactivated to generate a recall response. Limited data suggest that this includes both the early (non-proliferative) and later (proliferative) phases of the recall response. However, there are some notable exceptions to this rule that provide tremendous insight into the factors that regulate the specificity of T-cell responses. One particularly interesting and instructive example is the $CD8^+$ T-cell response to influenza virus in C57BL/6 mice, which exhibits different specificity profiles in the primary and recall responses [67]. Whereas T cells specific for both the $NP_{366-374}/D^b$ and $PA_{224-233}/D^b$ epitopes are present in equivalent numbers following primary infection, T cells specific for the $NP_{366-374}/D^b$ epitope dominate the secondary response (Figure 9.4). Several laboratories have been aggressively exploring the basis for this shift in immunodominance. One intriguing possibility is that it is due to differences in cross-presentation of the two antigens (e.g., the more abundant nucleoprotein may be preferentially cross-presented) [68]. The problem with this hypothesis is that it is hard to understand why the pattern would change between a primary and secondary infection. This could be attributed to the presence of anti-nucleoprotein or anti-acidic polymerase antibodies generated during the primary response, but the same patterns of immunodominance are seen in mice that specifically lack antibody responses. Another possibility is that the shift in immunodominance following secondary infection is mediated by differential antigen presentation. We have shown that whereas the $NP_{366-374}/D^b$ epitope is processed and efficiently presented by many cells types, the $PA_{224-233}/D^b$ epitope is efficiently presented only on dendritic cells [69]. Because dendritic cells drive the primary T-cell response, T cells to both epitopes would be efficiently presented. In contrast, it is believed that memory cell activation can be mediated by both dendritic and non-dendritic cells. Thus, $NP_{366-374}/D^b$-specific T cells would respond to antigen on a wide array of cells, whereas $PA_{224-233}/D^b$-specific T cells would only respond to antigen on dendritic cells. The more widespread expression of the $NP_{366-374}/D^b$ epitope would give $NP_{366-374}/D^b$-specific memory CD8 T cells a competitive advantage following secondary infection[2]. A key feature of this hypothesis is that it does not require a change in the pattern of antigen presentation between the primary and secondary infection. Instead, it depends on the differential capacity of naïve and memory T cells to perceive antigen presented on different cell types. In support of this general idea, we have shown that concurrent naïve and memory T-cell responses in a single mouse (where antigen presentation is identical for both sets of T cells) faithfully replicate the $NP_{366-374}/D^b$ and $PA_{224-233}/D^b$ immunodominance patterns characteristic of primary and secondary infections [69]. Taken together, it is clear that there are several possible explanations for the regulation of immunodominance patterns during recall responses in the lung, and further study will be required to identify the underlying mechanism. An understanding of this mechanism should reveal important insights into other aspects of T-cell immunodominance.

2) Consistent with this idea, the change in immunodominance during a secondary infection appears to reflect a greater expansion of $NP_{366-374}/D^b$-specific T cells, rather than a reduction in the magnitude of the $PA_{224-233}/D^b$-specific T-cell response. As illustrated in Figure 9.4, the magnitude of the $PA_{224-233}/D^b$-specific T-cell response does not change between a primary and secondary infection.

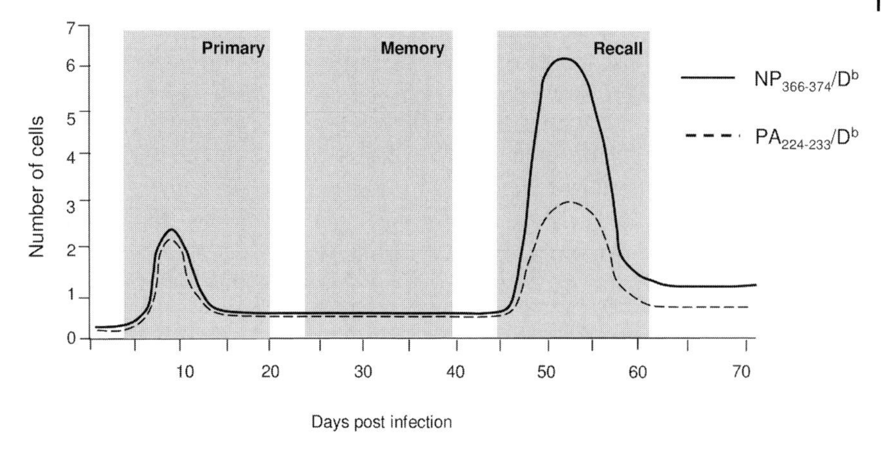

Figure 9.4 Kinetics of primary and secondary responses in the lung airways following influenza virus infection. T cells specific for the $NP_{366-374}/D^b$ and the $PA_{224-233}/D^b$ epitopes are generated in equivalent numbers during the primary response to influenza virus infection in C57BL/6 mice. These cells also persist in equal numbers the memory T-cell pool that is established after virus is cleared. Following secondary viral challenge, $NP_{366-374}/D^b$-specific T cells dominate the recall response with four to five times as many $NP_{366-374}/D^b$-specific T cells than $PA_{224-233}/D^b$-specific T cells.

It is currently unclear whether the changing patterns of immunodominance among $NP_{366-374}/D^b$– and $PA_{224-233}/D^b$-specific T cells represent a unique situation or are representative of a class of epitopes. Recently, Zhong et al. identified an array of previously unidentified $CD8^+$ T-cell epitopes in the PR8/8/34 strain of influenza virus and examined the ability of epitope-specific CD8 T cells to lyse peptide-pulsed or virally infected EL4 target cells [19]. Interestingly, the study identified two new epitopes that, like the $PA_{224-233}/D^b$ epitope, do not appear to be presented by virally infected EL4 cells. Thus, there is an indication that differential antigen processing and presentation may occur with other influenza virus epitopes. It remains unclear whether there is differential antigen processing and presentation of class II epitopes following influenza virus infection. With the recent identification of new class II epitopes, investigations regarding these questions are currently underway.

9.8 Modification of Immunodominance Hierarchies by Vaccination

An understanding of T-cell immunodominance is important when considering epitopes for inclusion in a vaccine designed to promote T-cell immunity in the lung. A key question has been whether T cells specific for subdominant epitopes can be protective if elicited in advance by vaccination. The answer to this is a clear yes. Studies by Oukka et al. have shown that vaccination with peptides that elicit T-cell responses to subdominant influenza virus epitopes is highly protective [70,

71]. We have made similar observations in the Sendai virus system [72]. In both of these cases, the response to subsequent viral challenge was dominated by T cells specific for the antigen that was used in the vaccination. Moreover, these T cells were capable of migrating to the lungs following infection and exhibited potent cytolytic activity and cytokine production. In other words, vaccination resulted in a complete switch in immunodominance. In addition, it appears that there is no fundamental defect in T cells specific for subdominant T cells. The switch in immunodominance following vaccination is only temporary, as the dominance patterns tend to revert back to the normal state on subsequent infections (D.L. Woodland, unpublished data).

While T cells specific for subdominant epitopes were clearly protective in the previous examples, there are cases in which T-cell responses to specific epitopes can be non-protective. An interesting example of this comes from the influenza virus model [73, 74]. As discussed earlier, the $PA_{224-233}/D^b$ epitope, unlike the $NP_{366-374}/D^b$ epitope, appears to be poorly expressed in cells such as lung epithelial cells that are the primary targets of influenza virus infection. Vaccination of mice with the $NP_{366-374}$ peptide was highly protective, whereas vaccination of mice with the $PA_{224-233}$ peptide resulted in higher viral titers and actually delayed viral clearance compared to normal unvaccinated mice [73]. Both vaccinations resulted in increased numbers of epitope-specific cytotoxic $CD8^+$ T cells capable of migrating to the lung airways and producing IFN-γ and TNF-α following infection. One explanation for these findings is that the $PA_{224-233}/D^b$ epitope is poorly expressed on lung epithelial cells and that T cells specific for this epitope not only fail to clear virus but also may also crowd out the protective $NP_{366-374}/D^b$-specific T cells. However, it has also been noted that cells specific for these two epitopes differ in their production of antiviral cytokines. T cells specific for the $PA_{224-233}/D^b$ epitope typically produce much higher levels of TNF-α in response to antigen-loaded dendritic cells in the lung, resulting in increased immunopathology and impaired control of the virus [73, 75]. These possibilities are currently under investigation.

9.9 Conclusions

The T-cell response to pulmonary pathogens is a highly complex process that involves the activation of distinct subsets of T cells, the trafficking of T cells to different sites, the elaboration of a variety of effector functions, and the development of distinct pools of memory T cells. A key aspect of T-cell immunity is the specificity for different antigens produced by the pathogen. As discussed in this review, multiple factors influence the specificity of the responses at different stages of the response. Depending on the circumstances, patterns of immunodominance established in primary responses may either stay the same or significantly change in memory T-cell pools and recall responses. These differences in specificity have a profound impact on the efficacy of the response and have significant implications for vaccine design. Therefore, developing a complete understanding of the mechanisms underlying the specificity of T-cell responses under different circumstances is an important goal for the future.

Acknowledgments

The authors would like to thank Drs. Marcy Blackman, Ken Ely, and Jake Kohlmeier for help in preparing this manuscript. Much of the work from the Woodland laboratory outlined in this review was funded by the PHS and the Trudeau Institute.

References

1 Blackman, M.A., J.W. Kappler, and P. Marrack. 1988. T-cell specificity and repertoire. *Immunol Rev* 101:5–19.
2 Yewdell, J.W., and J.R. Bennink. 1999. Immunodominance in major histocompatibility complex class I-restricted T lymphocyte responses. *Annu Rev Immunol* 17:51–88.
3 Chen, W., L.C. Anton, J.R. Bennink, and J.W. Yewdell. 2000. Dissecting the multifactorial causes of immunodominance in class I-restricted T cell responses to viruses. *Immunity* 12:83–93.
4 Horimoto, T., and Y. Kawaoka. 2001. Pandemic threat posed by avian influenza A viruses. *Clin Microbiol Rev* 14:129–149.
5 Palese, P. 2004. Influenza: old and new threats. *Nat Med* 10:S82–87.
6 Woodland, D.L. 2003. Cell-mediated immunity to respiratory virus infections. *Curr Opin Immunol* 15:430–435.
7 Doherty, P.C., D.J. Topham, R.A. Tripp, R.D. Cardin, J.W. Brooks, and P.G. Stevenson. 1997. Effector $CD4^+$ and $CD8^+$ T-cell mechanisms in the control of respiratory virus infections. *Immunol Rev* 159:105–117.
8 Flynn, K.J., G.T. Belz, J.D. Altman, R. Ahmed, D.L. Woodland, and P.C. Doherty. 1998. Virus-specific $CD8^+$ T cells in primary and secondary influenza pneumonia. *Immunity* 8:683–691.
9 Cole, G.A., T.L. Hogg, and D.L. Woodland. 1994. The MHC class I-restricted T cell response to Sendai virus infection in C57BL/6 mice: a single immunodominant epitope elicits an extremely diverse repertoire of T cells. *Int Immunol* 6:1767–1775.
10 Cole, G.A., J.M. Katz, T.L. Hogg, K.W. Ryan, A. Portner, and D.L. Woodland. 1994. Analysis of the primary T-cell response to Sendai virus infection in C57BL/6 mice: $CD4^+$ T-cell recognition is directed predominantly to the hemagglutinin-neuraminidase glycoprotein. *J Virol* 68:6863–6870.
11 Eichelberger, M., W. Allan, M. Zijlstra, R. Jaenisch, and P.C. Doherty. 1991. Clearance of influenza virus respiratory infection in mice lacking class I major histocompatibility complex-restricted $CD8^+$ T cells. *J Exp Med* 174:875–880.
12 Legge, K.L., and T.J. Braciale. 2003. Accelerated migration of respiratory dendritic cells to the regional lymph nodes is limited to the early phase of pulmonary infection. *Immunity* 18:265–277.
13 Norbury, C.C., D. Malide, J.S. Gibbs, J.R. Bennink, and J.W. Yewdell. 2002. Visualizing priming of virus-specific $CD8^+$ T cells by infected dendritic cells in vivo. *Nat Immunol* 3:265–271.
14 Topham, D.J., R.A. Tripp, and P.C. Doherty. 1997. $CD8^+$ T cells clear influenza virus by perforin or Fas-dependent processes. *J Immunol* 159:5197–5200.
15 Roman, E., E. Miller, A. Harmsen, J. Wiley, U.H. Von Andrian, G. Huston, and S.L. Swain. 2002. CD4 effector T cell subsets in the response to influenza: heterogeneity, migration, and function. *J Exp Med* 196:957–968.

16 Zhong, W., A.D. Roberts, and D.L. Woodland. 2001. Antibody-independent antiviral function of memory $CD4^+$ T cells in vivo requires regulatory signals from $CD8^+$ effector T cells. *J Immunol* 167:1379–1386.

17 Graham, M.B., and T.J. Braciale. 1997. Resistance to and recovery from lethal influenza virus infection in B lymphocyte-deficient mice. *J Exp Med* 186:2063–2068.

18 Yewdell, J.W., and S.M. Haeryfar. 2005. Understanding presentation of viral antigens to $CD8^+$ t cells in vivo: The key to rational vaccine design. *Annu Rev Immunol* 23:651–682.

19 Zhong, W., P.A. Reche, C.C. Lai, B. Reinhold, and E.L. Reinherz. 2003. Genome-wide characterization of a viral cytotoxic T lymphocyte epitope repertoire. *J Biol Chem* 278:45135–45144.

20 Stevenson, P.G., G.T. Belz, J.D. Altman, and P.C. Doherty. 1999. Changing patterns of dominance in the $CD8^+$ T cell response during acute and persistent murine gamma-herpesvirus infection. *Eur J Immunol* 29:1059–1067.

21 Crowe, S.R., S.C. Miller, D.M. Brown, P.S. Adams, R.W. Dutton, A.G. Harmsen, F.E. Lund, T.D. Randall, S.L. Swain, and D.L. Woodland. 2005. Uneven distribution of MHC class II epitopes within the influenza virus. *Vaccine* (in press).

22 Tynan, F.E., N.A. Borg, J.J. Miles, T. Beddoe, D. El-Hassen, S.L. Silins, W.J. van Zuylen, A.W. Purcell, L. Kjer-Nielsen, J. McCluskey, S.R. Burrows, and J. Rossjohn. 2005. The high resolution structures of highly bulged viral epitopes bound to the major histocompatability class I: Implications for T-cell receptor engagement and T-cell immunodominance. *J Biol Chem.*

23 Turner, S.J., K. Kedzierska, H. Komodromou, N.L. La Gruta, M.A. Dunstone, A.I. Webb, R. Webby, H. Walden, W. Xie, J. McCluskey, A.W. Purcell, J. Rossjohn, and P.C. Doherty. 2005. Lack of prominent peptide-major histocompatibility complex features limits repertoire diversity in virus-specific $CD8^+$ T cell populations. *Nat Immunol* 6:382–389.

24 Turner, S.J., K. Kedzierska, N.L. La Gruta, R. Webby, and P.C. Doherty. 2004. Characterization of CD8+ T cell repertoire diversity and persistence in the influenza A virus model of localized, transient infection. *Semin Immunol* 16:179–184.

25 Meijers, R., C.C. Lai, Y. Yang, J.H. Liu, W. Zhong, J.H. Wang, and E.L. Reinherz. 2005. Crystal structures of murine MHC Class I H-2 D^b and K^b molecules in complex with CTL epitopes from influenza A virus: implications for TCR repertoire selection and immunodominance. *J Mol Biol* 345:1099–1110.

26 Cole, G.A., T.L. Hogg, and D.L. Woodland. 1995. T cell recognition of the immunodominant Sendai virus NP324–332/Kb epitope is focused on the center of the peptide. *J Immunol* 155:2841–2848.

27 Nanda, N.K., R. Apple, and E. Sercarz. 1991. Limitations in plasticity of the T-cell receptor repertoire. *Proc Natl Acad Sci U S A* 88:9503–9507.

28 Perkins, D.L., Y.S. Wang, D. Fruman, J.G. Seidman, and I.J. Rimm. 1991. Immunodominance is altered in T cell receptor (beta-chain) transgenic mice without the generation of a hole in the repertoire. *J Immunol* 146:2960–2964.

29 Daly, K., P. Nguyen, D.L. Woodland, and M.A. Blackman. 1995. Immunodominance of major histocompatibility complex class I-restricted influenza virus epitopes can be influenced by the T-cell receptor repertoire. *J Virol* 69:7416–7422.

30 Welsh, R.M., and L.K. Selin. 2002. No one is naive: the significance of heterologous T-cell immunity. *Nat Rev Immunol* 2:417–426.

31 Brehm, M.A., A.K. Pinto, K.A. Daniels, J.P. Schneck, R.M. Welsh, and L.K. Selin. 2002. T cell immunodominance and maintenance of memory regulated by unexpectedly cross-reactive pathogens. *Nat Immunol* 3:627–634.

32 Rouse, B.T., and S. Suvas. 2004. Regulatory cells and infectious agents: detentes cordiale and contraire. *J Immunol* 173:2211–2215.

33 Rifa'i, M., Y. Kawamoto, I. Nakashima, and H. Suzuki. 2004. Essential roles of

CD8+CD122+ regulatory T cells in the maintenance of T cell homeostasis. *J Exp Med* 200:1123–1134.

34 Haeryfar, S.M., R.J. DiPaolo, D.C. Tscharke, J.R. Bennink, and J.W. Yewdell. 2005. Regulatory T cells suppress CD8$^+$ T cell responses induced by direct priming and cross-priming and moderate immunodominance disparities. *J Immunol* 174:3344–3351.

35 Doherty, P.C., W.E. Biddison, J.R. Bennink, and B.B. Knowles. 1978. Cytotoxic T-cell responses in mice infected with influenza and vaccinia viruses vary in magnitude with H-2 genotype. *J Exp Med* 148:534–543.

36 Belz, G.T., P.G. Stevenson, and P.C. Doherty. 2000. Contemporary analysis of MHC-related immunodominance hierarchies in the CD8$^+$ T cell response to influenza A viruses. *J Immunol* 165:2404–2409.

37 Cole, G.A., T.L. Hogg, M.A. Coppola, and D.L. Woodland. 1997. Efficient priming of CD8+ memory T cells specific for a subdominant epitope following Sendai virus infection. *J Immunol* 158:4301–4309.

38 Alexander-Miller, M.A. 2005. High-avidity CD8$^+$ T cells: optimal soldiers in the war against viruses and tumors. *Immunol Res* 31:13–24.

39 Seder, R.A., and R. Ahmed. 2003. Similarities and differences in CD4$^+$ and CD8$^+$ effector and memory T cell generation. *Nat Immunol* 4:835–842.

40 Moyron-Quiroz, J.E., J. Rangel-Moreno, K. Kusser, L. Hartson, F. Sprague, S. Goodrich, D.L. Woodland, F.E. Lund, and T.D. Randall. 2004. Role of inducible bronchus associated lymphoid tissue (iBALT) in respiratory immunity. *Nat Med* 10:927–934. Epub 2004 Aug 2015.

41 Hogan, R.J., W. Zhong, E.J. Usherwood, T. Cookenham, A.D. Roberts, and D.L. Woodland. 2001. Protection from respiratory virus infections can be mediated by antigen- specific CD4$^+$ T cells that persist in the lungs. *J Exp Med* 193:981–986.

42 Hogan, R.J., E.J. Usherwood, W. Zhong, A.A. Roberts, R.W. Dutton, A.G. Harmsen, and D.L. Woodland. 2001. Activated antigen-specific CD8$^+$ T cells persist in the lungs following recovery from respiratory virus infections. *J Immunol* 166:1813–1822.

43 Wiley, J.A., R.J. Hogan, D.L. Woodland, and A.G. Harmsen. 2001. Antigen-specific CD8$^+$ T cells persist in the upper respiratory tract following influenza virus infection. *J Immunol* 167:3293–3299.

44 Marshall, D.R., S.J. Turner, G.T. Belz, S. Wingo, S. Andreansky, M.Y. Sangster, J.M. Riberdy, T. Liu, M. Tan, and P.C. Doherty. 2001. Measuring the diaspora for virus-specific CD8$^+$ T cells. *Proc Natl Acad Sci U S A* 98:6313–6318.

45 Liang, S., K. Mozdzanowska, G. Palladino, and W. Gerhard. 1994. Heterosubtypic immunity to influenza type A virus in mice. Effector mechanisms and their longevity. *J Immunol* 152:1653–1661.

46 Sallusto, F., A. Langenkamp, J. Geginat, and A. Lanzavecchia. 2000. Functional subsets of memory T cells identified by CCR7 expression. *Curr Top Microbiol Immunol* 251:167–171.

47 Unsoeld, H., and H. Pircher. 2005. Complex memory T-cell phenotypes revealed by coexpression of CD62L and CCR7. *J Virol* 79:4510–4513.

48 Doherty, P.C., D.J. Topham, and R.A. Tripp. 1996. Establishment and persistence of virus-specific CD4$^+$ and CD8$^+$ T cell memory. *Immunol Rev* 150:23–44.

49 Topham, D.J., and P.C. Doherty. 1998. Longitudinal analysis of the acute Sendai virus-specific CD4$^+$ T cell response and memory. *J Immunol* 161:4530–4535.

50 Berard, M., K. Brandt, S.B. Paus, and D.F. Tough. 2003. IL-15 Promotes the Survival of Naive and Memory Phenotype CD8$^+$ T Cells. *J Immunol* 170:5018–5026.

51 Tan, J.T., B. Ernst, W.C. Kieper, E. LeRoy, J. Sprent, and C.D. Surh. 2002. Interleukin IL-15 and IL-7 jointly regulate homeostatic proliferation of memory phenotype CD8$^+$ cells but are not required for memory phenotype CD4$^+$ cells. *J Exp Med* 195:1523–1532.

52 Schluns, K.S., W.C. Kieper, S.C. Jameson, and L. Lefrancois. 2000. Interleukin-7 mediates the homeostasis

of naive and memory CD8 T cells in vivo. *Nat Immunol* 1:426–432.

53 Sun, J.C., and M.J. Bevan. 2003. Defective CD8 T cell memory following acute infection without CD4 T cell help. *Science* 300:339–342.

54 Shedlock, D.J., and H. Shen. 2003. Requirement for CD4 T cell help in generating functional CD8 T cell memory. *Science* 300:337–339.

55 Hou, S., L. Hyland, K.W. Ryan, A. Portner, and P.C. Doherty. 1994. Virus-specific CD8+ T-cell memory determined by clonal burst size. *Nature* 369:652–654.

56 Bouneaud, C., Z. Garcia, P. Kourilsky, and C. Pannetier. 2005. Lineage relationships, homeostasis, and recall capacities of central- and effector-memory CD8 T cells in vivo. *J Exp Med* 201:579–590.

57 Badovinac, V.P., B.B. Porter, and J.T. Harty. 2004. $CD8^+$ T cell contraction is controlled by early inflammation. *Nat Immunol* 5:809–817.

58 Liu, L., E. Flano, E.J. Usherwood, S. Surman, M.A. Blackman, and D.L. Woodland. 1999. Lytic cycle T cell epitopes are expressed in two distinct phases during MHV-68 infection. *J Immunol* 163:868–874.

59 van der Most, R.G., K. Murali-Krishna, J.G. Lanier, E.J. Wherry, M.T. Puglielli, J.N. Blattman, A. Sette, and R. Ahmed. 2003. Changing immunodominance patterns in antiviral CD8 T-cell responses after loss of epitope presentation or chronic antigenic stimulation. *Virology* 315:93–102.

60 Probst, H.C., K. Tschannen, A. Gallimore, M. Martinic, M. Basler, T. Dumrese, E. Jones, and M.F. van den Broek. 2003. Immunodominance of an antiviral cytotoxic T cell response is shaped by the kinetics of viral protein expression. *J Immunol* 171:5415–5422.

61 Ku, C.C., B. Kotzin, J. Kappler, and P. Marrack. 1997. $CD8^+$ T-cell clones in old mice. *Immunol Rev* 160:139–144.

62 LeMaoult, J., I. Messaoudi, J.S. Manavalan, H. Potvin, D. Nikolich-Zugich, R. Dyall, P. Szabo, M.E. Weksler, and J. Nikolich-Zugich. 2000. Age-related dysregulation in CD8 T cell homeostasis: kinetics of a diversity loss. *J Immunol* 165:2367–2373.

63 Messaoudi, I., J. Lemaoult, J.A. Guevara-Patino, B.M. Metzner, and J. Nikolich-Zugich. 2004. Age-related CD8 T cell clonal expansions constrict CD8 T cell repertoire and have the potential to impair immune defense. *J Exp Med* 200:1347–1358.

64 Roberts, A.D., and D.L. Woodland. 2004. Cutting edge: effector memory $CD8^+$ T cells play a prominent role in recall responses to secondary viral infection in the lung. *J Immunol* 172:6533–6537.

65 Ely, K.H., L.S. Cauley, A.D. Roberts, J.W. Brennan, T. Cookenham, and D.L. Woodland. 2003. Nonspecific recruitment of memory $CD8^+$ T cells to the lung airways during respiratory virus infections. *J Immunol* 170:1423–1429.

66 Woodland, D.L., and T.D. Randall. 2004. Anatomical features of anti-viral immunity in the respiratory tract. *Semin Immunol* 16:163–170.

67 Belz, G.T., W. Xie, J.D. Altman, and P.C. Doherty. 2000. A previously unrecognized $H\text{-}2D^b$-restricted peptide prominent in the primary influenza A virus-specific $CD8^+$ T-cell response is much less apparent following secondary challenge. *J Virol* 74:3486–3493.

68 Chen, W., K. Pang, K.A. Masterman, G. Kennedy, S. Basta, N. Dimopoulos, F. Hornung, M. Smyth, J.R. Bennink, and J.W. Yewdell. 2004. Reversal in the immunodominance hierarchy in secondary $CD8^+$ T cell responses to influenza A virus: roles for cross-presentation and lysis-independent immunodomination. *J Immunol* 173:5021–5027.

69 Crowe, S.R., S.J. Turner, S.C. Miller, A.D. Roberts, R.A. Rappolo, P.C. Doherty, K.H. Ely, and D.L. Woodland. 2003. Differential antigen presentation regulates the changing patterns of $CD8^+$ T cell immunodominance in primary and secondary influenza virus infections. *J Exp Med* 198:399–410.

70 Oukka, M., N. Riche, and K. Kosmatopoulos. 1994. A nonimmunodominant nucleoprotein-derived peptide is presented by influenza A virus-infected

H-2^{b} cells. *J.Immunol.* 152:4843–4851.

71 Oukka, M., J.C. Manuguerra, N. Livaditis, S. Tourdot, N. Riche, I. Vergnon, P. Cordopatis, and K. Kosmatopoulos. 1996. Protection against lethal viral infection by vaccination with nonimmunodominant peptides. *J Immunol* 157:3039–3045.

72 Chen, Y., R.G. Webster, and D.L. Woodland. 1998. Induction of $CD8^{+}$ T cell responses to dominant and subdominant epitopes and protective immunity to Sendai virus infection by DNA vaccination. *J Immunol* 160:2425–2432.

73 Crowe, S.R., S.C. Miller, R.M. Shenyo, and D.L. Woodland. 2005. Vaccination with an acidic polymerase epitope of influenza virus elicits a potent antiviral T cell response but delayed clearance of an influenza virus challenge. *J Immunol* 174:696–701.

74 Crowe, S.R., S.C. Miller, and D.L. Woodland. 2005. Identification of protective and non-protective T cell epitopes in influenza. *Vaccine* (in press).

75 Turner, S.J., N.L. La Gruta, J. Stambas, G. Diaz, and P.C. Doherty. 2004. Differential tumor necrosis factor receptor 2-mediated editing of virus-specific CD8+ effector T cells. *Proc Natl Acad Sci U S A* 101:3545–3550.

10
Effects of Pathogens on the Immune Response: HIV

Masafumi Takiguchi

10.1
Introduction

HIV-1-specific CD8$^+$ T cells were first implicated in suppressing HIV-1 replication by studies showing that the reduction in viremia in acute infection is temporally associated with the appearance of these T cells [1, 2]. The role of CD8$^+$ T cells was further indicated by studies using the SIV macaque model, which showed that SIV replication is not suppressed after the appearance of cytotoxic T lymphocytes (CTLs) was blocked *in vivo* by using anti-CD8 mAb [3–5]. Further studies demonstrated that HIV-1 replication is suppressed by killing HIV-1-infected cells and/or by factors secreted to prevent entry of HIV-1 or to directly suppress HIV-1 replication [6–8]. Although HIV-1-specific CTLs are induced in the host after a primary infection, they fail to completely suppress HIV-1 replication. These observations suggest that HIV-1 can escape from HIV-1-specific CTLs. There are several escape mechanisms proposed, including escape mutations of immunodominant epitopes [9, 10], reduction in the number of HIV-1-specific CTLs by apoptosis of CD8$^+$ T cells via Fas and TNF [11], skewed maturation of HIV-1-specific CD8$^+$ T cells [12], and impaired cytolytic activity of HIV-1-specific CTLs toward HIV-1-infected CD4$^+$ T cells by Nef-mediated downregulation of HLA class I molecules [13].

There are several reports of HLA association with the progression of AIDS. A study analyzing the large cohorts showed that HIV-1-infected individuals who are homozygous for one or more HLA class I loci progress to AIDS much more rapidly than those who are fully heterozygous at all three loci [14]. This finding indicates that the diversity of HIV-1-specific CTL epitopes is an important part of the successful control of HIV-1. In addition, a recent study using South African cohort showed that the HLA-B locus is more influential toward the outcome of HIV disease than are HLA-A or HLA-C [15]. There are many studies showing the association of each HLA allele with the progression of AIDS [16]. HLA-B*57/58, -B*27, and -B*51 are alleles associated with slow progression to AIDS, whereas the HLA-B*35 allele is associated with rapid progression [14]. More detailed analysis of the association of HLA-B*35 with AIDS progression showed that three B*35 subtypes, B*3502, B*3503, and B*3504, have a much stronger association with

Immunodominance: The Choice of the Immune System. Edited by Jeffrey A. Frelinger

ISBN: 3-527-31274-9

AIDS progression than does B*3501 [17]. The effect of each HLA class I allele on AIDS progression in the cohort of 600 Caucasians and in the South African cohort is presented in Figures 10.1 and 10.2, respectively [15]. These data strongly suggest that analysis of the HIV-1 dominant epitopes presented by these HLA alleles is very important for clarifying how these HLA alleles influence AIDS progression.

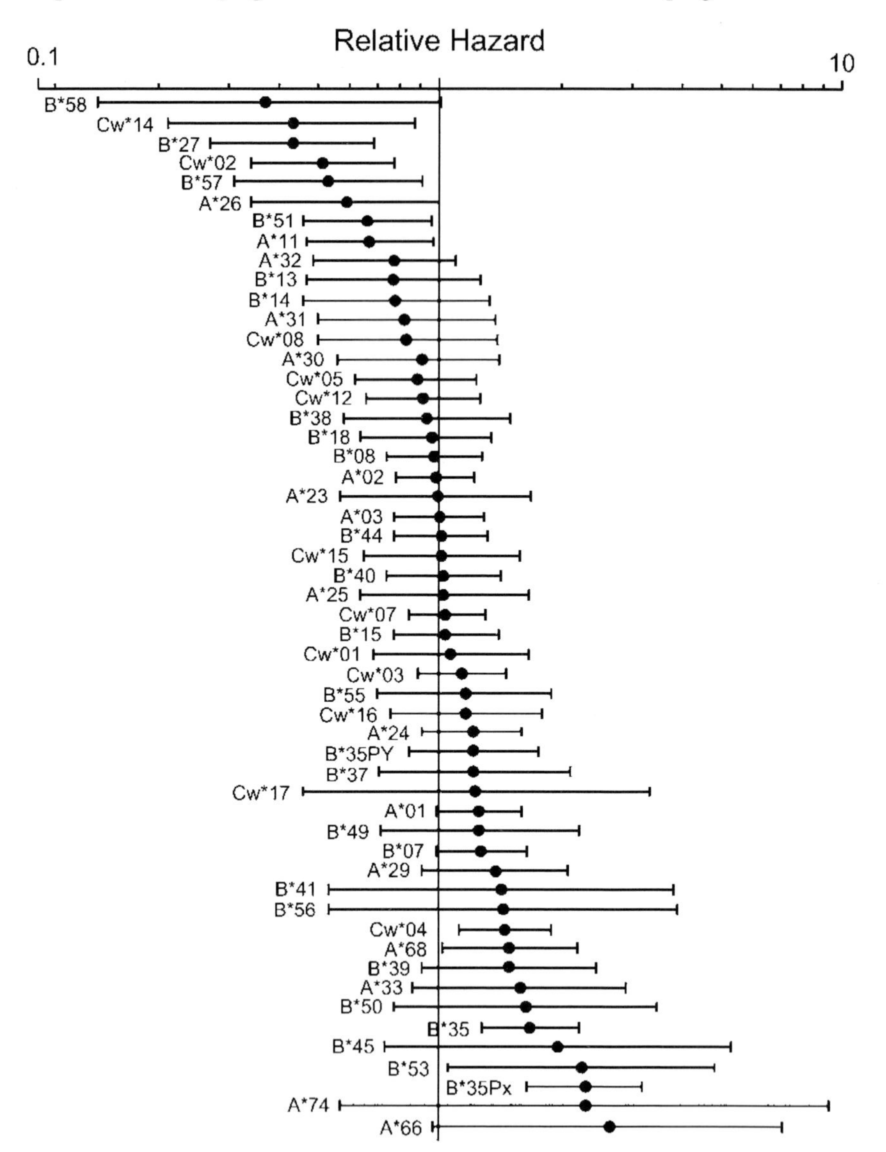

Figure 10.1 Association of HLA class I alleles with progression to AIDS. Relative hazard (RH) values were determined for each of the 54 HLA types in a Caucasian cohort including 600 seroconverter patients [89].

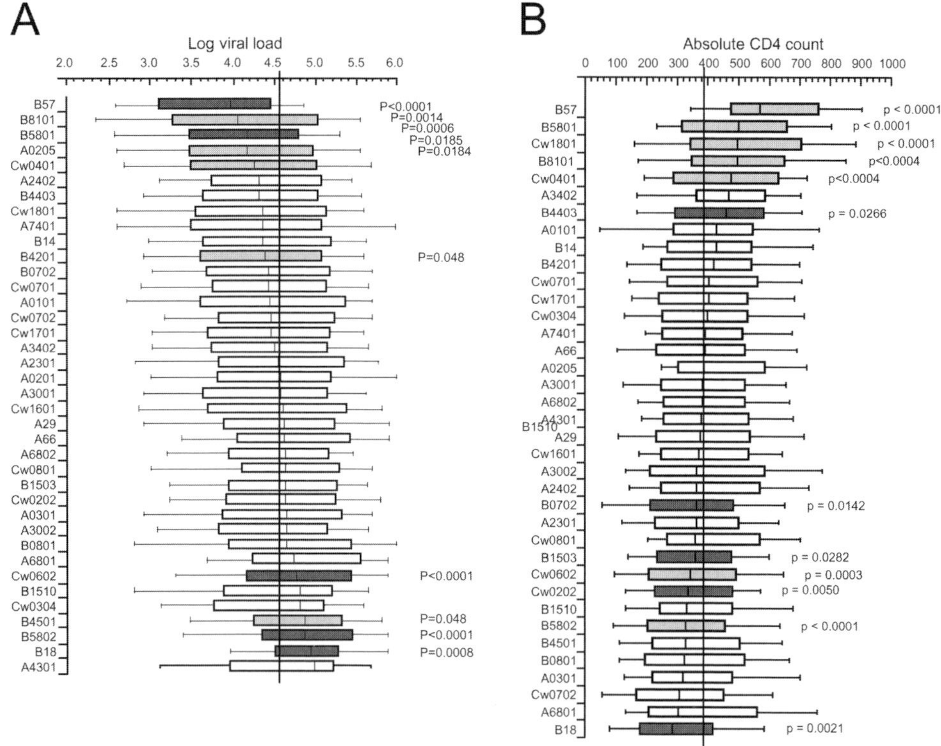

Figure 10.2 Association of HLA class I alleles with viral load and CD4 count. The influence of individual HLA alleles on the level of plasma viremia and CD4 count was investigated in a South African cohort [15].

10.2 Identification of HIV-1 CTL Epitopes

Identification of HIV-1 epitopes recognized by CTL is necessary for studies to investigate the role of HIV-1-specific CTLs in HIV-1 infections. This identification has not been easy because the epitope had to be determined by the following steps using HIV-1-specific CTL clones or lines. First, the target HIV-1 protein of the HIV-1-specific CTLs was identified by testing whether HIV-1-specific CTLs could kill target cells infected with vaccinia recombinant virus containing an HIV-1 gene encoding one HIV-1 protein or a part of one. Second, the epitope was identified by using overlapping synthetic peptides covering a part of one HIV-1 protein. Recently, a method directly using overlapping synthetic peptides or a technique called reverse immunogenetics has been used. Although many HIV-1 epitopes have been identified, the number of identified HIV-1 CTL epitopes still remains restricted.

10.2.1
Identification of HIV-1 CTL Epitopes by a Strategy Using Overlapping Peptides

HLA class I molecules present 8- to 11-mer peptides to T cells [19]. Peptides longer than 11-mer can bind to HLA class I molecules and are recognized by epitope-specific T cells if the peptide includes the T-cell epitope. These observations suggested that CTL epitopes could be identified by a strategy using overlapping peptides of more than 11-mer in length. A typical method using 15-mer overlapping peptides is shown in Figure 10.3. 15-mer HIV-1 sequences are selected from consensus amino acid sequences of each HIV-1 protein. 15-mers with a four-amino-acid shift in each starting from the N-terminus of the HIV-1 protein are synthesized. Peripheral blood mononuclear cells (PBMCs) are isolated from HIV-1-infected individuals and are then stimulated with each 15-mer or cocktails of these 15-mer peptides. PBMCs stimulated with the peptides are cultured for 7–14 days and are then tested for IFN-γ production or CTL activity. Peptides exhibiting a positive response in the CTL assay or producing IFN-γ are analyzed further to define the minimum length of the peptide for functioning as a CTL epitope. Peptides with more than a 7-mer length deleted from the N-terminus or C-terminus of the 15-mer peptides are synthesized. IFN-γ production by or CTL activity of the specific $CD8^+$ T cells is then further tested for these shorter peptides. The minimum length of the epitopes is determined based on the results of the response of specific $CD8^+$ T cells to these peptides. HLA restriction is identified by using a panel of cells sharing HLA class I alleles with HIV-1-infected individuals in whom the specific CTLs or $CD8^+$ T cells are induced. If HLA class I gene-trans-

Figure 10.3 Identification of HIV-1 CTL epitopes by a method using overlapping HIV-1 peptides. (1) 15-mer HIV-1 sequences are selected from consensus amino acid sequences of each HIV-1 protein. The 15-mer peptides, each with a four-amino-acid shift from the N-terminus of the HIV-1 protein, are synthesized. (2) PBMCs from HIV-1-infected individuals are isolated and are then stimulated with each 15-mer or cocktails of these 15-mer peptides. PBMCs stimulated with the peptides are cultured for 7–14 days and are then tested for IFN-γ production or CTL activity. Peptides (e.g., Gag-2 and Gag-3) showing a positive response in the CTL assay or eliciting IFN-γ production are selected for further analysis to define the minimum length of the epitope. Because the 11-mer RSLYNTVATLY is found in both Gag-2 and Gag-3, this sequence was selected for further analysis. (3) Peptides of more than 7-mer length with amino acids deleted from the N-terminus or C-terminus of the 11-mer RSLYNTVATLY are synthesized. IFN-γ production or CTL activity of specific $CD8^+$ T cells is further tested for these shorter peptides. The minimum length of the epitope (SLYNTVATL) is determined based on results of the responses of specific $CD8^+$ T cells to these peptides. (4) HLA restriction molecules are identified by using a panel of cells sharing HLA class I alleles with those of HIV-1-infected individuals in whom specific CTLs or $CD8^+$ T cells are induced. HLA-A*0201 is found to be a restriction molecule because the specific CTLs showed a positive response only to cells carrying HLA-A*0201. (5) Finally, to determine whether the peptide is recognized as a naturally occurring peptide by specific CTLs or $CD8^+$ T cells, cells infected with HIV-1 or HIV-1 recombinant vaccinia virus are tested for IFN-γ production or CTL activity.

▶

1. Synthesis of 15-mer overlapping peptides

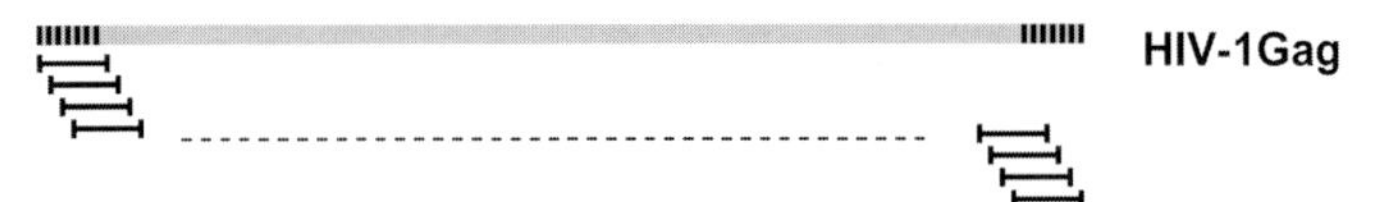

2. Induction of IFN-γ-producing CD8⁺T cells or CTL

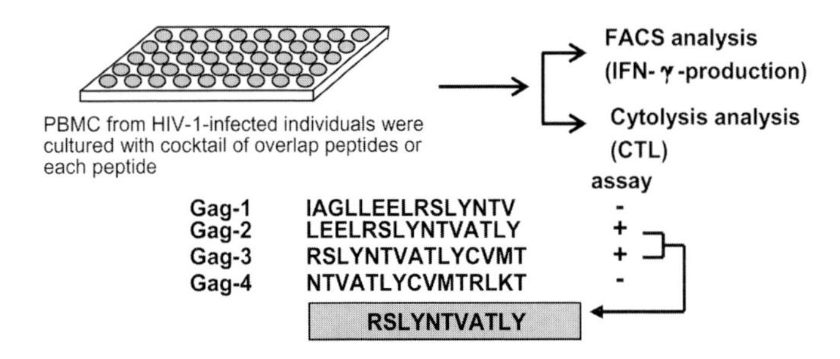

3. Definition of epitopes carrying optimal length

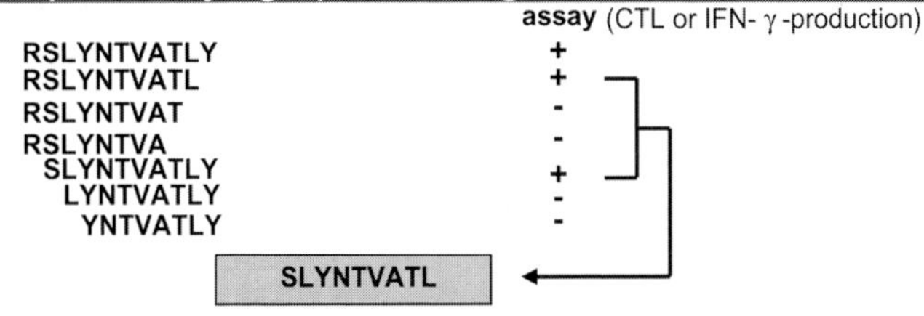

4. Identification of restricted HLA molecules

HLA alleles of HIV-1-infected individuals in whom specific CD8⁺T cells were detected: HLA-A*0201, -A*1101, -B*3501, -B*5101, -Cw0301, and Cw0401

use of HLA class I panel cells carrying one or more of these alleles for target (CTL assay) or stimulator (IFN-γ-production)

		assay
A	A*0301, A*1101, -B*4403, -B*5201, -Cw0302, Cw0702	-
B	A*2601, A*3303, -B*3501, -B*0702, Cw0401, Cw0702	-
C	A*0201, A*1101, -B*3901, -B*5301, -Cw0302, Cw0702	+
D	A*0201, A*2601, -B*5201, -B*5401, -Cw0702, Cw0801	+
E	A*1101, A*3101, -B*5101, -B*5201, -Cw0302, Cw0702	-

HLA restriction: A*0201

5. Confirmation of HIV-1 epitopes

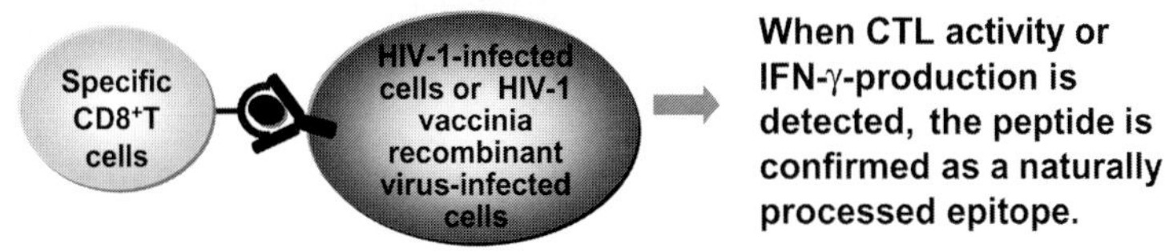

When CTL activity or IFN-γ-production is detected, the peptide is confirmed as a naturally processed epitope.

fected cells are available, they can be used to identify the HLA class I restriction. Finally, to determine whether the peptide is recognized as a naturally occurring peptide by specific CTLs or $CD8^+$ T cells, IFN-γ production by or CTL activity of specific T cells is tested for cells infected with HIV-1 or HIV-1 recombinant vaccinia virus.

1. Identification of motif of HLA-B*3501-binding peptides

Purification of HLA-B*3501 molecules → Isolation of B*3501 binding peptides → Analysis of amino acid sequences →

position

1	2	3	4	5	6	7	8	9
	P							Y
								F
								M
								L
								I

2. Synthesis of HIV-1 8-mer to 10-mer peptides carring B*3501 binding peptide motif

M D L P G R W K P V Y I G G R I K I R Q Y D Q I P I M D I C G H K F G P T P V N I I G R N L V T Q I

L P G R W K P V Y

I P I M D I C G H K F

T P V N I I G R N L

G P T P V N I I

3. Identification of B*3501-binding HIV-1 peptides by a peptide binding assay

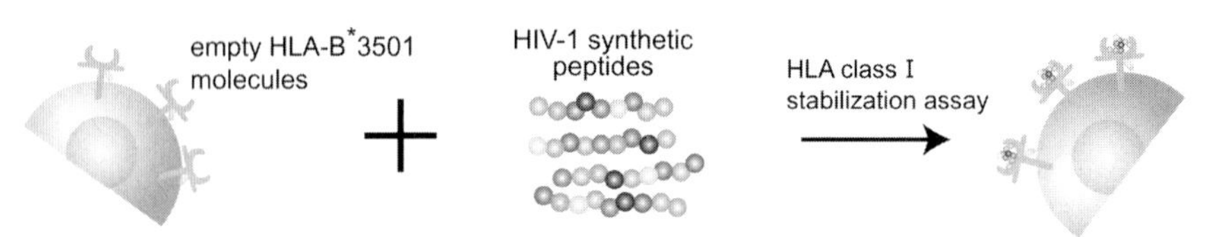

4. Induction of HIV-1 peptide-specific CTLs in PBMC from HIV-1-infected, HLA-B*3501+ Patients

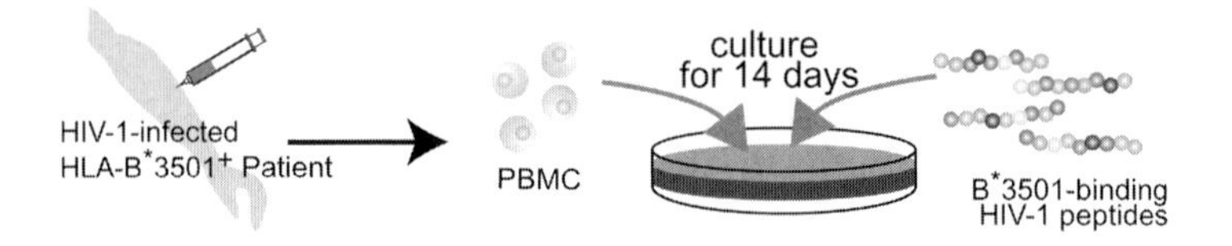

5. Killing of HIV-1-infected cells or HIV-1-recombinant vaccinia-infected cells by peptide-specific CTLs

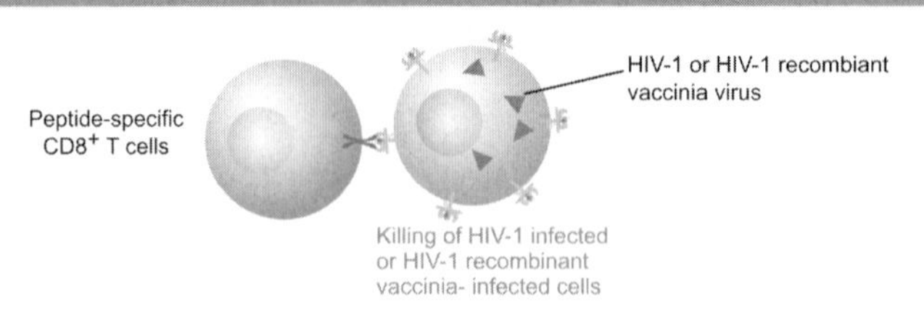

10.2.2
Identification of HIV-1 CTL Epitopes by the Strategy of Reverse Immunogenetics

HLA class I–binding peptides carry an allele-specific motif [19]. For example, HLA-B*3501-binding peptide possesses Pro at position 2 and the aromatic hydrophobic residues Tyr, Phe, Met, Leu, and Ile at its C-terminus, while HLA-A*2402 has Tyr at position 2 and Ile, Leu, and Phe at its C-terminus. This suggests that CTL epitopes also carry an allele-specific motif. A method employing this characteristic of HLA class I–binding peptides, which is called reverse immunogenetics, was used to identify CTL epitopes. A typical example of the identification of CTL epitopes by using reverse immunogenetics is shown in Figure 10.4. In the first step, the motif of a given HLA class I–binding peptide is identified. For example, HLA-B*3501 molecules are isolated from cell lines transfected with the HLA-B*3501 gene. Peptides are eluted from isolated HLA-B*3501 molecules, and then the eluted peptides are sequenced to determine their motif. In the second step, 8-mer to 11-mer sequences carrying the anchor residues are selected from the sequences of each HIV-1 protein. Synthesized HIV-1 peptides are then tested for binding to HLA-B*3501 by a peptide-binding assay such as the HLA class I stabilization assay [20]. In the third step, HLA-B*3501-binding peptides are further used to induce peptide-specific CTLs from PBMCs from HLA-B*3501-positive, HIV-1-infeceted individuals. The PBMCs are stimulated with each HLA-B*3501-binding peptide or cocktails of the peptides. Peptide-stimulated PBMCs are cultured for approximately 14 days. Peptide-specific CTLs and $CD8^+$ T cells are then identified

◀ **Figure 10.4** Identification of HLA-B*3501-restricted HIV-1 CTL epitopes by a method using reverse immunogenetics. (1) HLA-B*3501 molecules are isolated from HLA-negative cell lines transfected with the HLA-B*3501 gene. Peptides are eluted from isolated HLA-B*3501 molecules, and then the eluted peptides are sequenced to determine Immunodominant HIV-1 Epitopes Presented by HLA Allelesthe motif of HLA-B*3501-binding peptides. HLA-B*3501-binding peptide possessed Pro at position 2 and hydrophobic residues Tyr, Phe, Met, Leu, and Ile at the C-terminus. (2) 8-mer to 11-mer HIV-1 sequences carrying the HLA-B*3501 anchor residues at position 2 and the C-terminus are selected and synthesized. (3) Synthesized HIV-1 peptides are tested for binding to HLA-B*3501 by a peptide-binding assay such as the HLA class I stabilization assay. (4) HLA-B*3501-binding peptides are further used to induce peptide-specific CTLs from PBMCs of HLA-B*3501-positive, HIV-1-infected individuals. The PBMCs are stimulated with each HLA-B*3501-binding peptide or cocktails of the peptides. Peptide-stimulated PBMCs are cultured for approximately 14 days. Peptide-specific CTLs or $CD8^+$ T cells are identified by measuring the cytotoxic activity of cultured PBMCs toward peptide-pulsed cells or by measuring the production of IFN-γ by $CD8^+$ T cells in cultured PBMCs stimulated with peptide-pulsed cells. The peptides showing a positive response are considered epitope candidates. (5) To clarify whether the peptides are recognized as naturally occurring peptides by specific CTLs or $CD8^+$ T cells, cells infected with HIV-1 or HIV-1 recombinant vaccinia virus are used to stimulate IFN-γ production or for CTL activity. When peptide-specific T-cell clones or lines kill target cells infected with HIV-1 or HIV-1 recombinant vaccinia virus or produce IFN-γ after being stimulated with cells infected with HIV-1 or HIV-1 recombinant vaccinia virus, the peptides that these T cells recognize are concluded to be naturally occurring HIV-1 epitope peptides. (This figure also appears with the color plates.)

by measuring the cytotoxic activity of the cultured PBMCs toward peptide-pulsed cells or by measuring the production of IFN-γ by CD8$^+$ T cells in cultured PBMCs stimulated with peptide-pulsed cells, respectively. In the fourth step, to clarify whether the peptides are recognized as naturally occurring peptides by specific CTLs or CD8$^+$ T cells, IFN-γ production by or CTL activity of specific T cells is examined by using cells infected with HIV-1 or HIV-1 recombinant vaccinia virus. Peptide-specific T-cell clones or lines are established from the peptide-stimulated bulk cultured cells. These T-cell clones or lines are then tested for their ability to kill the target cells infected with HIV-1 or HIV-1 recombinant vaccinia virus or to produce IFN-γ after stimulation with cells infected with HIV-1 or HIV-1 recombinant vaccinia virus. When the specific T cells kill target cells infected with HIV-1 or HIV-1 recombinant vaccinia virus or produce IFN-γ that has been stimulated with cells infected with HIV-1 or HIV-1 recombinant vaccinia virus, the peptides that these T cells recognize are concluded to be naturally occurring HIV-1 epitope peptides. This method has been used to identify many HIV-1 epitope peptides presented by several HLA class I alleles such as HLA-B*3501, -A*2402, -B*5101, -A*1101, -B*3303, and -A*2601 [21–27].

10.3 Immunodominant HIV-1 Epitopes Presented by HLA Alleles Associated With Slow Progression to AIDS and Their Escape Mutants

It is well known that three HLA class I alleles, HLA-B57, -B27, and -B51, are associated with a slow progression to AIDS [28]. It is therefore speculated that CTLs specific for HIV-1 epitopes presented by these HLA molecules effectively suppress HIV-1 replication in long-term non-progressors (LTNPs) and slow progressors carrying these alleles. However, the majority of HIV-1-infected individuals with these alleles are progressors. Analysis of CTLs specific for epitopes presented by these HLA molecules not only in LTNPs and slow progressors but also in progressors is expected to clarify the mechanisms as to how HIV-1 replication is effectively suppressed in LTNPs and slow progressors and how HIV-1 escapes from HIV-1-specific CTLs in progressors.

10.3.1 HLA-B*57-restricted Immunodominant Epitopes and Their Escape Mutants

TW10 (TSTLQRQIAW: Gag HXB2 residues 240–249) is one of the well-known immunodominant Gag epitopes presented by HLA-B*57. CD8$^+$ T cells specific for this epitope were predominantly detected in acute HIV-1-infected HLA-B*57$^+$ individuals [29]. This epitope is also presented by HLA-B*58, which has a very similar structure. The accumulation of mutations within this epitope was detected in HIV-1-infected HLA-B*57$^+$ and -B*58$^+$ individuals, indicating a positive selection for these mutations by HLA-B*57- or -B*58-restricted CTLs [30]. The mutation of Thr to Asn at residue 242 (T242N) was predominantly found in these patients with chronic HIV-1 infection. This mutation significantly reduced the recognition of the specific T cells for the epitope. If these HIV-1 mutants are transmitted to individuals who do not carry HLA-B*57 or -B*58, the mutation at residue 242 reverts to the wild type [30].

IW9 (ISPRTLNAW: Gag HXB2 residues 147–155) is also an HLA-B*57-restricted immunodominant Gag epitope [31]. CD8$^+$ T cells specific for it were detected in half of the acute HIV-1-infected HLA-B*57$^+$ individuals examined [29]. Two major mutations were found at residues 146 and 147 in chronically HIV-1-infected HLA-B*57$^+$ individuals [32]. Sixty-one percent of HLA-B*57$^+$ individuals with a chronic HIV-1 infection had an A146P mutation in the flanking region, whereas only 14% of non-HLA-B*57 individuals chronically infected with HIV-1 had the same mutation. The IW9-specific CTL clones failed to recognize CD4$^+$ T cells infected with HIV-1 carrying this mutation [32], indicating that the mutation in the flanking region can alter the antigen processing of this epitope. Thus, IW9-specific immune selection pressure leads to a mutation in the flanking region of the epitope affecting antigen presentation.

KF11 (KAFSPEVIPMF: Gag residues 30–40) is a third HLA-B*57-restricted Gag epitope. The KF11-specific CD8$^+$ T cells were identified in LTNPs [33, 34], being predominantly detected in acute HIV-1-infected HLA-B*57$^+$ individuals [29]. Most KF11-specific CTL clones can effectively recognize the short peptide KF8 (KAFSPEVI), which is a part of the KF11 sequence. On the other hand, the existence of KF11-specific or KF8-specific CTL clones that fail to recognize the other epitope indicates that both epitopes can be independently recognized by T cells [35]. This epitope is conserved in HIV-1 isolates from both LTNPs and progressors [36, 37], suggesting the weak immune pressure by CTLs specific for this epitope. After stimulation with the epitope peptide, KF11-specific CD8$^+$ T cells produced a greater amount of IFN-γ than did two HLA-A*02-restricted and HLA-B*08-restricted, Gag-specific CTLs [35], implying that high responsiveness of KF11-specific CD8$^+$ T cells may contribute to the protective effect of HLA-B57 in HIV-1 infection.

QW9 (QASQEVKNW: Gag residues 308–316) is a fourth Gag epitope. QW9-specific CD8$^+$ T cells have been rarely detectable in acute HIV-1-infected HLA-B*57$^+$ individuals [29]. In contrast, they have been detected frequently in chronic patients, being more detectable in LTNPs than in progressors [36]. A glutamic acid-to-aspartic acid mutation at residue 312 (E312D) has been found more frequently in progressors than in LTNPs. In two of four progressors who had only HIV-1 carrying the E312D sequence in their plasma, their CD8$^+$ T cells produced IFN-γ after stimulation with QW9 peptide, but not after stimulation with E312D mutant peptide. In the two other progressors, the CD8$^+$ T cells produced IFN-γ after having been stimulated with both QW9 peptide and E312D mutant peptide. These findings indicate that E312D is an escape mutant in some patients but not in others [35].

Because Gag-specific CD8$^+$ T cells are much more frequently detected in HLA-B*57$^+$ patients with chronic HIV-1 infection than are CD8$^+$ T cells specific for other proteins, such HIV-1 protein-specific CD8$^+$ T cells have not been analyzed in detail. IW9 (IVLPEKDSW: RT residues 244–252)-specific and HQ10 (HTQGYFPDWQ)-specific CD8$^+$ T cells are frequently detectable in acute HIV-1-infected HLA-B*57$^+$ individuals [29]. $CD8^+$ T cells specific for several epitopes of Nef, Rev, Vif, and Vpr are also detectable in acute HIV-1-infected HLA-B*57^+ individuals.

Both the Caucasian allele HLA-B*5701 and the African allele HLA-B*5703 are associated with a slow progression to AIDS. The KF11 (KAFSPEVIPMF) epitope from subtype B has been well characterized in HLA-B*5701^+ Caucasian patients. Several mutants were detected in subtype A and C viruses. They include 2G4N (KGFNPEVIPMF), 3L (KALSPEVIPMF), 10V (KAFSPEVIPMV), 7I (KAF-SPEIIPMF), 5Q (KAFSQEVIPMF), and 2N (KNFSPEVIPMF). KF11-specific, B*5701- or B*5703-restricted CTL clones effectively recognized 2G4N, 3L, 10V, 7I, and 2N variants [38]. These mutant peptides also induced specific $CD8^+$ T cells in HLA-B*5703 donors. Thus, both HLA-B*57 alleles can effectively present KF11 and its mutants.

10.3.2 HLA-B*27-restricted Immunodominant Epitopes

Gag p24 epitope KK10 (KRWIILGLNK, residues 131–140) is a well-known HLA-B*27-restricted immunodominant epitope [39]. This epitope is presented by HLA-B*2705 [40]. KK10-specific CTLs were detected in primary HIV-1-infected patients [40] and chronically infected patients [10]. Several mutations of this epitope have been reported in HIV-1 isolates from HIV-1-infected donors carrying HLA-B*27 [41]. K2M6 (KKWILMGLNK) is a well-known mutant. HIV-1 variants carrying this mutation were observed in two chronically HIV-1-infected hemophiliacs 9–12 years after the primary infection. When the viral load became very high, the proportion of the variant virus increased to 100% [10]. K2M6 is an escape mutant because KK10-specific CTLs failed to recognize K2M6 peptide. The G2 (KGWIILGLNK) mutant of T2 epitope (KTWILMGLNK) has been reported, and it is also known as an escape mutant [42]. This mutant was detected in a perinatally HIV-1-infected child when the viral load was increased at 9.4 years of age. These peptides poorly bound to HLA-B*2705 [43], suggesting that the poor binding of the peptides to HLA-B*2705 results in loss of the specific T-cell recognition.

10.3.3 HLA-B*51-restricted Immunodominant Epitopes

HLA-B*51 is also associated with a slow progression to AIDS [28]. Three HLA-B*5101 epitopes (Pol 283–8: TAFTIPSI; Pol 743–9: LPPVVAKEI; and Gag 327–9: NANPDCKTI) were suggested as immunodominant epitopes because the specific CTLs were frequently detected in patients with chronic HIV-1 infection [23]. Pol 283–8-specific and Pol 743–9-specific CTLs more effectively suppress HIV-1 replication than do Gag 327–9-specific ones [44], suggesting that Pol 283–8 and Pol 743–9 are much stronger epitopes than is Gag 327–9. Pol 743–9 and Gag 327–9 were conserved among HIV-1 clade B isolates, whereas several mutants of Pol 283–8 were observed in clade B isolates (HIV sequence database, Los Alamos National Laboratory). The sequence analysis of these epitopes in HIV-1 isolates from HLA-B*5101^{+} and HLA-B*5101^{-} HIV-1-infected individuals will clarify whether escape mutants appear in HLA-B*5101^{+} donors.

10.4 Immunodominant HIV-1 Epitopes Presented by HLA Alleles Associated With Rapid Progression to AIDS

HLA-B*35 is well known as an HLA allele associated with a rapid progression to AIDS [14]. A previous study revealed that HLA-B*3501 can present more than nine epitopes [18], suggesting that this association is not due to a failure of HIV-1 epitope presentation by HLA-B*3501. Three B*35 subtypes, B*3502, B*3503, and B*3504, are much more strongly associated with AIDS progression than is B*3501 [17]. The only difference between B*3501 and the other three subtypes was in residue 116 on the floor of the F-pocket. HLA-B*3501 and -B*3502 carry Ser and Tyr, respectively, at residue 116. The Tyr substitution at this residue critically affected the binding of peptides carrying Tyr at the C-terminus [45]. Pol 587–596 (EPIVGAETFY), Pol 273–282 (VPLDKDFRKY), and Nef 75–85 (DPNPQEVVL) of the HLA-B*3501 epitopes are thought to be immunodominant because CTLs specific for them were detected in approximately half of the chronically HIV-1-infected individuals tested [18]. It is therefore speculated that HLA-B*3502 may fail to present two immunodominant epitopes: Pol 587–596 and Pol 273–282 epitopes. However, the mechanism underlying the stronger association of these three HLA-B*35 subtypes with a rapid progression still remains unknown.

10.5 Immunodominant HIV-1 Epitopes Presented by Other HLA Alleles

10.5.1 HLA-A*02-restricted Immunodominant Epitopes

SL9 (Gag 77–85: SLYNTVATL) was previously identified as an HLA-A*0201-restricted immunodominant epitope [46, 47]. SL9-specific CTLs were induced in approximately 60–75% of the chronically HIV-1-infected HLA-A*02^{+} individuals examined [48–50], whereas they were not induced in HLA-A*02^{+} individuals with an acute HIV-1 infection [51]. The fact that none of the 13 HIV vaccine recipients with documented anti-Gag CTL reactivity had SL9-specific responses may also support the data showing that SL9-specific CTLs cannot be induced in acute HIV-1 infection [52]. On the other hand, the magnitude of the HLA-A*0201-restricted, SL2-specific response has been negatively associated with viral load [53]. The escape mutations of this epitope and at the flanking region of the epitope arise during the course of chronic infection [49, 54]. These findings suggest that the

SL9-specific CTL response is one with some selection pressure on the virus. Since the mutations in this epitope were found only in the chronic phase, the immune pressure may be weak.

10.5.2
HLA-B*08-restricted Immunodominant Epitopes

HLA-B*0801 is associated with neither a slow nor a rapid progression to AIDS. However, FL8 (Nef 90–97: FLKEKGGL) is thought to be an immunodominant epitope because FL-8-specific CTLs were detected in 100% of HLA-B*0801^{+} slow progressors and LTNPs [55]. Analysis of FL8-specific CTLs from LTNPs showed that the specific CD8^{+} T cells express Vβ13.2 carrying long β-chain CDR3 regions in and that they are resistant to apoptosis to contrast with FL8-specific CTLs expressing other TCRs in progressors [56]. These findings suggest that these CTLs may be linked with a better clinical outcome. But, so far the T cells possessing such a TCR have been found only among FL8-specific CTLs. Analysis of TCRs in other dominant epitope–specific CTLs will be required to confirm this phenomenon.

10.6
Escape Mutations and Viral Fitness

It is well known that escape mutations can be observed in both acute and chronic phases of an HIV-1 infection. However, the escape mutations were most commonly detectable late in the course of infection in both humans and the SIV-macaque models. For example, HLA-B*27-restricted, KK10-specific CTLs are associated with effective control of viremia, whereas the escape mutant of this epitope, which appears late in the course of infection, seems to be coincident with loss of this immune-mediated control [10]. This fact implies that the escape mutants affecting effective immune control do not appear in early HIV-1 infection. A similar finding was also reported for the Mamu-A*01-restricted Gag epitope CM9 [57].

There are several mechanisms of mutational escape of HIV-1 epitopes in the pathway of antigen presentation (Figure 10.5). A mutation at the C-terminus of the HLA-A*03-restricted epitope may affect proteasomal processing [58]. An Ala-to-Pro mutation just before the N-terminus of HLA-B*57-restricted epitope ISW9 (ISPRTLNAW) prevented the aminopeptidase ERAAP from cleaving the glutamine residue from the 11-mer precursor QP ISPRTLNAW [32]. Loss of peptide binding to restriction HLA molecules and loss of TCR recognition are commonly detected in escape mutants.

In the case of the KK10 epitope, the most common escape mutations are the substitutions of arginine with lysine at residue 264 (R264K) and of leucine with methionine at residue 268 (L268M). The R264K mutation follows the L268M one in every case [10]. Because the former critically affects the binding to HLA-B*27, this mutation should be critical for T-cell recognition. On the other hand, the L268M mutation seems to occupy a crucial position in the capsid protein, suggest-

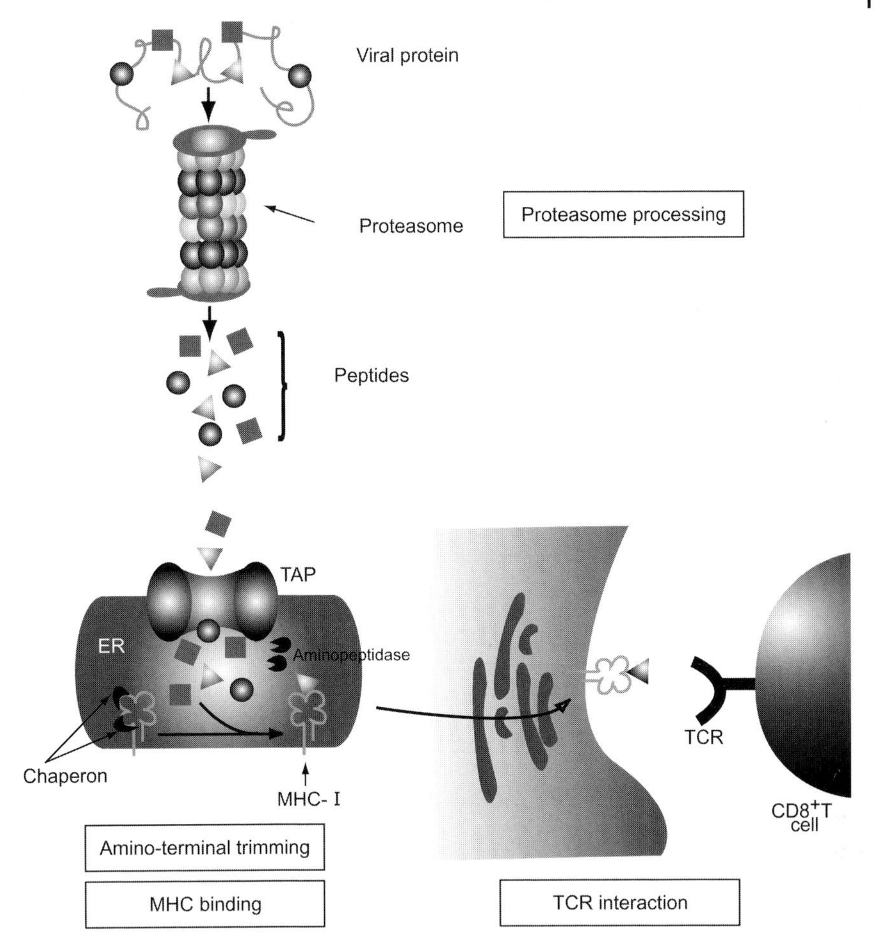

Figure 10.5 Mechanisms of mutational escape of HIV-1 epitopes in the antigen presentation pathway. Escape mutants affect the four steps of antigen presentation: proteasome processing, amino-terminal trimming, MHC binding, and TCR recognition.

ing that this mutant has a marked effect on the ability of the virus to replicate and is involved in the fitness cost. This compensatory mutation may explain why escape mutations appear late in the infection.

The reversion of an escape mutant to the wild type is known in HLA mismatched hosts (Figure 10.6). In 75% of HLA-B*5701$^+$ or B*5801$^+$ individuals, escape mutant T242N of the HLA-B*57-restricted Tw10 epitope was detected in the infection. In approximately 30% of HLA-B*5701$^+$ or B*5801$^+$ individuals, a mutation of glycine with alanine (G248A) was also detected at position 9 of the epitope, either alone or in combination with the T242N mutant [30]. Interestingly, when these escape mutants were transmitted to non-HLA-B*5701/B*5801 indi-

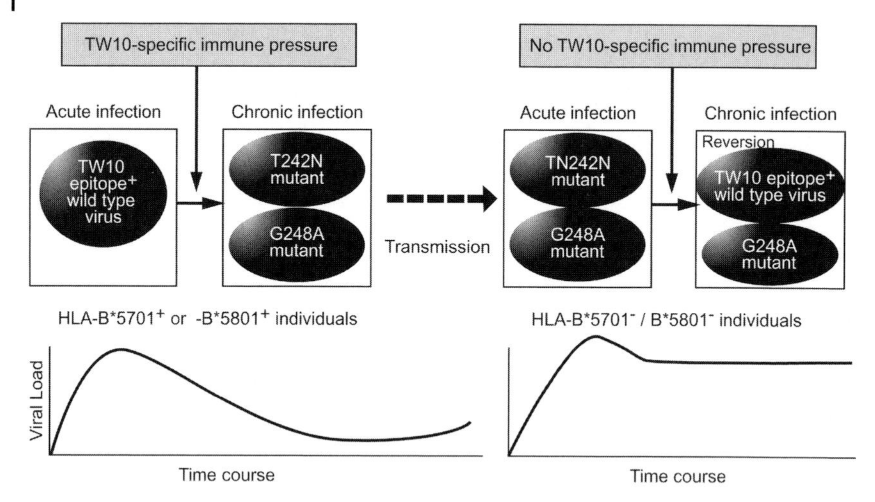

Figure 10.6 Appearance of an escape mutant and reversion of an escape mutant. The escape mutants T242N and G248A of HLA-B*5701/B*5801 epitope TW10 were found in HLA-B*5701 or HLA-B*5801 donors with chronic infection. Reversion of the T242N mutation was found when these mutants were transmitted to non-HLA-B*5701/B*5801donors, whereas G248A mutation was maintained after the transmission.

viduals, the reversion of T242N to the wild type was observed, but the G248A mutant persisted [30]. Thus, T242N is an example of an escape mutant that has a fitness cost for the virus.

10.7 Effect of Nef-mediated HLA Class I Downregulation on Recognition of HIV-1-infected CD4⁺ T Cells by HIV-1-specific CD8⁺ T Cells

It is well known that the surface expression of HLA class I molecules on HIV-1-infected cells is downregulated [59–61]. This downregulation is mediated by the function of several HIV-1 proteins, i.e., Nef, Tat, and Vpu, with the effect of Nef protein being much stronger than that of the others [62–64]. The exact mechanism of Nef-mediated disruption of the surface expression of HLA class I molecules remains unclear. The C-terminus of HLA class I molecules is involved in Nef-mediated HLA class I downregulation. A study using HIV-1 carrying a one-amino-acid Nef mutant demonstrated that residue 20 of Nef is critical for Nef-mediated HLA class I downregulation but not for other Nef functions [65], indicating that the part of the Nef protein functionally involved in the HLA class I downregulation is different from that involved in other Nef functions. These studies suggest that Nef-mediated HLA class I downregulation affects HIV-1-specific CTL recognition of HIV-1-infected cells. It has been demonstrated that HIV-1-specific CTLs effectively killed CD4⁺ T cells infected with Nef-defective HIV-1 but failed to kill the cells infected with Nef-positive HIV-1, indicating that the ability of HIV-1-specfic $CD8^+$ T cells to kill HIV-1-infected cells is affected by Nef-mediated HLA class I downregulation [13]. These studies support the finding *in vivo* that a patient infected with a Nef-defective HIV-1 mutant showed a good clinical course [66–68]. HLA-A and HLA-B allele molecules are downregulated in HIV-1-infected cells, whereas HLA-C and HLA-E molecules are not [69]. However, because HLA-C allele molecules are expressed on cell surface at a much lower density than HLA-A and HLA-B allele molecules [70], it is speculated that HLA-C molecules have a much weaker ability to present epitope peptides to T cells than do HLA-A and HLA-B molecules. On the other hand, the expression of HLA-C molecules on HIV-1-infected cells prevents NK cells from killing HIV-1-infected cells [69]. Thus, HIV-1-infected cells can escape from HIV-1-specific $CD8^+$ T cells and NK cells.

The above study showing that HLA-A*0201-restricted, HIV-1-specific CTL clones failed to kill HIV-1-infected $CD4^+$ T cells suggests that Nef-mediated HLA class I downregulation affects the recognition of HIV-1-infected $CD4^+$ T cells by HIV-1-specific CTLs [13]. The following studies using two HLA-B*3501-restricted, two HLA-A*3303-restricted, and two HLA-B*5101-restricted CTL clones confirmed the effect of Nef-mediated HLA class I downregulation on killing activity for HIV-1-infected $CD4^+$T cells and further showed that HIV-1-specific CTLs can partially suppress HIV-1 replication when they are cultured with HIV-1-infected $CD4^+$ T cells [44, 70]. They also showed that these CTL clones can produce cytokines after they are cultured with HIV-1-infected $CD4^+$ T cells for 6 hours [44]. These findings indicate that HIV-1-specific CTLs can recognize epitopes expressed on HIV-1-infected $CD4^+$ T cells, although they fail to kill these cells. Nef-mediated HLA class I downregulation affects the ability of HIV-1-infected CTLs to kill HIV-1-infected $CD4^+$ T cells *in vitro*, but it remains unknown whether these cells also failed to kill HIV-1-infected $CD4^+$ T cells *in vivo*. It is thought that HIV-1 replication is partially suppressed in chronically HIV-1-infected individuals who have less than a 10^5 viral load, because most AIDS patients have more than a 10^5 viral load. Partial suppression of HIV-1 replication by HIV-1-specific $CD8^+$ T cells found *in vitro* may account for the partial suppression of HIV-1 *in vivo*.

A recent study showed the existence of HIV-1-specific CTLs possessing a strong ability to kill HIV-1-infected $CD4^+$ T cells and to suppress HIV-1 replication [70]. Two Pol-specific, HLA-B*5101-restricted CTLs effectively killed HIV-1-infected $CD4^+$ T cells and suppressed HIV-1 replication, although the surface expression of HLA-B*5101 is downregulated by Nef in HIV-1-infected $CD4^+$ T cells. In contrast, Gag-specific and Rev-specific HLA-B*5101-restricted CTLs partially or hardly suppressed HIV-1 replication, respectively, and failed to kill HIV-1-infected $CD4^+$ T cells. Thus, the effect of Nef-mediated HLA class I downregulation seems to be epitope-dependent (Figure 10.7). This may be explained by the difference in expression level between HLA-B*5101 molecules carrying these epitope peptides (Figure 10.8). Another explanation is that these Pol-specific CTLs have TCRs with high affinity, which is advantageous for recognition of a low number of epitopes. The existence of HIV-1-specific CTLs possessing a strong ability to kill HIV-1-infected $CD4^+$ T cells and to suppress HIV-1 replication may explain the HLA-B*5101 association with a slow progression to AIDS.

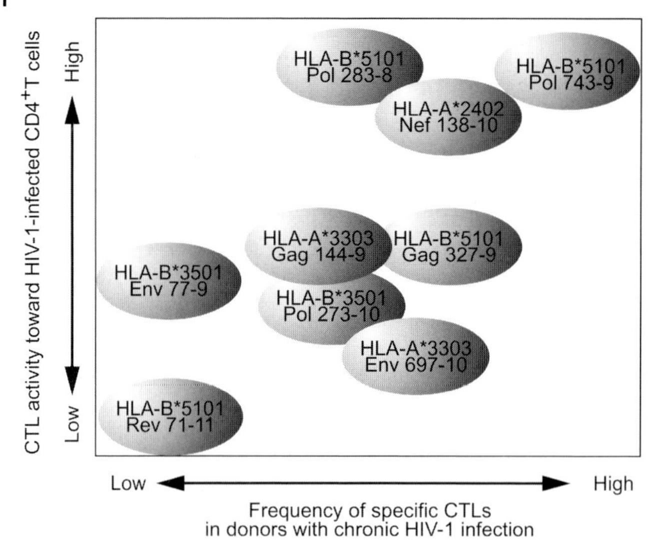

Figure 10.7 Ability of CTLs to recognize HIV-1-infected CD4⁺ T cells versus frequency of CTLs induced in individuals with chronic HIV-1 infection. The relationship between CTL activity toward HIV-1-infected CD4⁺ T cells and the frequency of specific CTLs in donors with chronic HIV-1 infection is shown. Pol 743–9-, Nef 138–10-, and Pol 283–8-specific CD8⁺ T cells exhibited high CTL activity and high frequency in donors with chronic HIV-1 infection. CTL activity was evaluated by measuring killing activity for HIV-1-infected CD4⁺ T cells and suppression activity of HIV-1 replication.

10.8
Skewed Maturation of HIV-1-specific CD8⁺ T Cells

HLA class I tetramers are a very useful tool for detecting antigen-specific CD8⁺ T cells by using flow cytometry [53, 71–73]. *Ex vivo* analysis using the tetramers demonstrated that human cytomegalovirus (HCMV)-specific CD8⁺ T cells, which can effectively kill target cells, in both healthy and chronically HIV-1-infected individuals expressed a high level of perforin [12, 74, 75]. In contrast, HIV-1-specific CD8⁺ T cells showed a low level of perforin expression [12, 74, 76]. These observations suggest skewed maturation of HIV-1-specific CD8⁺ T cells in the HIV-1-infected host.

Many studies have attempted to phenotypically classify human CD8⁺ T cells [75, 77–84]. Particular expression patterns of the costimulatory receptors CD27 and CD28 as well as CD45RA or CD45RO are associated with the naïve, memory, and effector function of human CD8⁺ T cells. Effector and memory/effector CD8⁺ T cells have a $CD28^-CD45RA^+$ or $CD27^-CD45RA^+$ and $CD28^-CD45RA^-$ or $CD27^-CD45RA^-$ phenotype, respectively, whereas naïve and memory CD8⁺ T cells have a $CD28^+CD45RA^+$ or $CD27^+CD45RA^+$ and $CD28^+CD45RA^-$ or $CD27^+CD45RA^-$ phenotype, respectively [77, 84]. In addition, a recent study on human CD8⁺ T cells by use of multicolor flow cytometric analysis showed that $CD27^+CD28^-CD45RA^-CD8^+$ T cells have cytotoxic activity and can effectively produce

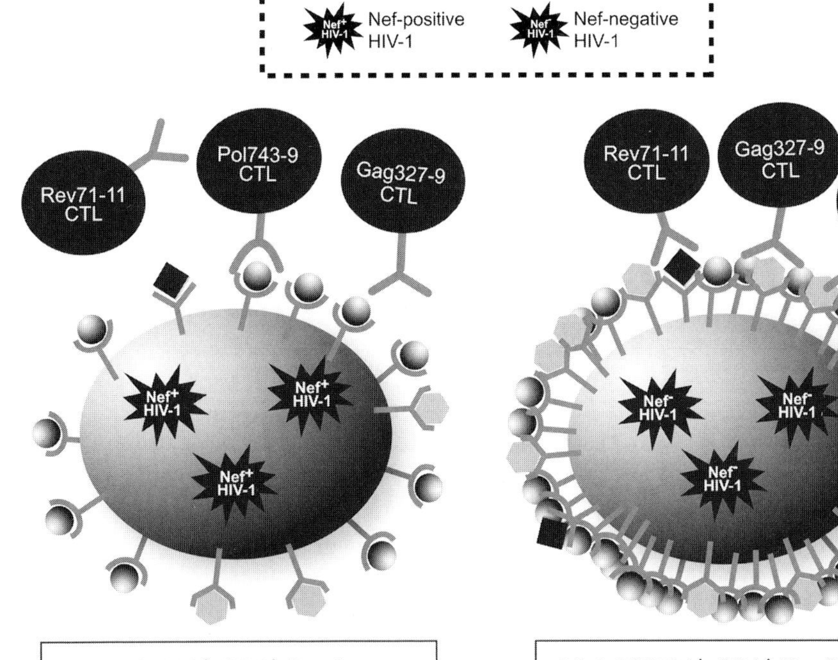

HLA-B*5101⁺ CD4⁺ T cells infected with Nef-positive HIV-1

		Activity for suppression of HIV-1 replication
Pol743-9-specific CTL	:	strong
Gag327-9-specific CTL	:	weak
Rev71-11-specific CTL	:	no

HLA-B*5101⁺ CD4⁺ T cells infected with Nef-negative HIV-1

		Activity for suppression of HIV-1 replication
Pol743-9-specific CTL	:	strong
Gag327-9-specific CTL	:	strong
Rev71-11-specific CTL	:	weak

Figure 10.8 Downregulation of HLA-A and HLA-B molecules affects the recognition of HIV-1-specific $CD8^+$ T cells. HLA-A and HLA-B molecules are downregulated by HIV-1 Nef. Nef-mediated HLA class I downregulation affects the ability of HIV-1-specific CTLs to kill Nef-positive, HIV-1-infected $CD4^+$T cells and to suppress Nef-positive HIV-1 replication. In three HLA-B*5101-restricted CTLs, Pol 743–9-specific and Gag 327–9-specific CTLs effectively suppressed Nef-negative HIV-1 replication, whereas Rev 71–11-specific CTLs weakly suppressed it. In contrast, Pol 743–9-specific CTLs effectively suppressed Nef-positive HIV-1 replication, whereas Gag 327–9-specific CTLs partially suppressed it and Rev 71–11-specific CTLs failed to suppress it. Comparison analysis of the ability of these peptides to bind HLA-B*5101 suggests that the expression level of these peptides on HIV-1-infected $CD4^+$ T cells may determine the cytotoxic activity of the CTLs [70].

cytokines, suggesting that this phenotype is that of memory/effector $CD8^+$ T cells [75]. In another study, human $CD8^+$ T cells were classified into four major subsets according to CD27 and CD28 expression: naïve ($CD27^+CD28^+$), early ($CD27^+CD28^+$), intermediate ($CD27^+CD28^-$), and late ($CD27^-CD28^-$); however, naïve and early subsets were mostly discriminated by the expression of CD45RA and CCR7 [74].

The chemokine receptor CCR7 is very useful for discriminating naïve and memory $CD8^+$ T cells from effector-memory and effector $CD8^+$ T cells [85, 86]. CCR7 functions as a homing receptor and is expressed in naïve $CD8^+$ T cells and in a subset of memory $CD8^+$ T cells. These studies showed the following classification of $CD8^+$ T cells using CCR7 and CD45RA: naïve, $CCR7^+CD45RA^+$; central/memory, $CCR7^+CD45RA^-$; effector/memory, $CCR7^-CD45RA^{+/-}$. CCR5 is also a useful chemokine receptor for discriminating naïve $CD8^+$ T cells from other $CD8^+$ T cells. Recent studies showed that the CCR5 chemokine receptor, whose ligands are RANTES, MIP-1α, and MIP-1β, is not expressed in naïve $CD8^+$ T cells but is expressed in memory, memory/effector, and effector $CD8^+$ T cells [77, 87], with the number of $CCR5^+CD8^+$ T cells decreasing during differentiation of $CD27^+CD28^+CD45RA^-$ T cells to $CD27^-CD28^-CD45RA^-$ T cells [77].

Phenotypes of HIV-1-specific, EBV-specific, and HCMV-specific $CD8^+$ T cells were compared to investigate the maturation stages of these $CD8^+$ T cells. Phenotypical analysis of EBV-specific $CD8^+$ T cells exhibited that they expressed the phenotype of $CD27^+CD28^+CD45RA^-CCR7^-CCR5^+$ in healthy and HIV-1-infected indi-

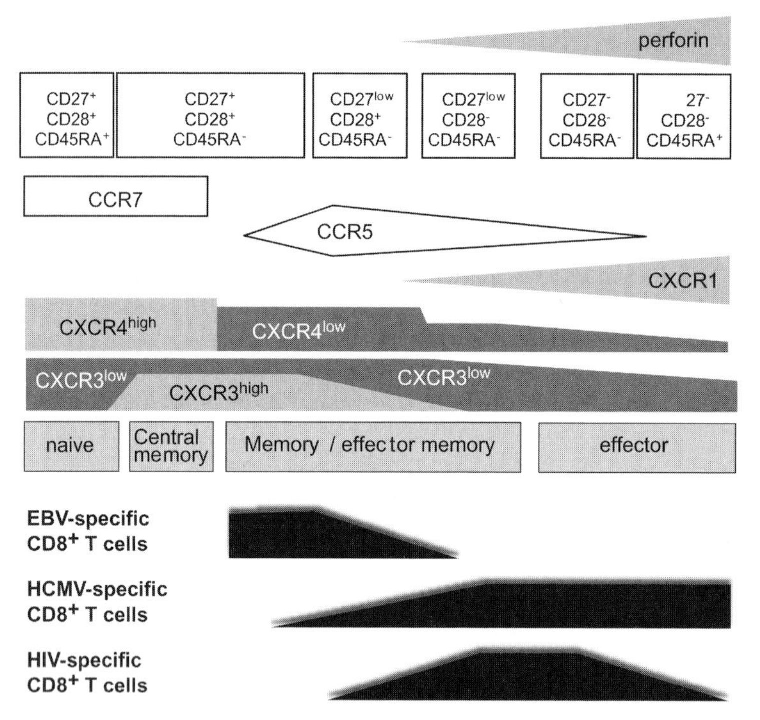

Figure 10.9 Differentiation pathway of human $CD8^+$ T cells and skewed maturation of HIV-1-specific $CD8^+$ T cells. In healthy and HIV-1-infected individuals, EBV-specific $CD8^+$ T cells exhibit the $CD27^+CD28^+CD45RA^-CCR7^-CCR5^+$ memory phenotype, whereas HCMV-specific $CD8^+$ T cells show the $CD27^-CD28^-CD45RA^{+/-}CCR7^-CCR5^{+/-}$ effector phenotype. In donors with a chronic HIV-1 infection, HIV-1-specific $CD8^+$ T cells predominantly exhibit the $CD27^{-/low}CD28^-CD45RA^{+/-}CCR7^-CCR5^{+/-}$ phenotype, suggesting that HIV-1-specific $CD8^+$ T cells are less differentiated than HCMV-specific $CD8^+$ T cells and undergo skewed maturation.

viduals [74, 77]. These CD8$^+$ T cells failed to kill the target cells [77]. These observations indicate that EBV-specific CD8$^+$ T cells are mostly memory T cells. Similar analysis of HCMV-specific CD8$^+$ T cells showed that they predominantly expressed the CD27$^-$CD28$^-$CD45RA$^{+/-}$CCR7$^-$CCR5$^{+/-}$ phenotypes in healthy and HIV-1-infected individuals [74, 75]. Isolated HCMV-specific CD8$^+$ T cells effectively killed the target cells, indicating them to be effector T cells. On the other hand, HIV-1-specific CD8$^+$ T cells predominantly expressed the phenotype of CD27$^{-/low-}$CD28$^-$CD45RA$^{+/-}$CCR7$^-$CCR5$^{+/-}$, although a small portion of these cells carry the CD27$^-$CD28$^-$CD45RA$^{+/-}$CCR7$^-$CCR5$^{+/-}$ effector phenotype [74, 86, 88]. These CD8$^+$ T cells had a weak ability to kill target cells [74, 86]. These observations suggest that maturation of HIV-1-specific CD8$^+$ T cells is impaired in chronically HIV-1-infected individuals (Figure 10.9).

References

1 Koup, R. A., Safrit, J. T., Cao, Y., Andrews, C. A., McLeod, G., Borkowsky, W., Farthing, C., Ho, D. D. *J. Virol.* **1994**, 68, 4650–4655.

2 Borrow, P., Lewicki, H., Hahn, B. H., Shaw, G. M., Oldstone, M.B. *J. Virol.* **1994**, 68, 6103–6110.

3 Matano, T., Shibata, R., Siemon, C., Connors, M., Lane, H. C., Martin, M. A. *J. Virol.* **1998**, 72, 164–169.

4 Schmitz, J. E., Kuroda, M. J., Santra, S., Sasseville, V. G., Simon, M. A., Lifton, M. A., Racz, P., Tenner-Racz., K, Dalesandro, M., Scallon, B. J., Ghrayeb, J., Forman, M. A., Montefiori, D. C., Rieber, E. P., Letvin, N. L., Reimann, K. A. *Science* **1999**, 283, 856–860.

5 Jin, X., Bauer, D. E., Tuttleton, S. E., Lewin, S., Gettie, A., Blanchard, J., Irwin, C. E., Safrit, J. T., Mittler, J., Weinberger, L., Kostrikis, L. G., Zhang, L., Perelson, A. S., Ho, D. D. *J. Exp. Med.* **1999**, 189, 991–998.

6 Cocchi, F., A. L. DeVico, A. Garzino-Demo, S. K. Arya, R. C. Gallo, and P. Lusso. *Science* **1995**, 270, 1811–1815.

7 Levy, J. A., Mackewicz, C. E., Barker, E. *Immunol. Today.* **1996**, 17, 217–224.

8 Yang, O. O., Kalams, S. A., Trocha, A., Cao, H., Luster, A., Johnson, R. P., B. Walker, D. *J. Virol.* **1997**, 7, 3120–3128.

9 Borrow, P., Lewicki, H., Wei, X., Horwitz, M. S., Peffer, N., Meyers, H., Nelson, J. A., Gairin, J. E., Hahn, B. H., Oldstone, M. B. A., Shaw, G. M. *Nat. Med.* **1997**, 3, 205–211.

10 Goulder, P. J., Phillips, R. E., Colbert, R. A., McAdam, S., Ogg, G., Nowak, M. A., Giangrande, P., Luzzi, G., Morgan, B., Edwards, A., McMichael, A. J., Rowland-Jones, S. *Nat. Med.* **1997**, 3, 212–217.

11 Xu, X. N., Laffert, B., Screaton, G. R., Kraft, M., Wolf, D., Kolanus, W., Mongkolsapay, J., McMichael, A. J., Baur, A. S. *J. Exp. Med.* **1999**, 199, 1489–1496.

12 Appay, V., Nixon, D. F., Donahoe, S. M., Gillespie, G. M. A., Dong, T., King, A., Ogg, G. S., Spiegel, H. M. L., Conlon, C., Spina, C. A., Havlir, D. V., Richman, D. D., Waters, A., Easterbrook, P., McMichael, A. J., Rowland-Jones, S. L. *J. Exp. Med.* **2000**, 192, 63–75.

13 Collins, K. L., Chen, B. K., Kalams, S. A., Walker, B. D., Baltimore, D. *Nature.* **1998**, 391, 397–401.

14 Carrington, M., Nelson, G. W., Martin, M. P., Kissner, T., Vlahov, D., Goedert, J. J., Kaslow, R., Buchbinder, S., Hoots, K., O'Brien, S. J. *Science* **1999**, 283, 1748–1752.

15 Kiepiela, P., Leslie, A. J., Honeyborne, I., Ramduth, D., Thobakgale, C., Chetty, S., Rathnavalu, P., Moore, C.,

Pfafferott, K. J., Hilton, L., Zimbwa, P., Moore, S., Allen, T., Brander, C., Addo, M. M., Altfeld, M., James, I., Mallal, S., Bunce, M., Barber, L. D., Szinger, J., Day, C., Klenerman, P., Mullins, J., Korber, B., Coovadia, H. M., Walker, B. D., Goulder, P. J. *Nature* **2004**, 432, 769–774.

16 Stephens, H. A. F. *Trends in Immunol.* **2005**, 26, 41–47.

17 Engl, N. *J. Med.* **2001**, 344, 1668–1675.

18 Tomiyama, H., Miwa, K., Shiga, H., Ikeda-Moore, Y., Oka, S., Iwamoto, A., Kaneko, Y., Takiguchi, M. *J. Immunol.* **1997**, 158, 5026–5034.

19 Falk, K., Rotzschke, O., Stevanovic, S., Jung, G., Rammensee, H.G. *Nature* **1991**, 351, 290–296.

20 Takamiya, Y., Schönbach, C., Nokihara, K., Ferrone, S., Yamaguchi, M., Kano, K., Egawa, K., Takiguchi, M. *Int. Immunol.* **1994**, 6, 255–261.

21 Shiga, H., Shioda, T., Tomiyama, H., Takamiya, Y., Oka, S., Kimura, S., Yamaguchi, Y., Gojoubori, T., Rammensee, H.-G., Miwa, K., Takiguchi, M. *AIDS* **1996**, 10, 1075–1083.

22 Ikeda-Moore, Y., Tomiyama, H., Miwa, K., Oka, S., Iwamoto, A., Kaneko, Y., Takiguchi, M. *J. Immunol.* **1997**, 159, 6242–6252.

23 Tomiyama, H., Sakaguchi, T., Miwa, K., Oka, S., Iwamoto, A., Kaneko, Y., Takiguchi, M. *Hum. Immunol.* **1999**, 60, 177–186.

24 Tomiyama, H., Chujoh, Y., Shioda, T., Miwa, K., Oka, S., Kaneko, Y., Takiguchi, M. *AIDS* **1999**, 13, 861–863.

25 Fukada, K., Chujoh, Y., Tomiyama, H, Miwa, K., Kaneko, Y., Oka, S., Takiguchi, M. *AIDS* **1999**, 13, 1413–1414.

26 Hossain, M. S., Tomiyama, H., Inagawa, T., Ida, S., Oka, S., Takiguchi, M. *AIDS Res. Huma.* Retroviruses. **2003**, 19, 503–510.

27 Satoh, M., Takamiya, Y., Oka, S., Tokunaga, K., Takiguchi, M. *Vaccine* **2005**, 23, 3783–3790.

28 Kaslow, R.A., Carrington, M., Apple, R., Park, L., Muñoz, A., Saah, A. J., Goedert, J.J., Winkler, C., O'Brien, S. J., Rinaldo, C., Detels, R., Blattner, W., Phair, J., Erlich H., Mann, D. L. *Nat. Med.* **1996**, 4, 405–411.

29 Altfeld, M., Addo, M. M., Rosenberg, E. S., Hecht, F. M., Lee, P. K., Vogel, M., Yu, X. G., Draenert, R., Johnston, M. N., Strick, D., Allen, T. M., Feeney, M.E., Kahn, J. O., Sekaly, R. P., Levy, J. A., Rockstroh, J. K., Goulder, P. J., Walker, B.D. *AIDS* **2003**, 17, 2581–91.

30 Leslie, A. J., Pfafferott, K. J., Chetty, P., Draenert, R., Addo, M. M., Feeney, M., Tang, Y., Holmes, E. C., Allen, T., Prado, J. G., Altfeld, M., Brander, C., Dixon, C., Ramduth, D., Jeena, P., Thomas, S. A., St John A., Roach, T. A., Kupfer, B., Luzzi, G., Edwards, A., Taylor, G., Lyall, H., Tudor-Williams, G., Novelli, V., Martinez-Picado, J., Kiepiela, P., Walker, B. D., Goulder, P. *J. Nat Med.* **2004**, 10, 282–289.

31 Klein, M.R., van der Burg, S. H., Hovenkamp, E., Holwerda, A. M., Drijfhout, J. W., Melief, C. J., Miedema, F. *J. Gen. Virol.* **1998**, 79 (Pt 9), 2191–2201.

32 Draenert, R., Le Gall, S., Pfafferott, K. J., Leslie, A. J., Chetty, P., Brander, C., Holmes, E. C., Chang, S. C., Feeney, M. E., Addo, M. M., Ruiz, L., Ramduth, D., Jeena, P., Altfeld, M., Thomas, S., Tang, Y., Verrill, C. L., Dixon, C., Prado, J. G., Kiepiela, P., Martinez-Picado, J., Walker, B. D., Goulder, P. J. *J Exp Med.* **2004**, 199, 905–915.

33 Goulder, P. J., Bunce, M., Krausa, P., McIntyre, K., Crowley, S., Morgan, B., Edwards, A., Giangrande, P., Phillips, R. E., McMichael, A. J. *AIDS Res. Hum. Retroviruses* **1996**, 12, 1691–1698.

34 Migueles, S. A., Sabbaghian, M. S., Shupert, W, L., Bettinotti, M. P, Marincola, F. M., Martino, L., Hallahan, C. W., Selig, S. M., Schwartz, D., Sullivan, J., Connors, M. *Proc Natl Acad Sci U S A.* **2000**, 97, 2709–2714.

35 Goulder, P. J., Tang, Y., Pelton, S. I., Walker, B.D. *J. Virol.* **2000**, 74, 5291–5299.

36 Migueles, S. A., Laborico, A. C., Imamichi, H., Shupert, W. L., Royce, C., McLaughlin, M., Ehler, L., Metcalf, J., Liu, S., Hallahan, C. W., Connors, M. *J . Virol.* **2003**, 77, 6889–6898.

37 Jansen, CA, Kostense, S, Vandenberghe, K, Nanlohy, NM,

De Cuyper, IM, Piriou, E, Manting, EH, Miedema, F, van Baarle, D. Eur. *J. Immunol.* **2005**, 35, 150–158.

38 Gillespie, G. M., Kaul, R., Dong, T., Yang, H. B., Rostron, T., Bwayo, J. J., Kiama, P., Peto, T., Plummer, F. A., McMichael, A. J., Rowland-Jones, S.L. *AIDS* **2002**, 16, 961–972.

39 Buseyne, F., McChesney, M., Porrot, F., Kovarik, S., Guy, B., Riviere, Y. *J. Virol.* **1993**, 67, 694–702.

40 Wilson, JD, Ogg, GS, Allen, RL, Davis, C, Shaunak, S, Downie, J, Dyer, W, Workman, C, Sullivan, S, McMichael, AJ, Rowland-Jones, SL. *AIDS* **2000**, 14, 225–233.

41 Nietfield, W., Bauer, M., Fevrier, M., Maier, R., Holzwarth, B., Frank, R., Maier, B., Riviere, Y., Meyerhans, A. *J. Immunol.* **1995**, 154, 2188–2197.

42 Feeney, M. E., Tang, Y., Roosevelt, K. A., Leslie, A. J., McIntosh, K., Karthas, N., Walker, B. D., Goulder, P. J. *J Virol.* **2004**, 78, 8927–8930.

43 Kelleher, A. D., Long, C., Holmes, E. C., Allen, R. L., Wilson, J., Conlon, C., Workman, C., Shaunak, S., Olson, K., Goulder, P., Brander, C., Ogg, G., Sullivan, J. S., Dyer, W., Jones, I., McMichael, A. J., Rowland-Jones, S., Phillips, R. E. *J. Exp. Med.* **2001**, 193, 375–385.

44 Tomiyama, H., Akari, H., Adachi, A., Takiguchi, M. *J. Virol.* **2002** , 76, 7535–7543.

45 Kubo, H, Ikeda-Moore, Y, Kikuchi, A, Miwa, K, Nokihara, K, Schonbach, C, Takiguchi, M. *Immunogenetics* **1998**, 47, 256–263.

46 Johnson, R. P., Trocha, A., Yang, L., Mazzara, G. P., Panicali, D. L., Buchanan, T. M., Walker, B. D. *J. Immunol.* **1991**, 147, 1512–1521.

47 Tsomides, T. J., Aldovini, A., Johnson, R. P., Walker, B. D., Young, R. A., Eisen, H. N. *J. Exp. Med.* **1994**, 180, 1283–1293.

48 Gray, C. M., Lawrence, J., Schapiro, J. M., Altman, J. D., Winters, M. A., Crompton, M., Loi, M., Kundu, S. K., Davis, M. M., Merigan, T. C. *J. Immunol.* **1999**, 162, 1780–1788.

49 Goulder, P. J. R., Sewell, A. K., Lalloo, D. G., Price, D. A., Whelan, J. A., Evans, J., Taylor, G. P., Luzzi, G., Giangrande, P., Phillips, R. E., McMichael, A. J. *J. Exp. Med.* **2004**, 185, 1423–1433.

50 Brander, C., Hartman, K. E., Trocha, A. K., Jones, N. G., Johnson, R. P., Korber, B., Wentworth, P., Buchbinder, S. P., Wolinsky, S., Walker, B. D., Kalams, S. A. *J. Clin. Invest.* **1998**, 101, 2559–2566.

51 Cao, J., McNevin, J., Holte, S., Fink, L., Corey, L., McElrath, M. J. *J. Virol.* **2003**, 77, 6867–6878.

52 Ferrari, G., Neal, W., Ottinger, J., Jones, A. M., Edwards, B. H., Goepfert, P., Betts, M. R., Koup, R. A., Buchbinder, S., McElrath, M. J., Tartaglia, J., Weinhold, K. J. *J. Immunol.* **2004**, 173, 2126–2133.

53 Altman, J. D., Moss, P. A., Goulder, P. J., Barouch, D. H., McHeyzer-Williams, M. G., Bell, J. I., McMichael, A. J., Davis, M. M. *Science.* **1996**, 274, 94–96.

54 Jamieson, B. D., Yang, O. O., Hultin, L., Hausner, M. A., Hultin, P., Matud, J., Kunstman, K., Killian, S., Altman, J., Kommander, K., Korber, B., Giorgi, J., Wolinsky, S. *J. Immunol.* **2003**, 171, 5372–5379.

55 Papagno, L., Appay, V., Sutton, J., Rostron, T., Gillespie, G. M., Ogg, G. S., King, A., Makadzanhge, A. T., Waters, A., Balotta, C., Vyakarnam, A., Easterbrook, P. J., Rowland-Jones, S. L. *Clin. Exp. Immunol.* **2002**, 130, 509–517.

56 Dong, T., Stewart-Jones, G., Chen, N., Easterbrook, P., Xu, X., Papagno, L., Appay, V., Weekes, M., Conlon, C., Spina, C., Little, S., Screaton, G., van der Merwe, A., Richman, D. D., McMichael, A. J., Jones, E. Y., Rowland-Jones, S. L. *J. Exp. Med.* **2004**, 200, 1547–1557.

57 Barouch, D.H., Kunstman, J., Kuroda, M. J., Schmitz, J. E., Santra, S., Peyerl, F.W., Krivulka G. R., Beaudry, K., Lifton, M.A., Gorgone, D. A., Montefiori, D. C., Lewis, M. G., Wolinsky, S. M., Letvin, N. L. *Nature* **2002**, 415, 335–339.

58 Allen, T. M., Altfeld, M., Yu, X. G., O'Sullivan, K. M., Lichterfeld, M., Le Gall, S., John, M., Mothe, B. R., Lee, P. K., Kalife, E. T., Cohen, D. E., Freedberg, K. A., Strick, D. A., Johnston, M. N., Sette, A.,

Rosenberg, E. S., Mallal, S. A., Goulder, P. J., Brander, C., Walker, B. D. *J Virol.* **2004**, 78, 7069–7078.

59 Kerkau, T., R. Schmitt-Landgraf, A. Schimpl, and E. Wecker. *AIDS Res. Hum. Retrovir.* **1989**, 5, 613–620.

60 Noraz, N., Verrier, B., Fraisier, C., Desgranges, C. *AIDS Res. Hum. Retrovir.* **1995**, 11, 145–154.

61 Scheppler, J. A., Nicholson, J. K., Swan, D. C., Ahmed-Ansari, A., McDougal, J. S. *J. Immunol.* **1989**, 143, 2858–2866.

62 Kerkau, T., Bacik, I., Bennink, J. R., Yewdell, J. W., Hunig, T., Schimpl, A., Schubers, U. *J. Exp. Med.* **1997**, 185,1295–1305.

63 Matsui, M., Warburton, R. J., Cogswell, P. C., Baldwin Jr. A. S., Frelinger, J. A. *J. Acquir. Immune Defic. Syndr. Hum. Retrovirol.* **1996**, 11, 233–240.

64 Schwartz, O., Marechal, V., Le Gall, S., Lemonnier, F., Heard. J. M. *Nat. Med.* **1996**, 2, 338–342.

65 Akari, H., Arold, S., Fukumori, T., Okazaki, T., Strebel, K., Adachi, A. *J. Virol.* **2000**, 74, 2907–2912.

66 Deacon, N. J., Tyskin, A., Solomon, A., Smith, K., Ludford-Menting, M., Hooker, D.J., McPhee, D.A., Greenway, A.L., Ellett, A., Chatfield, C., Lawson, V. A., Crowe, S., Maerz, A., Sonza, S., Learmont, J., Sullivan, J. S., Cunningham, A., Dwyer, D., Dowton, D., Mills, J. *Science* 1995, 270, 988–991.

67 Kirchhoff, F., Greenough, T.C., Brettler, D.B., Sullivan, J.L., Desrosiers, R.C. *N. Engl. J. Med.* **1995**, 332, 228–232.

68 Mariani, R., Kirchhoff, F., Greenough, T. C., Sullivan, J. L., Desrosiers, R. C., Skowronski, J. *J. Virol.* **1996**, 70, 7752–7764.

69 Cohen, G., Gandhi, R., Davis, D., Mandelboim, O., Chen, B., Strominger, J., Baltimore, D. *Immunity*, **1999**, 10, 661–671

70 Tomiyama, H., Fujiwara, M., Oka, S., Takiguchi, M. *J. Immunol.* **2005**, 74, 36–40.

71 Ogg, G. S., McMichael, AJ. *Curr Opin Immunol.* **1998**, 10, 393–396.

72 Ogg ,G. S, Jin, X., Bonhoeffer, S., Dunbar, P. R., Nowak, M. A., Monard, S., Segal, J. P., Cao, Y., Rowland-Jones, S. L., Cerundolo, V., Hurley, A., Markowitz, M., Ho, D. D., Nixon, D. F., McMichael, A. J. *Science.* **1998**, 279, 2103–2106.

73 Bodinier, M., Peyrat, M. A., Tournay, C., Davodeau, F., Romagne, F., Bonneville, M., Lang, F. *Nat Med.* **2000**, 6, 707–710.

74 Appay, V., Dunbar, P. R., Callan, M., Klenerman, P., Gillespie, G. M., Papagno, L., Ogg, G. S., King, A., Lechner, F., Spina, C. A., Little, S., Havlir, D. V., Richman, D. D., Gruener, N., Pape, G., Waters, A., Easterbrook, P., Salio, M., Cerundolo, V., McMichael, A. J., Rowland-Jones, S. L. *Nat Med.* **2002**, 8, 379–85.

75 Tomiyama, H., Takata, H., Matsuda, T., Takiguchi, M. *Eur J Immunol.* **2004**, 34, 999–1010.

76 Andersson, J., Behbahani, H., Lieberman, J., Connick, E., Landay, A., Patterson, B., Sonnerborg, A., Lore, K., Uccini, S., Fehniger, T. E. *AIDS.* **1999**, 13, 1295–303.

77 Tomiyama, H., Matsuda, T., Takiguchi, M. *J. Immunol.* 2002, **168**, 5538–5550.

78 Kaech, S., Hemby, M. S., Kersh, E., Ahmed, R. *Cell.* 2002. **111:** 837–851.

79 Posnett, D. N., Edinger, J. W., Manavalan, J. S., Irwin, C., Marodon, G. *Int. Immunol.* **1999**, 11, 229–241.

80 Nociari, M. M., Telford, W., Russo, C. *J. Immunol.* **1999**, 162, 3327–3335.

81 Kern, F., Khatamzas, E., Surel, I., Frommel, C., Reinke, P., Waldrop, S. L., Picker, L. J., Volk, H. D. *Eur. J. Immunol.* **1999**, 29, 2908–2915.

82 Wills, M. R., Carmichael, A. J., Weekes, M. P., Mynard, K., Okecha, G., Hicks, R., Sissons, J. G. *J. Immunol.* **1999**, 165, 7080–7087.

83 Weekes, M. P., Carmichael, A. J., Wills, M. R., Mynard, K., Sissons, J. G. *J. Immunol.* **1999**, 162, 7569–7577.

84 Hamann, D., Roos, M. T., van Lier, R. A. *Immunol. Today.* **1999**, 20, 177–180.

85 Sallusto, F., Lenig, D., Forster, R., Lipp, M. and Lanzavecchia, A. *Nature* **1999**, 401, 708–712.

86 Champagne, P., Ogg, G. S., King, A. S., Knabenhans, C., Ellefsen, K., Nobile, M., Appay, V., Rizzardi, G. P., Fleury, S., Lipp, M., Förster, R., Rowland-Jones, S., Sékaly, R. P., McMichael, A. J., Pantaleo, G. *Nature* **2001**, 410, 106–111.

87 Fukada, K., Sobao, Y., Tomiyama, H., Oka, S., Takiguchi, M. *J. Immunol.* **2002**, 168, 2225–32.

88 Tomiyama, H., Oka, S., Ogg, G. S., Ida, S., McMichael, A. J., Takiguchi, M. *AIDS.* **2000**, 14, 2049–51.

89 O'Brien, S. J., Gao, X., Carrington, M. *Trends Mol Med.* **2001**, 7, 379–81.

11
The Effects of Pathogens on the Immune System: Viral Hepatitis

Mala Maini and Antonio Bertoletti

11.1
Introduction

Hepatitis B virus (HBV) and hepatitis C virus (HCV) are the two major causes of chronic liver inflammation worldwide [1, 2]. Although the two viruses have distinct virological features, both are preferentially hepatotropic, are not directly cytopathic, and elicit liver diseases that share a similar natural history.

11.2
The Viruses and the Disease

11.2.1
Genomic Organization

HBV, a member of the hepadnavirus family, is a DNA virus that uses a reverse transcriptase for replication from a pre-genomic RNA template (Figure 11.1a). The genome of HBV is approximately 3200 nucleotides long and contains four open reading frames encoding the viral envelope, nucleocapsid, polymerase, and X proteins [3]. HCV is a single-stranded RNA virus of positive polarity with a genome of 9500 nucleotides (Figure 11.1b). It contains a large open reading frame encoding a precursor polyprotein of approximately 3000 amino acids, which is processed to produce structural (core, envelope 1 and 2) and at least six nonstructural proteins (NS2, NS3, NS4A, NS4B, NS5A and NS5B). HCV shows a genetic organization similar to that of the Flaviviridae (e.g., dengue, yellow fever) and shares some sequence homology with members of this family [4].

11.2.2
Prevalence

Despite the presence of an effective prophylactic vaccine, 300 million people remain HBV carriers, with a particularly high prevalence in Asia and Africa [1].

Immunodominance: The Choice of the Immune System. Edited by Jeffrey A. Frelinger

ISBN: 3-527-31274-9

a) HBV:

Genome: DNA(3200 nucleotides)

Viral proteins: envelope, core, polymerase, X

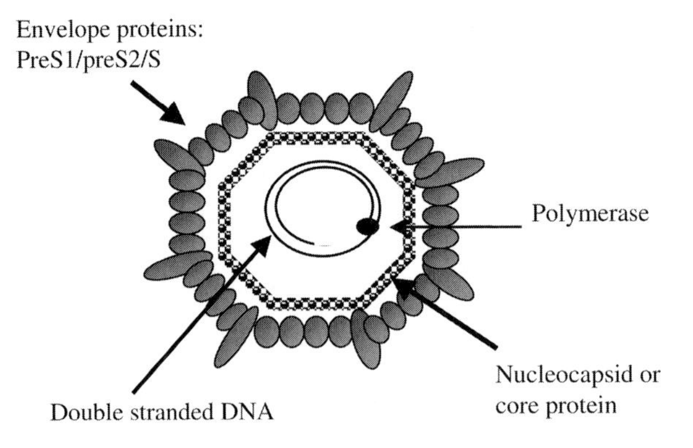

Epidemiology: ~300 million chronically infected
(mostly in Asia and Africa)

b) HCV:

Genome: RNA(9500 nucleotides)
Viral proteins: envelope, core, 6 non-structurals
Morphology:unknown

HCV genome: single stranded, RNA

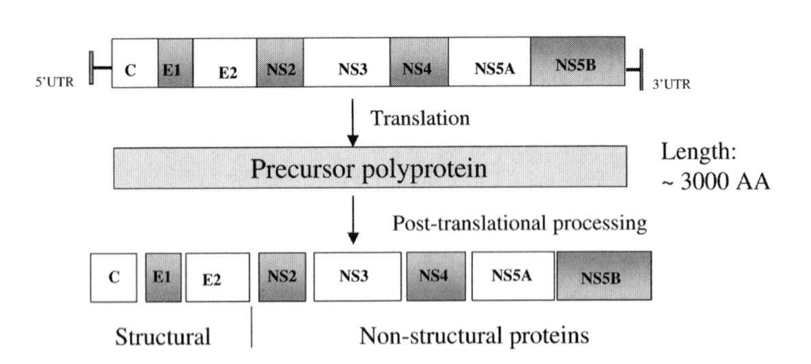

Epidemiology: ~200 million chronically infected
(major agent of liver disease in USA)

Figure 11.1 (a) HBV structure (b) HCV genomic organization.

HCV has infected around 200 million people and is a major causative agent of liver disease in the West, with an estimated prevalence of HCV infection in the general U.S. population of around 1–2%[2].

11.2.3
Hepatic Disease and Chronicity After Infection

HBV and HCV elicit liver diseases that share a similar natural history. In both infections, disease severity varies greatly from person to person. In some subjects the immune system is able to control infection efficiently and clears the virus from the bloodstream either without clinically evident liver disease or with an acute inflammation of the liver (acute hepatitis) that can resolve without long-term clinical sequelae. Other patients fail to clear the virus and develop chronic infection. Most chronically infected patients remain largely asymptomatic without life-threatening liver disease, but about 10–30% develop liver cirrhosis with possible progression to liver cancer [1, 5]. It is assumed that 75–80% of subjects infected with HCV develop a chronic infection [5]. The rate of HBV chronicity seems much lower (10% of adult subjects), but age and route of infection influence the outcome, and exposure in neonatal life usually leads to HBV persistence [1]. It is likely that the rate of HCV chronicity has been overestimated. While HBV and HCV infections can have a completely asymptomatic course, antiviral antibodies, used to define the prevalence of infected people, have a different rate of disappearance in the two infections. HCV-specific antibodies can disappear soon after infection [6], thus potentially underestimating the number of subjects exposed. In line with this possibility, a higher rate of spontaneous HCV clearance has been reported particularly among women and children (45% recovery) [7, 8]. Moreover, HCV-specific $CD8^+$ T cell are frequently detectable, without evidence of specific antibodies, in healthy family members exposed to chronically infected HCV patients [9, 10]. Thus, the risk of chronicity is still difficult to define and likely is dependent on host (age and route of infection, genetic background) [11–13] and virus (dose and strain of infectious virus) [14–18] variables.

11.3
Importance of CD4 and CD8 T Cells in HBV and HCV Control

CD4 and CD8 T-cell responses are crucial and complimentary mediators of control of HBV and HCV infection and protection from persistent infection. A clear dichotomy in the adaptive immunity of patients with chronic or resolved HBV and HCV infections has been shown in several studies. Multispecific antiviral CD4 and CD8 responses with a type 1 profile of cytokine production are usually detectable in subjects with a favorable outcome [19–32]. These responses are quantitatively stronger than those found in patients with chronic infection, who are instead characterized by a virus-specific T-cell response that is difficult to detect and is Th2 oriented [33–38]. These observations have led to the concept that the ability to mount an efficient cellular immune response is the main mechanism responsible for HBV and HCV control, while a defect in this response leads to chronicity. Recent studies in animal models have been able to directly demonstrate that the associations reported in human studies have a causative effect. In chimpanzees infected with HCV, expansion of a multispecific and sustained HCV-specific, CD8-mediated T-cell response was observed in two out of two animals that cleared the virus, but not in the four animals that developed chronic infection [39]. More recent work has confirmed these initial results and shown that intrahepatic production of IFN-γ by T cell strongly correlates with HCV control [40]. In addition, depletion of $CD8^+$ or $CD4^+$ T cell from chimpanzees that have previously recovered from HCV abrogates protective immunity and results in persistent HCV infection upon rechallenge with HCV [41, 42].

The importance of CD8 and CD4 T-cell responses has also recently been reported in animal models of HBV infection, where a reduced early expansion of virus-specific T cell was associated with virus persistence [43]. Furthermore, $CD8^+$ T-cell deletion experiments performed in HBV-infected chimpanzees have provided further support to the concept that CD8 cells are the main cellular subset responsible for viral clearance [40]. Similar association between T-cell response and viral control was also observed in patients studied during the incubation phase of acute HCV and HBV infections [26, 44]. In this setting, expansion of virus-specific IFN-γ + CD8 and CD4 cells preceded complete virus clearance and was present only in subjects who controlled the infections.

However, it must be remembered that similar associations can also be found for humoral responses. HBV and HCV clearance is associated with production of anti-envelope antibodies [45, 46], and sera with high levels of antiviral antibodies (specific for the viral envelope) can control HBV [47] and HCV [48, 49]. Therefore, it is likely that the integrated and coordinate activation of both cellular and humoral arms of the adaptive immune response ultimately allows for control of both infections. The different components of the adaptive immune system are so interconnected that the failure of one of them clearly affects the expansion and protective efficacy of the others [50]. Lack of CD4 help can impair CD8 cells and antibody production [51], while the inability to mount a virus-specific CD8 response results in a quantity of circulating virus that cannot be cleared by antibodies alone [52].

11.4 Limitations of Existing Data

Before considering the question of immunodominance within the immune response to HBV and HCV, it is important to consider the type of data we currently have available and how much weight can be placed on them. The low frequency of virus-specific $CD4^+$ and $CD8^+$ T cell in the periphery, the fact that HBV and HCV do not readily infect antigen-presenting cells *in vitro*, the difficulty in analyzing the T-cell response longitudinally from the time of infection to the development of disease, and the variability of HBV and HCV sequences represent major problems faced by HBV and HCV immunologists analyzing these T-cell responses.

11.4.1
Low Frequency of HBV- and HCV-specific T Cells

The frequency of HCV- and HBV-specific $CD8^+$ T cell in the peripheral blood is usually very low compared to that found in other viral infections. Frequencies of less than 1–2% of the total circulating population of $CD8^+$ T cell are usually found during the acute phase of HBV and HCV infections, well below frequencies detectable in other viral diseases such as EBV or HIV [53]. During HBV ad HCV persistence, the situation is even worse, with frequencies of circulating HBV- and HCV-specific T cell so low that they are difficult to detect *ex vivo* by using MHC–peptide tetramer staining, intracellular cytokine staining, and Elispots [25, 38, 54]. A single round of *in vitro* expansion is often used to overcome this problem, but expanded responses may differ in their hierarchy from those found *in vivo* [54]. The inability of HBV and HCV to infect antigen-presenting cells *in vitro* necessitates the use of pre-selected synthetic peptides or eukaryotic expression vectors that can also potentially bias the results.

Magnetic bead enrichment of cytokine-producing T cell has recently been applied to the *ex vivo* quantification of very low-frequency HCV-specific CD4 [55] and CD8 T cell [56], and this method will have to be applied to a wide range of T-cell specificities to directly measure T-cell hierarchy.

11.4.2
Pre-selection of Epitopes

The inability of HBV and HCV to infect antigen-presenting cells and the necessity of expanding T cell *in vitro* before analysis have directed the studies of HBV- and HCV-specific T-cell responses towards the use of synthetic peptides mimicking the processed antigen fragments presented in the context of HLA class I [57]. Initially, peptides were chosen at random from viral protein sequences [24, 58], but the identification of HLA class I–binding motifs and the development of HLA-binding assays that identify a significant association between peptide affinity and immunogenicity have led to the selection of higher-affinity peptides for HBV and HCV immunogenicity studies [59, 60]. Because the binding motifs for HLA-A2 molecules were the first to be discovered [61], most studies have been carried out using predefined peptides selected for their ability to bind the HLA-A2 molecule. This historical bias has resulted in many of the defined HBV and HCV epitopes being restricted by HLA-A2, and disproportionately few HCV and HBV epitopes have been identified for other class I alleles [62, 63]. Recently, the use of overlapping peptides spanning the whole genome, allowing detection of responses regardless of HLA restriction, has been implemented in both HBV [64] and HCV [54, 65, 66] and has revealed a more complex picture than that suggested by the analysis of selected responses. An alternative approach, allowing the detection of naturally processed epitopes, has been the use of eukaryotic expression vectors (EBO [67], vaccinia [68], or adenovirus [69]) to induce endogenous synthesis of all the HCV and HBV proteins.

11.4.3 Stage of Infection

HBV and HCV infections are characterized by a long asymptomatic incubation phase of infection (about 1 month [53]). Recruiting patients within this largely asymptomatic phase of infection has proved difficult, and information about the early T-cell response profile is very limited, particularly after HBV infection. Recent studies of patients with acute HCV infection have started to reveal a kinetic regulation of the hierarchy of HCV-specific T-cell responses: strong CD4 [26, 70] and CD8 [71] responses can be primed and expanded in acute infection but are then lost in association with viral persistence. In the same way that the magnitude of responses detected varies from acute to chronic infection, the specificity of responses detected seems dependent on the stage of viral infection studied. This phenomenon has been clearly demonstrated in the setting of HIV-1 infection, where many dominant epitopes are not conserved between acute and chronic stages [72]. Evidence that viral persistence alters the profile of T-cell immunodominance is also accumulating in chronic HBV [38, 73] and HCV infection [65].

11.4.4 Variations in Viral Inoculum

The dose and route of viral inoculum are rarely known in human HBV and HCV infection but could be a major source of the variability between patients studied. In addition, sequence variability in the infecting virus is a major concern in studies of T-cell responses. Not only are there many HBV and HCV genotypes circulating in the population, but the viruses are prone to frequent mutation and, in the case of HCV, exist within an individual as a quasispecies of related but distinct genomes [74]. This makes the selection of peptides for the study of T-cell specificity extremely difficult, because customizing them following sequencing of autologous virus is rarely practical. It also means that studies of viral escape mutations are limited by lack of knowledge of the sequence of the infecting strain. However, there are cohorts of patients who have all been infected with a common source of HCV [12]. Studies of chimpanzees have also overcome these shortcomings, allowing careful analysis of viral and T-cell evolution following infection with a known single source of HCV [39, 75].

11.5 Hierarchy of T-cell Responses During HBV Infection: Helper CD4 T-cell Response

HBV-specific, HLA class II–restricted CD4 T cell able to produce Th1 cytokines can be detected in patients with self-limited acute hepatitis [19, 20, 22] and can persist for decades after clinical recovery [21]. In contrast, such helper T-cell responses are generally weak in patients with persistent infection. Multiple epitopes within the nucleocapsid protein are targeted by helper T cell, whereas the response against envelope antigens is weak in subjects with both self-limited and chronic hepatitis B. Immunodominant core epitopes have been identified with the nucleocapsid protein, and a sequence covering region 50–69 has been shown to stimulate helper T cell in 90% of the patients tested, irrespective of their HLA class II profile [76]. This degenerate sequence could be used in the general population to increase the immunogenicity of the current HBV vaccines.

The demonstration that increased core-specific CD4 responses are detectable during exacerbations of chronic hepatitis B preceding HBeAg seroconversion (indicative of a reduced level of viral replication) [77] represents a further indication of the importance of the nucleocapsid-specific CD4 response in HBV control. It is likely that HBeAg, a secretory form of the nucleocapsid protein, is able to tolerize HBcAg-specific T cell [78] and contribute substantially to the low level of core-specific helper T-cell response detectable in HBeAg^{+} chronic hepatitis B patients. This scenario has been well characterized and identified in mice [79].

Even though most of the data generated in resolved or chronic HBV patients have identified nucleocapsid-specific CD4 T cell as the dominant helper response correlating with HBV recovery, other aspects need to be considered. In particular, the helper T-cell response specific for the polymerase and X antigens has not been sufficiently investigated, and only recently have polymerase epitopes able to elicit CD4 helper T-cell responses been identified [80]. These polymerase epitopes were conserved among the different HBV genomes and bound to the most common HLA-DR, and they induced in resolved acute hepatitis B patients a helper T-cell response that seemed comparable to that detected against core peptides.

A different scenario is instead present for the envelope-specific CD4 T-cell response. In contrast to the immunogenicity of core and polymerase antigens, the HBV envelope antigen does not seem to expand an equally strong helper T-cell response during HBV infection [19, 81]. The limited expansion of envelope-specific CD4 cells does not imply that the envelope is a generally weak immunogen. On the contrary, the HBV envelope antigen elicited a strong helper T-cell response in subjects vaccinated with a plasma-derived or recombinant form of this antigen [81–83]. The differential immunogenicity of envelope antigens in vaccine recipients and in patients with natural infection suggests that differences in antigen presentation and/or the presence of "natural" or synthetic adjuvants influence the immunogenicity of the responses in these two groups. The HBV envelope antigen is produced in large excess and in a soluble form during HBV replication. Particles composed of only envelope antigen are present in a 10^{3}– to 10^{6}-fold excess over whole virions [3]. These particles are not infectious, and it is very likely that the evolution of such impressive levels of synthetic effort by HBV may be linked to the apparent ability of the secreted envelope antigens to induce a state of low T-cell response in the infected patients. A further possibility is that an anti-envelope-specific T-cell response dominates the immune response during the initial phases of HBV infection. Although recent analysis of the dynamics of the T-cell response during the incubation phase of acute HBV infection has not confirmed this hierarchical shift of T-cell responses between nucleocapsid and envelope response [44] suggested by initial studies [84], the quantitative kinetics of helper T-cell responses against the different HBV antigens needs to be examined in more detail before it will be possible to draw a precise antigenic hierarchy of the helper response during HBV infection.

11.6 Hierarchy of T-cell Responses During HBV Infection: Cytotoxic T-cell Response

Analysis of the HLA class I–restricted cytotoxic T-cell response to HBV has been severely hampered by the inability of HBV to be propagated in cell culture [85]. Initial reports of the presence of cytotoxic T cell in chronic hepatitis B utilized hepatocytes of infected individuals as targets [86], but the first definitive characterization of HLA class I–restricted cytotoxic T cell specific for HBV derived from the understanding that the sequence of the processed viral antigens presented by HLA class I molecules can be mimicked by synthetic peptides [24, 58]. Thus, the first identification of cytotoxic T cell specific for several viral epitopes within core [24, 58, 87], envelope [88], polymerase [23], and X [89] proteins of HBV was achieved by using synthetic peptides to expand memory CTL *in vitro*. These initial studies demonstrated that the magnitude of the HBV-specific CD8 response is stronger in self-limited than chronic infection [24, 58], that the CTL response persists decades after clinical recovery of acute infection [90], and that it can also be demonstrated after resolution of chronicity [91]. As reported previously, the majority of these studies have been carried out using peptides able to bind the HLA-A2 molecule, with the result that a disproportionate number of known HBV epitopes are HLA-A2 restricted. However, HBV-specific cytotoxic epitopes restricted by different HLA class I molecules [92] have now been identified, and a map of the different HLA-class I–restricted epitopes specific for each HBV protein is represented in Figure 11.2.

The recent development of methods such as MHC–peptide tetramer staining, intracellular cytokine staining, and Elispots, which are able to quantify virus-specific CD8 cells directly *ex vivo*, has permitted a more accurate analysis of HBV-specific $CD8^+$ T cell during the different phases of HBV infection. The use of HLA-A2 tetramer complexes that are able to directly visualize $CD8^+$ T cell specific for the more commonly recognized core (core 18–27), envelope (env 183–191 and env 335–343), and polymerase (pol 455–463) HBV epitopes confirmed the quantitative differences between self-limited and chronic infection [93] and demonstrated that the quantity of HBV-specific $CD8^+$ cells correlated with HBV control and not with liver damage [94]. This work also revealed that an epitope hierarchy exists within the HBV-specific $CD8^+$ T-cell response that can be altered by viral persistence. In the few patients with acute hepatitis B that have been analyzed longitudinally from incubation time to recovery, direct *ex vivo* analysis of HBV-specific $CD8^+$ T cell showed that the dominant response present in the acute phase is maintained after recovery [38]. The frequency of dominant response is usually around

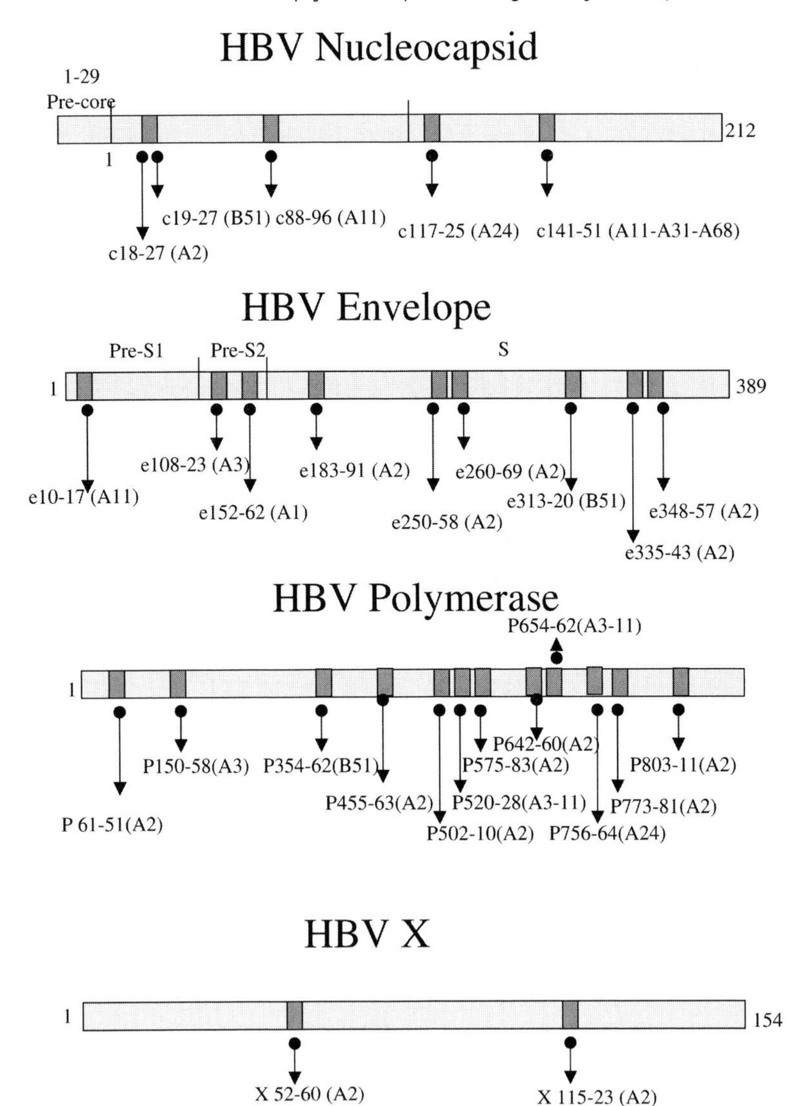

Figure 11.2 HBV-specific HLA class I–restricted epitopes.

1–2% during the acute clinical phase of infection, while subdominant responses are present a much lower frequency (0.2–0.3% of total $CD8^+$). The dominant response drops to a value of around 0.1–0.2% of total $CD8^+$ T cell in the recovery phase, and this frequency is then maintained even 2–3 years after resolution of the acute phase of infection. Subdominant $CD8^+$ responses instead become undetectable directly *ex vivo* after recovery and can be demonstrated only after *in vitro* expansion. Core 18–27–specific $CD8^+$ cells often represent the dominant response among the different A2-restricted epitopes tested in patients with acute hepatitis, but this is not absolute. In some patients, pol 455–463–specific, env 183–191–specific, or env 335–343–specific $CD8^+$ T cell were found to quantitatively dominate the $CD8^+$ T-cell response [38, 44]. The overall dominance of these three responses among the different HLA-A2-restricted epitopes within a patient is also maintained when immunodominance is defined as the most common responses among different patients [63]. The great majority of $A2^+$ patients with self-limited hepatitis B recognize the HBc 18–27, HBe 183–191, HBe 335–343, and HBp 455–463 epitopes. The cause of immunodominance of these sequences is likely linked to their good binding affinity to the HLA-A2 molecule. A direct relationship between immunogenicity and binding ability exists among different epitopes, and these epitopes represent good A2-binding peptides [59]. A further possible explanation of the dominance of these HLA-A2-restricted CD8 responses is the finding that some HLA class I epitopes are nested within helper T-cell epitopes. The well-characterized, often immunodominant HBc 18–27 epitope overlaps with an HLA class II–restricted epitope [95], and similar features have been described for new polymerase $CD8^+$ T-cell epitopes [80]. CD4 helper T cell are necessary for the maintenance of functional $CD8^+$ T cell, and the covalent linkage between helper and cytotoxic epitopes has been shown to be important for the induction of CTL responses [51]. However, it must be stressed that the overall hierarchy of CTL responses is still incomplete and there is no information available about competition among epitopes restricted by different HLA class I alleles. Overlapping peptides spanning the whole HBV genome, allowing the detection of responses regardless of HLA restriction, will have to be used to supplement our knowledge of the hierarchy of CTL responses in self-limited acute hepatitis B.

Although knowledge of immunodominance of different HBV proteins has been limited by the selected use of few peptides, detailed analysis of these HBV-specific $CD8^+$ responses has led to important information regarding the potential impact of the different CTL responses on HBV immunopathogenesis. Amino acid mutations within the core 18–27 region that are able to inhibit activation of the core 18–27–specific $CD8^+$ cells have been shown to occur in patients with chronic hepatitis B who are able to mount this specific $CD8^+$ T-cell response [96]. In contrast, mutations within polymerase and envelope epitopes are rare [97] and cannot be identified even in chronic patients who demonstrate the presence of envelope- and polymerase-specific $CD8^+$ cells [38], suggesting that the antiviral efficiency of the core 18–27–specific $CD8^+$ response is greater than the response against polymerase and envelope epitopes. Longitudinal analysis of HLA-A2-restricted HBV-specific $CD8^+$ T cell in resolved and chronic hepatitis B patients has also revealed that different epitope specificities are differentially regulated during chronic infection. Chronic hepatitis B is a heterogeneous disease, and patients can differ markedly in the levels of viral replication, liver disease activity, and humoral responses. The combined direct *ex vivo* and after *in vitro* expansion analysis of HBV-specific $CD8^+$ cells in chronic patients with different profiles of disease demonstrated that differences were detectable among $CD8^+$ T cell specific for different epitopes. Core 18–27–specific $CD8^+$ T cell (often immunodominant in self-limited hepatitis) cannot be detected in the circulation (either directly *ex vivo* or after *in vitro* expansion) when HBV-DNA levels are $>10^7$ copies per milliliter, and lack of detection within the circulatory compartment is not caused by preferential intrahepatic localization of core 18–27–specific $CD8^+$ T cell. On the contrary, the frequency of core 18–27–specific $CD8^+$ T cell within the liver is also inversely proportional to the level of HBV replication [38].

Envelope- and polymerase-specific $CD8^+$ T cell are instead the only specificities that can be demonstrated in chronic hepatitis B patients with concentrations of HBV-DNA $> 10^7$ copies per milliliter [38, 73]. Their ability to persist in the face of high levels of HBV replication is associated with an apparent inability to exert antiviral function. Envelope-specific $CD8^+$ cells are characterized by an altered phenotype (tetramer negative) [73], and their indifference to the dynamic fluctuations of HBV-DNA levels is suggestive of a tolerant state. The persistence of polymerase-specific $CD8^+$ T cell could instead be the result of the low quantity of polymerase epitopes expressed *in vivo* by infected hepatocytes, as suggested by results obtained in the transgenic mouse model of HBV infection [98].

The fact that these data have been generated using only selected HLA-A2 epitopes does not allow us to determine whether the different behavior of these distinct CD8 T-cell epitopes is due to their HBV protein derivation or to intrinsic features of the epitopes (such as affinity for HLA-A2 binding, efficiency of presentation, availability of a T-cell repertoire, cross-reactivity) [99–101]. Nevertheless accumulating evidence supports the former possibility. Firstly, the different behavior of HBV-specific $CD8^+$ T cell, according to their antigenic derivation, has been recently confirmed in transgenic mice [98]. Secondly, the persistence of envelope and polymerase $CD8^+$ T cell in patients with HBV $> 10^7$ copies per milliliter is not restricted to a single epitope but encompasses different epitopes present in the same HBV proteins [73]. Thirdly, $CD8^+$ T-cell responses specific for core epitopes, and restricted by non-HLA-A2 alleles, display behavior identical to that of HLA-A2-restricted, core 18–27–specific $CD8^+$ T cell. They are usually dominant in non-HLA-$A2^+$ patients with self-limited infection [63], and they are present only in chronic patients with low HBV-DNA levels [102].

11.7 Hierarchy of HCV Proteins

All the structural and nonstructural proteins of HCV (Figure 11.1b) are cleaved from one long polyprotein precursor, which is transcribed and translated from a single open reading frame of this small RNA virus [4]. It might therefore be expected that all HCV proteins would be equally and simultaneously available for class I processing and presentation. This differs from the situation in other viral infections in which the differing kinetics of viral protein expression has been shown to impact on the pattern of T-cell immunodominance. In SIV/HIV infection, Nef, Tat, and Rev are the earliest expressed viral proteins and have a prominent role as targets of the CD8 T-cell response in acute infection, which is attributable to the fact that they are presented before other HIV proteins [103,

104]. In EBV infection it has also recently been recognized that the hierarchy of immunodominance among lytic cycle antigens can be explained by the efficiency of presentation of these antigens in the target lytically infected cells [105].

Early reports also suggested some focusing of both CD4 and CD8 responses at the level of HCV proteins. Patients who had controlled HCV appeared to have dominant CD4 T-cell responses to the nonstructural protein NS3 [29]. However, subsequent data have revealed that the majority of individuals with self-limiting infections have helper responses targeted against most of the HCV proteins [6, 26, 28, 55]. Thus, the apparent immunodominance of the NS3 protein was probably merely a reflection of limitations in the methods of analysis.

Similarly for CD8 responses, when a comprehensive and unbiased analysis was carried out, it became apparent that all viral proteins could be targeted [69]. In this study, defective adenoviruses carrying the genes encoding the different HCV proteins were used to infect monocytes, allowing endogenous presentation of naturally processed epitopes. Although there was some evidence of preferential targeting of nonstructural proteins, all adenoviruses carrying different HCV genes were able to elicit significant levels of CTL activity among the patients studied.

Apart from the timing of presentation to the immune system, individual proteins might modulate immune responses through unique properties. An example of this is the ability of the HCV core protein (which may be released into the circulation as naked core protein [106]) to bind to the complement receptor gC1qR on T cell, directly inhibiting their proliferation [107]. Conversely, the E2 protein has been shown to bind to CD81 that provides a costimulatory signal and enhances T-cell proliferation [108]. Data from hepatocytes infected with HCV replicons suggest that the NS4 protein can contribute to downregulation of MHC class I molecules by inhibiting endoplasmic reticulum to Golgi transfer of glycoproteins [109]. However, all these HCV protein-specific effects would be expected to result in generalized alterations of T-cell responses rather than focusing the response on particular proteins.

11.8 Hierarchy of HCV Epitopes

In many viral infections, there are T-cell epitopes that exhibit a striking degree of immunodominance, both within an individual (the largest response) and within a population of HLA-matched individuals (the most common response). In HCV infection, the body of evidence to date suggests no obvious, consistent CD4 or CD8 epitope immunodominance in humans (or among the more limited number of responses mapped in chimpanzees) [62, 110]. The numerous CD8 epitopes identified in human HCV [62, 65] are illustrated in Figure 11.3. Patients who have resolved acute infection have multispecific CD4 responses directed to at least 4–14 epitopes, persisting for years after HCV RNA becomes undetectable [111]. Multispecific helper responses have also been detected during the acute phase of infection, both in patients who subsequently resolve and in those who fail to control

HCV, although responses may be more transient in the latter group [26, 112]. However, some evidence from the chimpanzee model suggests asynchronous evolution of these multispecific responses during the course of acute infection, with precise longitudinal sampling allowing identification of dominant epitopes temporally associated with viral control [113]. The HCV-specific CD8 response is similarly broadly directed during acute infection in humans [32, 65] and chimpanzees [39]. However, in patients [32] and chimpanzees [39, 40] who go on to develop chronic infection, there are some data pointing to the presence of a more narrowly focused CD8 T-cell response already present in acute infection.

Unlike the situation in HBV, in HCV infection no clear pattern emerges from the existing data of preferential focusing of the response on particular epitopes in association with resolution versus chronicity [65]. There have been descriptions of HCV epitopes that are dominant within particular patients [114] (see next section), but these are not "immunodominant" in the sense that they are not necessarily the commonest response in all HCV patients with the relevant HLA type. Conversely, a promiscuous HLA-DR-binding CD4 epitope ($NS3_{1248-1261}$) has been described [115, 116] that was the most frequently recognized within a population of HLA-DR11 individuals but was not immunodominant within the hierarchy of responses of any of those individuals [117]. Although the CD4 responses to this NS3 epitope were at very low frequency in all the patients studied, the virus showed extreme sequence conservation over this region, even between different genotypes. A model of the position of this epitope within NS3 shows it to be buried deeply within the structure where it is likely to be subject to structural constraints to mutation. A survey of other commonly recognized DR11-restricted CD4 epitopes in HCV suggested that sequence conservation might be a unifying feature of these frequent responses [117].

Because HCV is a rapidly replicating, error-prone RNA virus, the development of escape mutations is likely to be a major factor shaping the evolution of the CD8 T-cell response. As discussed above, T-cell responses directed to more structurally or functionally constrained epitopes are likely to be well conserved, and development of *de novo* CD8 responses to mutated epitopes may be limited by the inadequacy of CD4 help [51] and the phenomenon of original antigenic sin [118]. The association of HCV outcome with particular class II [12, 119] and, more recently, class I alleles [120] is in keeping with the hypothesis that certain T-cell specificities are more effective than others. A number of HCV mutations interfering with T-cell recognition have been described, in both humans and the well-characterized chimpanzee model. These include CD8 T-cell escape mutations [75, 121, 122] and partial agonists [122, 123]. Recently, it was also recognized that HCV mutations can impair the induction of CD8 T-cell responses by inhibiting proteasomal epitope processing [124–126]. These multiple mechanisms of viral escape from the HCV-specific T-cell response may partially explain the lack of consistent focusing of the response observed in the cohorts studied to date.

A number of other factors could contribute to the specificity of the T-cell response mounted in response to this virus. The affinity of MHC–peptide binding and stability and the abundance of these complexes on the surface of infected cells

a)

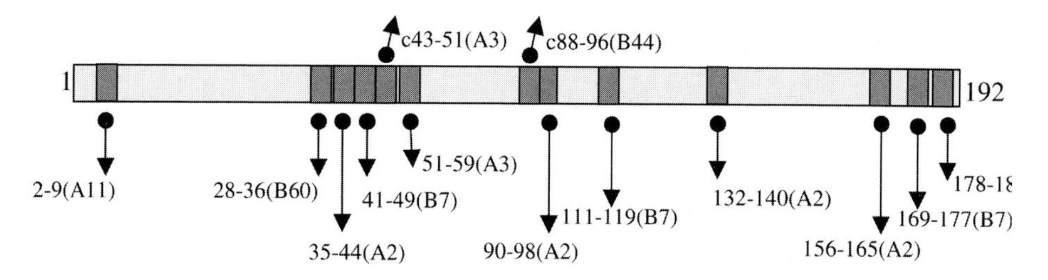

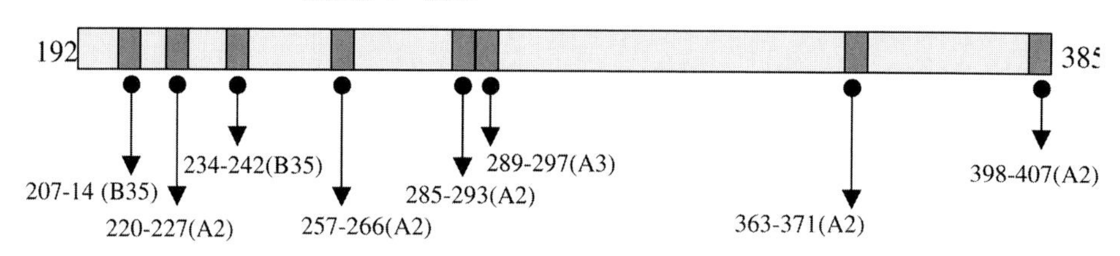

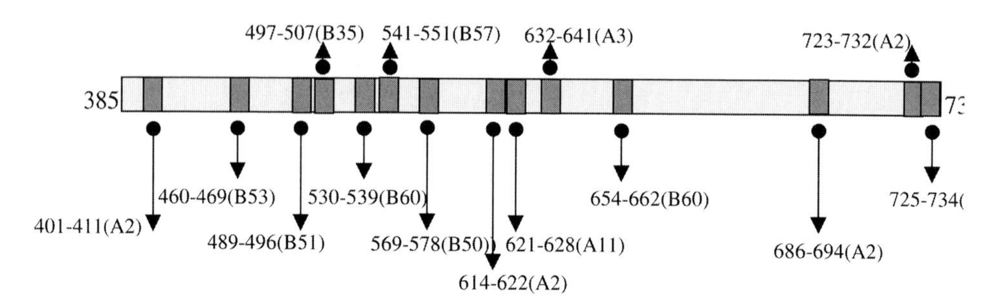

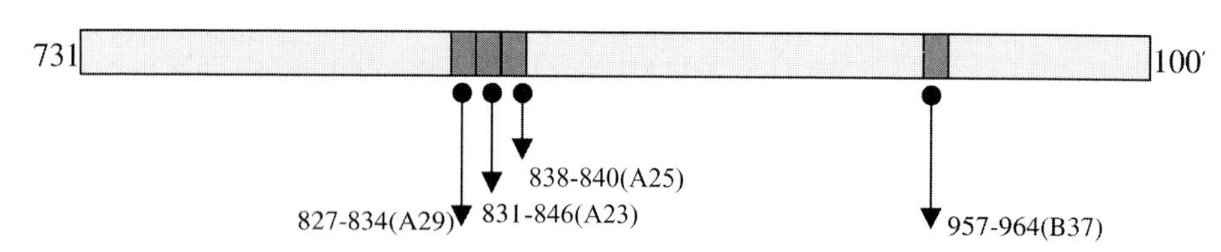

Figure 11.3 HCV-specific HLA class I–restricted epitopes. Epitopes are derived from Refs. [62] and [65] with kind permission from Paul Klenerman.

b)

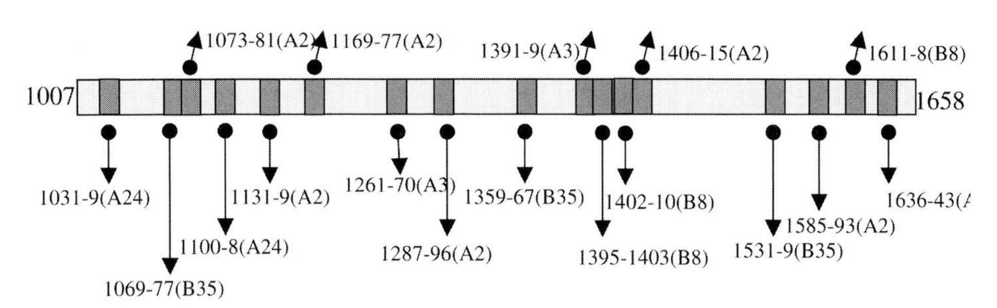
HCV NS3
1007
1658
1073-81(A2)
1169-77(A2)
1391-9(A3)
1406-15(A2)
1611-8(B8)
1031-9(A24)
1069-77(B35)
1100-8(A24)
1131-9(A2)
1261-70(A3)
1287-96(A2)
1359-67(B35)
1395-1403(B8)
1402-10(B8)
1531-9(B35)
1585-93(A2)
1636-43(

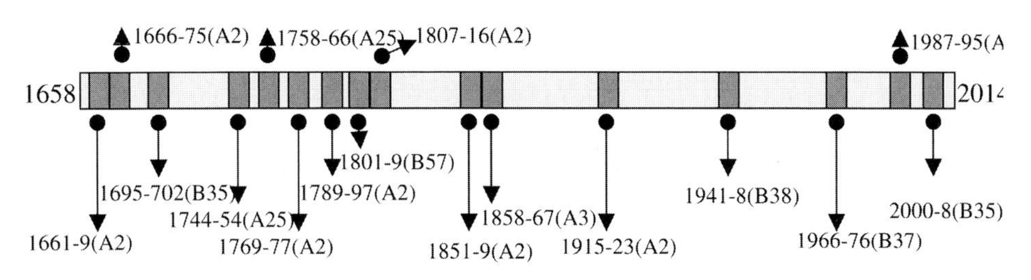
HCV NS4
1658
1666-75(A2)
1758-66(A25)
1807-16(A2)
1987-95(A
1661-9(A2)
1695-702(B35)
1744-54(A25)
1769-77(A2)
1789-97(A2)
1801-9(B57)
1851-9(A2)
1858-67(A3)
1915-23(A2)
1941-8(B38)
1966-76(B37)
2000-8(B35)

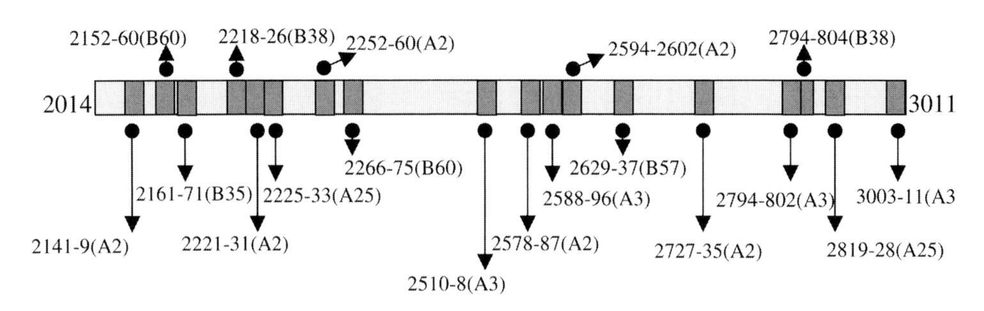
HCV NS5
2014
3011
2152-60(B60)
2218-26(B38)
2252-60(A2)
2594-2602(A2)
2794-804(B38)
2141-9(A2)
2161-71(B35)
2221-31(A2)
2225-33(A25)
2266-75(B60)
2510-8(A3)
2578-87(A2)
2588-96(A3)
2629-37(B57)
2727-35(A2)
2794-802(A3)
2819-28(A25)
3003-11(A3

may play a role, as may the avidity of the T cell. In an infection like HCV, where there is persistent high-level viral replication, inadequate damping down of viral load by the innate response may contribute to clonal deletion of the highest avidity responses. Accumulating data also point to the role of heterologous immunity in shaping antiviral T-cell responses [127], and new data described below have highlighted the relevance of this in the pathogenesis of severe acute HCV infection [114].

11.9 Immunodominance and Liver Pathology

T-cell immunodominance can be shaped by the presence of memory T cell, generated during an early infection, that can cross-react with a second, different infecting viral pathogen. This process, termed "heterologous immunity," was identified and characterized in studies of viral infection in mice, but evidence has accumulated that a similar phenomenon can occur during HCV infection. HCV/influenza virus cross-reactive cells have been described in healthy individuals who were not infected with HCV, and the presence of influenza memory–specific $CD8^+$ cells cross-reacting with HCV is the likely explanation of the marked immunodominance of NS3 1073–81–specific $CD8^+$ cells found in patients with severe acute hepatitis C. The recent work of Urbani et al. identified an unusually narrowly focused and high-frequency NS3 1073–81–specific $CD8^+$ T-cell response (36% and 12% of total $CD8^+$ cells) in two subjects with a rare fulminant onset of acute hepatitis C infection [114]. Patients with milder forms of disease displayed a broader repertoire of HCV-specific CD8 response, and, most importantly, the HCV-specific CD8 response in the fulminant patients cross-reacted with an influenza neuraminidase sequence. Thus, CD8 T-cell cross-reactivity might have a role in shaping the repertoire profile of the HCV-specific CD8 response and also may influence the severity of HCV-related liver pathology. It is interesting to note that despite the presence of a vigorous NS3-specific $CD8^+$ T-cell response, HCV was not controlled in the two patients with fulminant hepatitis. Both patients developed HCV chronicity and did not show any evidence of selection of HCV with mutations within the targeted NS3 1073–81 epitope.

The lack of antiviral efficiency of the NS3 1073–81–specific $CD8^+$ T cell may be a consequence of the intrinsic low protective value of this epitope. Alternatively, the absence of a concomitant multispecific $CD8^+$ T-cell response might be implicated in the inability to control HCV. The overall inadequacy of the single isolated, but strong, NS3 1073–81–specific $CD8^+$ response in controlling the virus may have contributed to the overall severity of liver disease. Intrahepatic activation of HCV-specific $CD8^+$ T cell with poor antiviral function may have sustained a massive recruitment of nonspecific immune cells, causing liver inflammation. These data cast a warning on the development of antiviral vaccines based on a few selected epitopes. Focusing the $CD8^+$ T-cell response on a single immunodominant epitope with poor antiviral efficacy can have severe pathological consequences.

11.10 Concluding Remarks

It is clear that a more extensive and unbiased analysis of the whole range of T-cell responses to HBV and HCV is required before definitive conclusions regarding immunodominance across different HLA types can be reached. However, it is important to keep in mind that immunodominance does not necessarily equate with protective capacity [128], and many responses identified may not exert the required effector functions. For example, HBV-specific $CD8^+$ T cell can be "tetramer-negative"[73] and HCV-specific T cell that bind tetramer often have a "stunned" phenotype [71], being unable to produce IFN-γ and therefore undetectable by IFN-γ-based assays. A combination of assays is therefore best employed to optimize the detection of these populations and to attempt to correlate them with viral control and clinical outcome of infection.

Ultimately the information gained could be applied to the tailoring of new immunotherapeutic approaches for treatment of these chronic viral infections. For example, it remains difficult to select the best HBV proteins for inclusion in a therapeutic vaccine. The fact that core-specific $CD8^+$ cells are associated, as shown both here and in other work [19, 129], with control of HBV replication suggests that core antigen should be included in a vaccine formulation. Nevertheless, association is not proof of a causative effect, and we should not forget that this association may be the consequence and not the cause of low HBV-DNA levels.

In HCV, there is no consensus supporting the inclusion of particular HCV proteins for boosting an effective T-cell response. Instead, the data highlight the importance of vaccines targeting multiple epitopes, preferably those identified as being less prone to undergo escape mutation.

References

1 Lok, A. S. & McMahon, B. J. (2001) *Hepatology* **34**, 1225–41.
2 Lauer, G. M. & Walker, B. D. (2001) *N Engl J Med* **345**, 41–52.
3 Seeger, C. & Mason, W. S. (2000) *Microbiol Mol Biol Rev* **64**, 51–68.
4 Houghton, M., Weiner, A., Han, J., Kuo, G. & Choo, Q.-L. (1991) *Hepatology* **14**, 381–88.
5 Alberti, A., Chemello, L. & Benvegnu, L. (1999) *J Hepatol* **31 Suppl 1**, 17–24.
6 Takaki, A., Wiese, M., Maertens, G., Depla, E., Seifert, U., Liebetrau, A., Miller, J., Manns, M. & Rehermann, B. (2000) *Nature Med.* **6**, 578–82.
7 Vogt, M., Lang, T., Frosner, G., Klingler, C., Sendl, A. F., Zeller, A., Wiebecke, B., Langer, B., Meisner, H. & Hess, J. (1999) *N Engl J Med* **341**, 866–70.
8 Kenny-Walsh, E. (1999) *N Engl J Med* **340**, 1228–33.
9 Koziel, M. J., Wong, D. K., Dudley, D., Houghton, M. & Walker, B. D. (1997) *J Infect Dis* **176**, 859–66.
10 Scognamiglio, P., Accapezzato, D., Casciaro, M. A., Cacciani, A., Artini, M., Bruno, G., Chircu, M. L., Sidney, J., Southwood, S., Abrignani, S., Sette, A. & Barnaba, V. (1999) *J Immunol* **162**, 6681–9.
11 Cote, P., Korba, B., Miller, R., Jacob, J., Baldwin, B., Hornbuckle, W., Purcell, R., Tennant, B. & Gerin, J. (2000) *Hepatology* **31**, 190–200.

12 Thursz, M., Yallop, R., Goldin, R., Trepo, C. & Thomas, H. C. (1999) *Lancet* **354**, 2119–24.
13 Thursz, M., Thomas, H., Greenwood, B. & Hill, A. (1997) *Nature Genet.* **17**, 11–12.
14 Pontisso, P., Gerotto, M., Chemello, L., Casarin, C., Tisminetzky, S., Baralle, F. & Alberti, A. (1995) *J Infect Dis* **171**, 760.
15 Simmonds, P. (1999) *J Hepatol* **31 Suppl 1**, 54–60.
16 Baumert, T. F., Rogers, S. A., Hasegawa, K. & Liang, T. J. (1996) *J Clin Invest* **98**, 2268–76.
17 Sterneck, M., Kalinina, T., Gunther, S., Fischer, L., Santantonio, T., Greten, H. & Will, H. (1998) *Hepatology* **28**, 1390–7.
18 Pollicino, T., Zanetti, A. R., Cacciola, I., Petit, M. A., Smedile, A., Campo, S., Sagliocca, L., Pasquali, M., Tanzi, E., Longo, G. & Raimondo, G. (1997) *Hepatology* **26**, 495–9.
19 Ferrari, C., Penna, A., Bertoletti, A., Valli, A., Degli Antoni, A., Giuberti, T., Cavalli, A., Petit, M. A. & Fiaccadori, F. (1990) *J. Immunol.* **145**, 3442–3449.
20 Jung, M., Spengler, U., Schraut, W., Hoffman, R., Zachoval, R., Eisemburg, J., Eichenlaub, D., Riethmuller, G., Paumgartner, G., Ziegler-Heitbrock, H. W. L. & Pape, G. R. (1991) *J. of Hepatol.* **13**, 310–317.
21 Penna, A., Artini, M., Cavalli, A., Levrero, M., Bertoletti, A., Pilli, M., Chisari, F. V., Rehermann, B., Del Prete, G., Fiaccadori, F. & Ferrari, C. (1996) *J-Clin-Invest* **98**, 1185–94 issn: 0021–9738.
22 Penna, A., Del Prete, G., Cavalli, A., Bertoletti, A., D'Elios, M. M., Sorrentino, R., D'Amato, M., Boni, C., Pilli, M., Fiaccadori, F. & Ferrari, C. (1997) *Hepatology* **25**, 1022–7 issn: 0270–9139.
23 Rehermann, B., Fowler, P., Sidney, J., Person, J., Redeker, A., Brown, M., Moss, B., Sette, A. & Chisari, F. V. (1995) *J. Exp. Med.* **181**, 1047–1058.
24 Penna, A., Chisari, F. V., Bertoletti, A., Missale, G., Fowler, P., Giuberti, T., Fiaccadori, F. & Ferrari, C. (1991) *J. Exp. Med.* **174**, 1565–1570.
25 Jung, M., Hartmann, B., Gerlach, J., Diepolder, H., Gruber, R., Schraut, W., Gruner, N., Zachoval, R., Hoffmann, R., Santantonio, T. & Pape, G. (1999) *Virology* **261**, 165–72.
26 Thimme, R., Oldach, D., Chang, K. M., Steiger, C., Ray, S. C. & Chisari, F. V. (2001) *J Exp Med* **194**, 1395–406.
27 Bottarelli, P., Brunetto, M., Minutello, M., Calvo, P., Unutmaz, D., Weiner, A., Choo, Q., Bonino, F., Houghton, M. & Abrignani, S. (1993) *Gastroenterology* **104**, 580–7.
28 Missale, G., Bertoni, R., Lamonaca, V., Valli, A., Massari, M., Mori, C., Rumi, M., Houghton, M., Fiaccadori, F. & Ferrari, C. (1996) *J. Clin. Invest.* **98**, 706–14.
29 Diepolder, H. M., Zachoval, R., Hoffmann, R. M., Wierenga, E., Santantonio, T., Jung, M. C., Eichenlaub, D. & Pape, G. R. (1995) *Lancet* **346**, 1006–1007.
30 Cramp, M., Carucci, P., Rossol, S., Chokshi, S., Maertens, G., Williams, R. & Naoumov, N. (1999) *Gut* **44**, 424–429.
31 Gr]uner, N. H., Gerlach, T. J., Jung, M. C., Diepolder, H. M., Schirren, C. A., Schraut, W. W., Hoffmann, R., Zachoval, R., Santantonio, T., Cucchiarini, M., Cerny, A. & Pape, G. R. (2000) *J Infect Dis* **181**, 1528–36.
32 Lechner, F., Wong, D., Dunbar, P., Chapman, R., Chung, R., Dohrenwend, P., Robbins, G., Phillips, R., Klenerman, P. & Walker, B. (2000) *J.Exp. Med.* **191**, 1499–512.
33 Lechmann, M., Ihlenfeldt, H. G., Braunschweiger, I., Giers, G., Jung, G., Matz, B., Kaiser, R., Sauerbruch, T. & Spengler, U. (1996) *Hepatology* **24**, 790–5.
34 Chang, K.-M., Thimme, R., Melpolder, J., Oldach, D., Pemberton, J., Moorhead-Lodis, J., McHutchison, J., Alter, H. & Chisari, F. (2001) *Hepatology* **33**, 267–76.
35 Tsai, S. L., Liaw, Y. F., Chen, M. H., Huang, C. Y. & Kuo, G. C. (1997) *Hepatology* **25**, 449–58.
36 Woitas, R. P., Lechmann, M., Jung, G., Kaiser, R., Sauerbruch, T. & Spengler, U. (1997) *J Immunol* **159**, 1012–8.
37 Kamal, S. M., Rasenack, J. W., Bianchi, L., Al Tawil, A., El Sayed Khalifa, K., Peter, T.,

Mansour, H., Ezzat, W. & Koziel, M. (2001) *Gastroenterology* **121**, 646–56.

38 Webster, G. J., Reignat, S., Brown, D., Ogg, G. S., Jones, L., Seneviratne, S. L., Williams, R., Dusheiko, G. & Bertoletti, A. (2004) *J Virol* **78**, 5707–19.

39 Cooper, S., Erickson, A., Adams, E., Kansopon, J., Weiner, A., Chien, D., Houghton, M., Parham, P. & Walker, C. (1999) *Immunity* **10**, 439–449.

40 Thimme, R., Bukh, J., Spangenberg, H. C., Wieland, S., Pemberton, J., Steiger, C., Govindarajan, S., Purcell, R. H. & Chisari, F. V. (2002) *Proc Natl Acad Sci U S A* **99**, 15661–15668.

41 Shoukry, N. H., Grakoui, A., Houghton, M., Chien, D. Y., Ghrayeb, J., Reimann, K. A. & Walker, C. M. (2003) *J Exp Med* **197**, 1645–55.

42 Grakoui, A., Shoukry, N. H., Woollard, D. J., Han, J. H., Hanson, H. L., Ghrayeb, J., Murthy, K. K., Rice, C. M. & Walker, C. M. (2003) *Science* **302**, 659–62.

43 Menne, S., Roneker, C. A., Roggendorf, M., Gerin, J. L., Cote, P. J. & Tennant, B. C. (2002) *J Virol* **76**, 1769–80.

44 Webster, G., Reignat, S., Maini, M., Whalley, S., Ogg, G., King, A., Brown, D., Amlot, P., Williams, R., Vergani, D., Dusheiko, G. & Bertoletti, A. (2000) *Hepatology* **32**, 1117–24.

45 Alberti, A., Diana, S., Sculard, G. H., Eddleston, A. L. & Williams, R. (1978) *Br Med J* **2**, 1056–8.

46 Zibert, A., Meisel, H., Kraas, W., Schulz, A., Jung, G. & Roggendorf, M. (1997) *Hepatology* **25**, 1245–9.

47 Grady, G. F., Lee, V. A., Prince, A. M., Gitnick, G. L., Fawaz, K. A., Vyas, G. N., Levitt, M. D., Senior, J. R., Galambos, J. T., Bynum, T. E., Singleton, J. W., Clowdus, B. F., Akdamar, K., Aach, R. D., Winkelman, E. I., Schiff, G. M. & Hersh, T. (1978) *J Infect Dis* **138**, 625–38.

48 Farci, P., Alter, H. J., Wong, D. C., Miller, R. H., Govindarajan, S., Engle, R., Shapiro, M. & Purcell, R. H. (1994) *Proc Natl Acad Sci U S A* **91**, 7792–6.

49 Farci, P., Shimoda, A., Wong, D., Cabezon, T., De Gioannis, D., Strazzera, A., Shimizu, Y., Shapiro, M., Alter, H. J. & Purcell, R. H. (1996) *Proc Natl Acad Sci U S A* **93**, 15394–9.

50 Klenerman, P., Lechner, F., Kantzanou, M., Ciurea, A., Hengartner, H. & Zinkernagel, R. (2000) *Science* **289**, 2003.

51 Kalams, S. A. & Walker, B. D. (1998) *J. Exp. Med.* **188**, 2199–2204.

52 Ciurea, A., Hunziker, L., Klenerman, P., Hengartner, H. & Zinkernagel, R. M. (2001) *J Exp Med* **193**, 297–305.

53 Bertoletti, A. & Ferrari, C. (2003) *Hepatology* **38**, 4–13.

54 Lauer, G. M., Ouchi, K., Chung, R. T., Nguyen, T. N., Day, C. L., Purkis, D. R., Reiser, M., Kim, A. Y., Lucas, M., Klenerman, P. & Walker, B. D. (2002) *J Virol* **76**, 6104–13.

55 Day, C. L., Seth, N. P., Lucas, M., Appel, H., Gauthier, L., Lauer, G. M., Robbins, G. K., Szczepiorkowski, Z. M., Casson, D. R., Chung, R. T., Bell, S., Harcourt, G., Walker, B. D., Klenerman, P. & Wucherpfennig, K. W. (2003) *J Clin Invest* **112**, 831–42.

56 Barnes, E., Ward, S. M., Kasprowicz, V. O., Dusheiko, G., Klenerman, P. & Lucas, M. (2004) *Eur J Immunol* **34**, 1570–7.

57 Chisari, F. (1997) *J. Clin. Invest.* **99**, 1472–1477.

58 Bertoletti, A., Ferrari, C., Fiaccadori, F., Penna, A., Margolskee, R., Schlicht, H., Fowler, P., Guilhot, S. & Chisari, F. (1991) *Proc. Natl. Acad. Sci. USA* **88**, 10445–10449.

59 Sette, A., Vitiello, A., Reherman, B., Fowler, P., Nayersina, R., Kast, W. M., Melief, C. J., Oseroff, C., Yuan, L., Ruppert, J. & et al. (1994) *J-Immunol* **153**, 5586–92.

60 Cerny, A., McHutchison, J. G., Pasquinelli, C., Brown, M. E., Brothers, M. A., Grabscheid, B., Fowler, P., Houghton, M. & Chisari, F. V. (1995) *J Clin Invest* **95**, 521–30.

61 Falk, K., Rotzschke, O., Stevanovic, S., Jung, G. & Rammensee, H. G. (1991) *Nature* **351**, 290–6.

62 Ward, S., Lauer, G., Isba, R., Walker, B. & Klenerman, P. (2002) *Clin Exp Immunol* **128**, 195–203.

63 Bertoni, R., Sidney, J., Fowler, P., Chesnut, R., Chisari, F. & Sette, A. (1997) *J. Clin. Invest.* **100**, 503–513.
64 Chang, J. J., Wightman, F., Bartholomeusz, A., Ayres, A., Kent, S. J., Sasadeusz, J. & Lewin, S. R. (2005) *J Virol* **79**, 3038–51.
65 Lauer, G. M., Barnes, E., Lucas, M., Timm, J., Ouchi, K., Kim, A. Y., Day, C. L., Robbins, G. K., Casson, D. R., Reiser, M., Dusheiko, G., Allen, T. M., Chung, R. T., Walker, B. D. & Klenerman, P. (2004) *Gastroenterology* **127**, 924–36.
66 Wertheimer, A. M., Miner, C., Lewinsohn, D. M., Sasaki, A. W., Kaufman, E. & Rosen, H. R. (2003) *Hepatology* **37**, 577–89.
67 Guilhot, S., Fowler, P., Portillo, G., Margolskee, R. F., Ferrari, C., Bertoletti, A. & Chisari, F. V. (1992) *J-Virol* **66**, 2670–8 issn: 0022–538x.
68 Wong, D. K., Dudley, D. D., Dohrenwend, P. B., Lauer, G. M., Chung, R. T., Thomas, D. L. & Walker, B. D. (2001) *J Virol* **75**, 1229–35.
69 Urbani, S., Uggeri, J., Matsuura, Y., Miyamura, T., Penna, A., Boni, C. & Ferrari, C. (2001) *Hepatology* **33**, 1533–43.
70 Gerlach, J., Diepolder, H. M., Jung, M. C., Gruener, N., Schraut, W., Zachoval, R., Hoffmann, R. M., Schirren, C., Santantonio, T. & Pape, G. (1999) *Gastroenterology* **117**, 933–941.
71 Gruener, N. H., Lechner, F., Jung, M. C., Diepolder, H., Gerlach, T., Lauer, G., Walker, B., Sullivan, J., Phillips, R., Pape, G. R. & Klenerman, P. (2001) *J Virol* **75**, 5550–8.
72 Goulder, P. J., Altfeld, M. A., Rosenberg, E. S., Nguyen, T., Tang, Y., Eldridge, R. L., Addo, M. M., He, S., Mukherjee, J. S., Phillips, M. N., Bunce, M., Kalams, S. A., Sekaly, R. P., Walker, B. D. & Brander, C. (2001) *J Exp Med* **193**, 181–94.
73 Reignat, S., Webster, G. J., Brown, D., Ogg, G. S., King, A., Seneviratne, S. L., Dusheiko, G., Williams, R., Maini, M. K. & Bertoletti, A. (2002) *J Exp Med* **195**, 1089–101.
74 Rehermann, B. & Nascimbeni, M. (2005) *Nat Rev Immunol* **5**, 215–29.
75 Erickson, A. L., Kimura, Y., Igarashi, S., Eichelberger, J., Houghton, M., Sidney, J., McKinney, D., Sette, A., Hughes, A. L. & Walker, C. M. (2001) *Immunity* **15**, 883–95.
76 Ferrari, C., Bertoletti, A., Penna, A., Cavalli, A., Valli, A., Missale, G., Pilli, M., Fowler, P., Giuberti, T., Chisari, F. V. & et al. (1991) *J-Clin-Invest* **88**, 214–22 issn: 0021–9738.
77 Tsai, S., Chen, M. & Yang, P. (1992) *J. Clin. Invest.* **98**, 1185–1194.
78 Milich, D. & Liang, T. J. (2003) *Hepatology* **38**, 1075–86.
79 Chen, M. T., Billaud, J. N., Sallberg, M., Guidotti, L. G., Chisari, F. V., Jones, J., Hughes, J. & Milich, D. R. (2004) *Proc Natl Acad Sci U S A* **101**, 14913–8.
80 Mizukoshi, E., Sidney, J., Livingston, B., Ghany, M., Hoofnagle, J. H., Sette, A. & Rehermann, B. (2004) *J Immunol* **173**, 5863–71.
81 Bocher, W., Herzog-Hauff, S., Schlaak, J., Meyer zum Buschenfelde, K. & Lohr, H. (1998) *Hepatology* **29**, 238–44.
82 Celis, E., Ou, D. & Otvos, L., Jr. (1988) *J Immunol* **140**, 1808–15.
83 Ferrari, C., Penna, A., Bertoletti, A., Cavalli, A., Valli, A., Schianchi, C. & Fiaccadori, F. (1989) *J Clin Invest* **84**, 1314–9.
84 Vento, S., Ranieri, S., Williams, R., Rondanelli, E., O'Brien, C. & Eddleston, A. (1987) *Lancet* **2(8511)**, 119–22.
85 Chisari, F. & Ferrari, C. (1995) *Ann. Rev. Immunol.* **13**, 29–60.
86 Mondelli, M., Vergani, G. M., Alberti, A., Vergani, D., Portmann, B., Eddleston, A. L. & Williams, R. (1982) *J Immunol* **129**, 2773–8.
87 Missale, G., Redeker, A., Person, J., Fowler, P., Guilhot, S., Schlicht, H. J., Ferrari, C. & Chisari, F. V. (1993) *J-Exp-Med* **177**, 751–62.
88 Nayersina, R., Fowler, P., Guilhot, S., Missale, G., Cerny, A., Schlicht, H. J., Vitiello, A., Chesnut, R., Person, J. L., Redeker, A. J. & Chisari, F. V. (1993) *J. Immunol.* **150**, 4659–4671.
89 Hwang, Y. K., Kim, N. K., Park, J. M., Lee, K., Han, W. K., Kim, H. I. & Cheong, H. S. (2002) *Vaccine* **20**, 3770–7.

90 Rehermann, B., Ferrari, C., Pasquinelli, C. & Chisari, F. V. (1996) *Nat. Med.* **2**, 1104–1108.
91 Rehermann, B., Lau, D., Hoofnagle, J. H. & Chisari, F. V. (1996) *J.Clin.Invest.* **97**, 1655–1665.
92 Sobao, Y., Sugi, K., Tomiyama, H., Saito, S., Fujiyama, S., Morimoto, M., Hasuike, S., Tsubouchi, H., Tanaka, K. & Takiguch, M. (2001) *J Hepatol* **34**, 922–9.
93 Maini, M. K., Boni, C., Ogg, G. S., King, A. S., Reignat, S., Lee, C. K., Larrubia, J. R., Webster, G. J. M., McMichael, A. J., Ferrari, C., Williams, R., Vergani, D. & Bertoletti, A. (1999) *Gastroenterology* **117**, 1386–96.
94 Maini, M. K., Boni, C., Lee, C. K., Larrubia, J. R., Reignat, S., Ogg, G. S., King, A. S., Herberg, J., Gilson, R., Alisa, A., Williams, R., Vergani, D., Naoumov, N. V., Ferrari, C. & Bertoletti, A. (2000) *J.Exp. Med.* **191**, 1269–80.
95 Bertoletti, A., Southwood, S., Chesnut, R., Sette, A., Falco, M., Ferrara, G., Penna, A., Boni, C., Fiaccadori, F. & Ferrari, C. (1997) *Hepatology* **26**, 1027–1034.
96 Bertoletti, A., Costanzo, A., Chisari, F. V., Levrero, M., Artini, M., Sette, A., Penna, A., Giuberti, T., Fiaccadori, F. & Ferrari, C. (1994) *J. Exp. Med.* **180**, 933–943.
97 Rehermann, B., Pasquinelli, C., Mosier, S. M. & Chisari, F. V. (1995) *J-Clin-Invest* **96**, 1527–1534.
98 Kakimi, K., Isogawa, M., Chung, J., Sette, A. & Chisari, F. V. (2002) *J Virol* **76**, 8609–20.
99 Levitsky, V., Zhang, Q. J., Levitskaya, J. & Masucci, M. G. (1996) *J-Exp-Med* **183**, 915–26.
100 Chen, W., Anton, L. C., Bennink, J. R. & Yewdell, J. W. (2000) *Immunity* **12**, 83–93.
101 Brehm, M. A., Pinto, A. K., Daniels, K. A., Schneck, J. P., Welsh, R. M. & Selin, L. K. (2002) *Nat Immunol* **3**, 627–34.
102 Sobao, Y., Tomiyama, H., Sugi, K., Tokunaga, M., Ueno, T., Saito, S., Fujiyama, S., Morimoto, M., Tanaka, K. & Takiguchi, M. (2002) *J Hepatol* **36**, 105–15.
103 Allen, T. M., O'Connor, D. H., Jing, P., Dzuris, J. L., Mothe, B. R., Vogel, T. U., Dunphy, E., Liebl, M. E., Emerson, C., Wilson, N., Kunstman, K. J., Wang, X., Allison, D. B., Hughes, A. L., Desrosiers, R. C., Altman, J. D., Wolinsky, S. M., Sette, A. & Watkins, D. I. (2000) *Nature* **407**, 386–90.
104 van Baalen, C. A., Guillon, C., van Baalen, M., Verschuren, E. J., Boers, P. H., Osterhaus, A. D. & Gruters, R. A. (2002) *Eur J Immunol* **32**, 2644–52.
105 Pudney, V. A., Leese, A. M., Rickinson, A. B. & Hislop, A. D. (2005) *J Exp Med* **201**, 349–60.
106 Kanto, T., Hayashi, N., Takehara, T., Hagiwara, H., Mita, E., Naito, M., Kasahara, A., Fusamoto, H. & Kamada, T. (1994) *Hepatology* **19**, 296–302.
107 Kittlesen, D. J., Chianese-Bullock, K. A., Yao, Z. Q., Braciale, T. J. & Hahn, Y. S. (2000) *J Clin Invest* **106**, 1239–49.
108 Wack, A., Soldaini, E., Tseng, C., Nuti, S., Klimpel, G. & Abrignani, S. (2001) *Eur J Immunol* **31**, 166–75.
109 Konan, K. V., Giddings, T. H., Jr., Ikeda, M., Li, K., Lemon, S. M. & Kirkegaard, K. (2003) *J Virol* **77**, 7843–55.
110 Shoukry, N. H., Cawthon, A. G. & Walker, C. M. (2004) *Annu Rev Microbiol* **58**, 391–424.
111 Day, C. L., Lauer, G. M., Robbins, G. K., McGovern, B., Wurcel, A. G., Gandhi, R. T., Chung, R. T. & Walker, B. D. (2002) *J Virol* **76**, 12584–95.
112 Lechner, F., Gruener, N. H., Urbani, S., Uggeri, J., Santantonio, T., Kammer, A. R., Cerny, A., Phillips, R., Ferrari, C., Pape, G. R. & Klenerman, P. (2000) *Eur J Immunol* **30**, 2479–87.
113 Shoukry, N. H., Sidney, J., Sette, A. & Walker, C. M. (2004) *J Immunol* **172**, 483–92.
114 Urbani, S., Amadei, B., Fisicaro, P., Pilli, M., Missale, G., Bertoletti, A. & Ferrari, C. (2005) *J Exp Med* **201**, 675–80.
115 Diepolder, H. M., Gerlach, J. T., Zachoval, R., Hoffmann, R. M., Jung, M. C., Wierenga, E. A., Scholz, S., Santantonio, T., Houghton, M., Southwood, S., Sette, A. & Pape, G. R. (1997) *J Virol* **71**, 6011–9.

116 Lamonaca, V., Missale, G., Urbani, S., Pilli, M., Boni, C., Mori, C., Sette, A., Massari, M., Southwood, S., Bertoni, R., Valli, A., Fiaccadori, F. & Ferrari, C. (1999) *Hepatology* **30**, 1088–98.

117 Harcourt, G. C., Lucas, M., Sheridan, I., Barnes, E., Phillips, R. & Klenerman, P. (2004) *J Viral Hepat* **11**, 324–31.

118 Klenerman, P. & Zinkernagel, R. M. (1998) *Nature* **394**, 482–485.

119 Fanning, L. J., Levis, J., Kenny-Walsh, E., Whelton, M., O'Sullivan, K. & Shanahan, F. (2001) *Hepatology* **33**, 224–30.

120 McKiernan, S. M., Hagan, R., Curry, M., McDonald, G. S., Kelly, A., Nolan, N., Walsh, A., Hegarty, J., Lawlor, E. & Kelleher, D. (2004) *Hepatology* **40**, 108–14.

121 Weiner, A., Erickson, A. L., Kansopon, J., Crawford, K., Muchmore, E., Hughes, A. L., Houghton, M. & Walker, C. M. (1995) *Proc-Natl-Acad-Sci-U-S-A* **92**, 2755–9 issn: 0027–8424.

122 Chang, K. M., Rehermann, B., McHutchison, J. G., Pasquinelli, C., Southwood, S., Sette, A. & Chisari, F. V. (1997) *J Clin Invest* **100**, 2376–85.

123 Tsai, S.-L., Chen, Y.-M., Chen, M.-H., Huang, C.-Y., Sheen, I.-S., Yeh, C.-T., Huang, J.-H., Kuo, G.-C. & Liaw, Y.-F. (1998) *Gastroenterology* **115**, 954–66.

124 Seifert, U., Liermann, H., Racanelli, V., Halenius, A., Wiese, M., Wedemeyer, H., Ruppert, T., Rispeter, K., Henklein, P., Sijts, A., Hengel, H., Kloetzel, P. M. & Rehermann, B. (2004) *J Clin Invest* **114**, 250–9.

125 Kimura, Y., Gushima, T., Rawale, S., Kaumaya, P. & Walker, C. M. (2005) *J Virol* **79**, 4870–6.

126 Timm, J., Lauer, G. M., Kavanagh, D. G., Sheridan, I., Kim, A. Y., Lucas, M., Pillay, T., Ouchi, K., Reyor, L. L., Zur Wiesch, J. S., Gandhi, R. T., Chung, R. T., Bhardwaj, N., Klenerman, P., Walker, B. D. & Allen, T. M. (2004) *J Exp Med* **200**, 1593–604.

127 Selin, L. K. & Welsh, R. M. (2004) *Immunity* **20**, 5–16.

128 Gallimore, A., Dumrese, T., Hengartner, H., Zinkernagel, R. M. & Rammensee, H. G. (1998) *J. Exp. Med.* **187**, 1647–1657.

129 Lau, G. K., Suri, D., Liang, R., Rigopoulou, E. I., Thomas, M. G., Mullerova, I., Nanji, A., Yuen, S. T., Williams, R. & Naoumov, N. V. (2002) *Gastroenterology* **122**, 614–24.

12
Immunodominance in the T-Cell Response to Herpesviruses

Michael W. Munks and Ann B. Hill

12.1
Introduction

Herpesviruses are ubiquitous companions of the adaptive immune system; most mammals are infected for life by at least one of these viruses. They are large, complex viruses with orchestrated gene expression cascades, and they encode a myriad of genes that enable them to manipulate and avoid host immune responses. Nevertheless, herpesvirus manipulation of the immune system is largely benign: rather than seeking short-term advantage, these viruses take the long view, and their evolutionary success can be attributed to the minimal cost they impose on host well-being. These viruses make use of the immune response to position themselves in the host: while the immunocompetent host remains an asymptomatic carrier of the virus, immunocompromise upsets the host–virus balance, and recrudescent herpesvirus infections can lead to death.

The T-cell response to herpesviruses is an essential part of the virus–host equilibrium: it controls but does not eradicate virus. In this chapter we review the immunodominance choices the immune system makes in responding to herpesvirus infections and consider how they fit into this scenario. Because the literature is much more extensive for CD8 than for CD4 T cells, we focus mostly on CD8 T cells. In addition, the CD8 T-cell response to herpesviruses provides two of the most extreme examples of immunodominance that have yet been described in the literature. While the basis of these is not yet understood, they are highlighted here because of the opportunity they provide to probe the finer decision-making processes that the immune system undertakes to determine immunodominance.

Immunodominance: The Choice of the Immune System. Edited by Jeffrey A. Frelinger

ISBN: 3-527-31274-9

12.2
General Considerations

12.2.1
Herpesviruses: A Brief Virological Primer

Herpesviruses are an ancient virus family with two broad lineages: one infecting birds and mammals and another that infects poikilothermic animals such as oysters [1]. Herpesviruses of birds and mammals are divided into three subfamilies: alpha, beta, and gamma. These three lineages were clearly established before the mammalian radiation 60–80 million years ago, and in general, herpesviruses are thought to have cospeciated with their hosts [1, 2]. Herpesvirus genomes contain genes that have clearly been acquired from their hosts, and it is apparent that there has been a close coevolution. Primordial herpesviruses may well have been with us as the adaptive immune system developed; at any rate, it is clear that this family of viruses has an intimate knowledge of the workings of our immune system and that it exploits that knowledge to maintain its impressive evolutionary stability.

Herpesviruses have very large genomes, encoding between 80 and 250 viral proteins. They have a conserved virion structure, and the mammalian and avian herpesviruses share a homologous set of core genes that control the basic program of virus replication [1]. These viruses also have a common lifestyle, establishing latent infection (Table 12.1) for the life of the host. The focus of this chapter is the nine human herpesviruses and their murine models (Table 12.2).

Table 12.1 Terms used to describe herpesvirus gene expression programs.

Productive infection/ lytic cycle	**Virus replication leading to production of new infectious virions; because this results in cell lysis, it is also referred to as the lytic cycle.**
Latency	Maintenance of the viral genome in a cell without production of infectious virions, with or without viral protein synthesis.
Reactivation	The conversion of latent virus infection to productive infection.
Abortive reactivation	Latent virus initiates the "reactivation" sequence of gene expression, but the process is aborted or the cell destroyed before infectious virions are produced.
Persistent infection	Continued virus infection with production of new virions after the acute phase of virus infection of the animal has passed.
Chronic infection	Many virologists use the term chronic infection interchangeably with persistent infection, implying continued lytic cycle virus replication. We use the term here to refer to an animal that continues to harbor the herpesvirus after the acute infection has been resolved. The infection may be latent or persistent; very likely both genetic programs occur simultaneously in the animal, with foci of virus replication alongside many cells and tissues in which the virus is maintained in a true latent state.

Table 12.2 The nine known human herpesviruses and their mouse models.

Subfamily	Human virus	Disease	Mouse model	Comment on the mouse model
Alpha	Herpes simplex virus (HSV) 1	Herpes labialis	HSV	Human virus used to infect mice
	HSV-2	Genital herpes		
	Varicella-zoster virus (VZV)	Varicella (chicken pox) and herpes zoster (shingles)	None	
Beta	Human cyto-megalovirus (HCMV)	Congenital malformations; disease in the immunocom-promised	Murine CMV (MCMV)	Natural pathogen of laboratory mice (*Mus musculis* species)
	HHV-6A and HHV-6B	Roseola infantum	None	
	HHV-7		None	
Gamma-1	Esptein-Barr virus	Infectious mononucleosis, Burkitt's lym-phoma, and naso-pharyngeal lym-phoma	MHV-68	Actually a gamma-2 herpesvirus, a natural pathogen of wood mice (genus *Apodemus*) that is able to infect mice of the *Mus* genus
Gamma-2	Kaposi's sarcoma–associated herpes-virus (KSHV)/HHV-8	Kaposi's sarcoma	MHV-68	See above

The lytic cycle of herpesvirus infection leads to production of new infectious virions. For all herpesviruses, lytic cycle gene expression is characterized by a regulated, ordered sequence of gene expression (Figure 12.1), from immediate early (IE) to early (E) to late (L) genes. The programs of latency differ for the three herpesvirus families. The gammaherpesviruses encode sets of genes that are uniquely expressed during latency and are responsible for maintenance and propagation of the latent virus genome. Alphaherpesviruses express latency-specific transcripts, but no latent proteins are known. For betaherpesviruses, the program of gene expression during latency, if any, is not known.

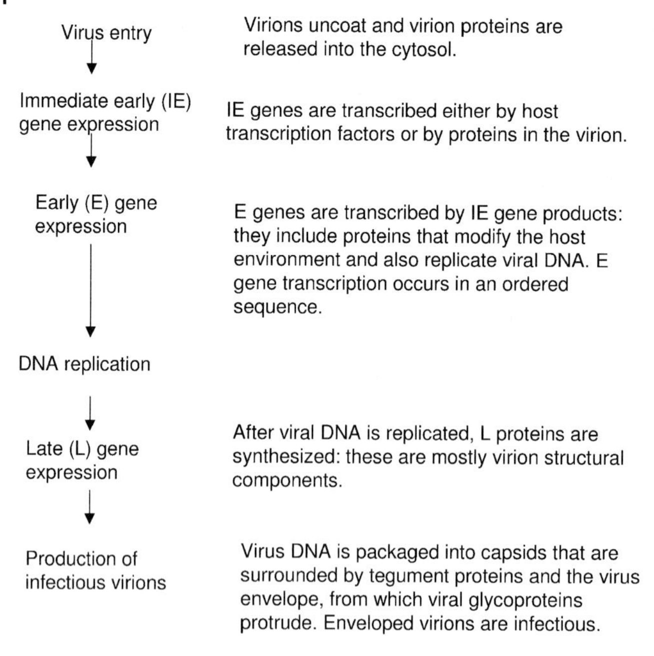

Figure 12.1 Scheme of herpesvirus replication showing the kinetic classes of viral gene transcription.

In choosing immunodominant epitopes for herpesvirus infections, the immune system operates within the parameters that the earlier chapters of this book attempt to elucidate. Overlying this, there are several herpesvirus-specific considerations that need to be taken into account. These considerations have led us to the following general framework, which will serve as a scaffold for the discussion of immunodominance patterns in the T-cell response to each of the three subfamilies of herpesvirus.

12.2.2
A General Framework for Thinking About Immunodominance in the T-Cell Response to Herpesvirus Infections

We can conceptualize three broad classes of explanation for immunodominance in herpesvirus infections, which we call virus centric, APC centric, and T cell centric.

Virus-centric explanations are based on the availability of viral antigen: the more abundant a parental antigen is, the more substrate there is available for processing and presentation, and the more likely an antigen is to be immunodominant. In herpesvirus infections, antigen abundance is affected by the program of viral gene expression.

During lytic cycle infection, herpesvirus gene expression occurs in a regulated cascade from immediate early (IE) to early (E) to late (L) proteins (Figure 12.1). Even before any viral gene activity occurs, virion structural proteins may be delivered into the cytosol with the infective virus inoculum. If most virus-infected cells are destroyed (by the immune system) during the early phases of infection, then IE genes, or possibly virion structural proteins, will be more abundantly presented than genes expressed later during infection.

The second herpesvirus-specific aspect of abundance of gene expression that needs to be considered is the nature of viral latency. As described above, all herpesviruses establish latent infection, during which the viral genome is maintained without infectious virions being produced. For most of the life of the host, most herpesvirus infections are clinically latent, i.e., there is no evidence of productive infection. Therefore, if proteins are expressed during latency, these are likely to provoke immunodominant responses during the lifelong latent state. In order to maintain their genomes during latency, gammaherpesviruses express a unique set of latency-specific proteins that are not expressed during lytic infection. These genes can lead to transformation and uncontrolled replication of the host cell, presumably as a means of propagating the virus genome. Thus, viral antigen abundance for gammaherpesviruses will depend on whether latent- or lytic-phase infection is dominant. The alpha- and betaherpesviruses are not known to encode separate proteins that are expressed only during latency. For these viruses, the proteins that are first expressed upon reactivation from latency, such as the IE1 protein of cytomegalovirus (CMV), may have a viral protein abundance advantage during latency, especially if such reactivation is frequently abortive.

Another virus-centric explanation of particular relevance to herpesvirus infections is the impact of viral genes that interfere with antigen presentation (VIPRs) [3], which all herpesviruses seem to encode. VIPRs impair or prevent presentation in the infected cell. If directly infected cells are important for priming the CD8 T-cell response, VIPRs should have a profound impact on the immune system's choice of immunodominant CD8 T-cell epitopes.

APC-centric explanations are concerned with the processing and presentation of viral epitopes. Obviously, to be immunodominant a peptide must be able to bind to MHC, and the availability of peptides with high affinity for the available MHC types is likely to be the single most important factor influencing immunodominance [4, 5]. Similarly, the epitope must be processed and presented by the relevant antigen-presenting cell (APC) type. The relevant APC is presumed to be dendritic cells (DCs), either directly presenting or cross-presenting, for the initial priming of T cells. Whether other APC types can then go on to maintain (save from activation-induced cell death) or expand T-cell populations is not clear, although we note with interest that DCs seem uniquely able to promote activation even of memory CD8 T cells [6]. On the other hand, for the B cell–trophic gammaherpresviruses, antigen presentation by B cells seems likely to play a prominent role. (Interference with antigen presentation by VIPRs is, of course, another APC-centric consideration.)

T cell–centric explanations are based on the assumption that some T-cell receptors (TCRs) are better than others. All other things being equal, T cells bearing those receptors win out over other T cells. The concept of immunodomination, by which a dominant T-cell response actually suppresses other responses, is an extension of this explanation [4, 5]. One issue in T cell–centric explanations is that we don't really understand what makes one TCR better than another: it is not clear what the best relationship between TCR and MHC–peptide is, in terms of affinity, occupancy time, and conformational change induced by the interaction. Even within responses to some peptide–MHC complexes, certain T-cell clones come to dominate. We need to introduce the concept of TCR immunodominance (dominance of one TCR within the response to a given peptide–MHC ligand) to add to the more general concept of epitope immunodominance (dominance of a response to one peptide over others). The two most striking examples of TCR immunodominance in the literature are provided by herpesvirus immunology: one in the murine model of herpes simplex virus (HSV) infection and the other in human Epstein-Barr virus (EBV) infection. Although the basis for these is not understood, it can be hoped that the experimental clarity offered by this extreme immunodominance may lead in time to elucidation of the principles that underlie the more frequent examples of epitope immunodominance.

12.3 Immunodominance in the CD8 T-Cell Response to the Three Classes of Herpesvirus

12.3.1 Alphaherpesviruses

12.3.1.1 Human Studies: Immunodominance of Structural Virion Proteins that can be Presented in the Face of Immune Evasion

The alphaherpesviruses HSV and varicella-zoster virus (VZV) are neurotrophic viruses that cause recognizable clinical symptoms in both acute infection and upon reactivation (Table 12.1). VZV cannot be readily propagated *in vitro*, with the result that little is known about its immunobiology. HSV, on the other hand, grows easily in tissue culture and has been studied for some time in both mouse models and humans. HSV-2 causes genital herpes and is also the main cause of fatal HSV encephalitis in newborns. Because of its clinical importance, much of the best human HSV immunology has been performed for HSV-2. The clearance of virus from the lesions and the resolution of clinical symptoms in genital herpes correlates with infiltration of CD8 T cells into the lesions [7, 8]. This suggests that CD8 T cells are able to detect and control HSV infection *in vivo*.

Several HSV antigens that are recognized by CD8 T cells from infected people have been identified. Theses are mostly structural virion proteins: glycoproteins and tegument proteins [9, 10]. In addition, IE antigens are also targets for CD8 T cells [9, 11]. A careful analysis of a large panel of clones that could recognize HSV-2-infected cells confirmed the immunodominance of responses to structural virion proteins and to IE proteins [10].

Koelle et al. have used drug blockade and viral mutants to investigate when in the infectious cycle the structural viral proteins, which are all encoded by late genes, can be recognized by CD8 T cells [9]. If a high multiplicity of infection (MOI) is used, these structural virion proteins can be processed and presented without requiring synthesis of new viral proteins. Similarly, the immediate early protein ICP0 is presented very shortly after viral entry. HSV-2 encodes two powerful genes that inhibit the MHC class I antigen-processing pathway. ICP47 is an IE protein (the only one of HSV's five IE proteins that is not a transcription factor) that potently inhibits the TAP transporter [12, 13]. The virion host shutoff (vhs) protein encoded by UL41 is a potent inhibitor of host class I proteins, including HLA class I. vhs is a virion protein that begins to shut down host protein synthesis within the first hour after virion entry [14]. Antigens that are present in virions in large quantities, or that are synthesized in abundance from IE genes, would have the greatest chance of being presented before these two powerful immune evasion functions prohibit presentation of other epitopes. HSV can spread directly from an infected cell to its neighbor. When this happens, a large number of virions are delivered into the cell, analogous to infection *in vitro* with a high MOI. Hence, processing of antigens from input structural virions is physiologically feasible, and epitopes presented from this source may be the target of a protective CD8 T-cell response. Thus, Koelle et al. postulate that structural virion proteins and IE proteins are immunodominant in the CD8 T-cell response to HSV because of their ability to escape immune evasion.

However, are these responses truly immunodominant? Before accepting this conclusion, it is worthwhile to examine the way the data were generated. Koelle et al. generated CD8 T-cell clones from T cells that were infiltrating genital herpes lesions [9] or from peripheral blood CD8 T cells bearing a tissue-homing receptor [10]. Clones that could recognize HSV-2-infected cells were selected for antigen identification, and this may have introduced a major bias into their system. In the CMV and EBV systems, it is clear that CD8 T-cell responses are readily generated (we believe by cross-priming) that are unable to recognize infected cells (this is described later in the chapter). If that were also the case for HSV, these authors would not have detected them. The antigens identified by Koelle et al. are extremely important, because cytotoxic T lymphocytes (CTLs) recognizing them are actually able to recognize infected cells and hence would be useful vaccine candidates. However, we do not believe that studies have yet been performed that would enable us to determine whether or not these antigens are truly immunodominant.

12.3.1.2 Mouse Studies of HSV

Mice are not a natural host for HSV, and we are not aware of an alphaherpesvirus that is a natural mouse pathogen. Although mice can be efficiently infected by HSV, usually by footpad injection or corneal scarification, the natural course of infection is different from that seen in humans. In sublethal infection, after acute infection is resolved, virus does persist in a latent state. However, observable disease resulting from spontaneous reactivation does not occur. This may be due to virological reasons or to more efficient control of latent HSV by the adaptive immune system. HSV-specific CD8 T cells surround infected cells in the trigeminal ganglia [15]. The main HSV VIPR, ICP47, has only weak activity against mouse TAP [16] and has not been shown to inhibit antigen presentation. Mouse studies have mostly been carried out with HSV-1 strains, which have a weak ability to impair host protein synthesis that is not sufficient to impair antigen presentation. As a result, CD8 T cells are likely to be much more efficient in controlling HSV in mice. Thus, when the immune system "chooses" its immunodominant CD8 T-cell epitopes in HSV-infected mice, it is able to do so free from any constraints imposed by immune evasion of the class I pathway.

In the mouse system, in parallel to the human, attention has focused on structural viral glycoproteins and IE antigens, and evidence of responses to both has been found, with different immunodominance patterns found for different mouse strains. gB was identified as immunodominant in C57BL/6 mice, whereas gC [17] and gD [18] were implicated as antigens in mice that expressed the H-2^k MHC haplotype. The IE protein ICP4 is an antigen for H-2^k mice [19], whereas ICP27 is antigenic for H-2^d and H-2^b mice. In BALB/c (H-2^d), ICP27 is relatively immunodominant: limiting dilution assay indicated that 25% of clones that could recognize HSV were able to recognize a recombinant vaccinia encoding ICP27 [20].

12.3.1.3 The Remarkable Immunodominance of gB-SSIEFARL in B6 Mice

In C57BL/6 mice, a K^b-restricted epitope (SSIEFARL) from gB is immunodominant. Two other epitopes have been described, one from ICP6, the large subunit of the viral ribonucleotide reductase (RR) [21], and one from ICP27 [22]. The gB epitope is profoundly immunodominant: CD8 T cells recognizing gB comprise 60–90% of HSV-reactive CD8 T cells in draining lymph nodes at the height of the acute response [23]. This extraordinary immunodominance appears to be based on a very limited TCR usage and thus is likely to represent an extreme example of T cell–centric immunodominance. There is some disagreement in the literature about the extent of response to the other two epitopes. However, the massive immunodominance of gB has been confirmed by other groups, particularly when newer methodologies rather than limiting dilution analysis have been used.

The response to K^b-SSIEFARL in C57BL/6 mice is probably the most extreme example of immunodominance in the literature. Because gB is not particularly immunodominant in other mice or humans, its immunodominance in C57BL/6 mice is likely to be due to some unique features of the TCRs that interact with K^b-SSIEFARL. Despite its impressive immunodominance, the K^b-SSIEFARL response is not marked by high affinity, nor is it especially protective [24]. The Carbone group performed an interesting experiment that may shed some light on the question of how the immune system chooses which TCRs to use in response to a given determinant [25]. They generated a TCR β chain transgenic mouse bearing the TCR β chain from a K^b-SSIEFARL-specific T-cell clone. While a fairly diverse TCR repertoire was generated in these mice, the choice of TCR α chains was nevertheless constrained, and approximately 25% of all CD8 T cells in the spleen could bind K^b-SSIEFARL tetramers. When the mice were infected with HSV, only a subset of these tetramer-binding T cells expanded. The subset that expanded did not appear to be of higher affinity than those that did not expand, based rather roughly on tetramer binding. Thus, only a subset of TCRs that can bind a given ligand were able to transmit an activation signal to the T cell. While the authors propose that this select subset of TCRs included those that underwent a conformational change upon ligand binding, there is no experimental evidence to support this claim.

A crystal structure of the immunodominant TCR alone and complexed to K^b-SSIEFARL will be required to determine whether this TCR does undergo a marked conformational change upon ligand binding. To our knowledge, no other hypothesis has been proposed to explain this extraordinary TCR immunodominance, and at present it remains an unsolved mystery.

12.3.2 Betaherpesviruses

There are three known members of the betaherpesvirus family: the cytomegaloviruses, HHV-6, and HHV-7. As yet, no T-cell antigens have been described for HHV-6 or HHV-7, so these will be discussed only briefly in the section on T-cell cross-reactivity. However, a substantial body of literature on CD8 T-cell immunodominance to murine cytomegalovirus (MCMV) exists, and on both CD4 and CD8 T-cell immunodominance to human cytomegalovirus (HCMV), and these will be the focus of this section.

Each cytomegalovirus is specific for only one mammalian species. Both HCMV and MCMV are highly prevalent, with 60–90% of individuals infected. CMV infections are normally asymptomatic, with innate immunity (particularly NK cells) and all facets of adaptive immunity (CD4 T cells, CD8 T cells, and antibody) playing an important role in immune control.

12.3.2.1 The Acute CD8 T-Cell Response to MCMV

Because HCMV infection is normally asymptomatic and goes undetected, there is scant literature on acute T-cell responses to HCMV. In contrast, the acute CD8 T-cell response to MCMV has been characterized in great detail in both the BALB/c ($H\text{-}2^d$) and C57BL/6 ($H\text{-}2^b$) strains of mice.

Because more MCMV pathology is seen in the sensitive BALB/c mouse strain, many studies of MCMV immunity have focused on this model. The ability to restrict herpesvirus gene expression to the *IE* class was used to determine that an *IE* gene was a major CD8 T-cell antigen in BALB/c ($H\text{-}2^d$) mice [26], and in 1989 a peptide from IE1/pp89 became the first CD8 T-cell epitope from any herpesvirus infection to be identified [27]. Further epitopes from early genes and structural proteins were identified by motif-based epitope predictions or by analogy with the human response [28, 29].

Eight days after footpad infection, there is a codominant CD8 T-cell response to IE1 and to the E gene m164 in the popliteal lymph node, as measured by ELISPOT [30], with subdominant responses to m18 and M45 also detectable directly *ex vivo*. Responses to the other identified epitopes were detectable only after *in vitro* expansion of low-frequency memory CD8 T cells. Together, these epitopes account for most of the CD8 T cells that are capable of responding to anti-CD3 stimulation, which is used as an estimate of the total MCMV response.

The CD8 T-cell response in C57BL/6 mice is both larger and more diverse than that of BALB/c mice. Our laboratory first used an expression library of genomic MCMV DNA fragments to identify the E antigen M45-D^b as immunodominant in C57BL/6 mice [31]. In a separate venture, we cloned and expressed each open reading frame (ORF) from MCMV; by screening this "ORF library" with CD8 T cells taken directly from infected mice, we were able to identify a total of 27 $H\text{-}2^b$-restricted antigens that are recognized during the acute response, from which we have identified 24 peptide epitopes (M. Munks, unpublished data). These epitopes account for the majority of the CD8 T-cell response to MCMV. The M45 epitope still reigns as the most immunodominant, but it constitutes less than one-third of the virus-specific CD8 T-cell response. Most of these epitopes are encoded by E genes. Interestingly, no IE epitope was identified during the acute response to MCMV in C57BL/6 mice.

Thus, in BALB/c mice, the CD8 T-cell response is dominated by two epitopes, but responses against a total of four epitopes are detectable directly *ex vivo* and comprise most of the MCMV-specific response. The majority of CD8 T-cell epitopes are also identified for C57BL/6 mice, with 24 known epitopes, but no single epitope constitutes more than one-third of the total response.

12.3.2.2 The Memory CD8 T-Cell Response to MCMV Becomes More Focused Over Time

In MCMV, there is a growing body of evidence about how the CD8 T-cell response evolves from acute to chronic infection. In humans, where it is not possible to experimentally control the time of infection, information is more limited. But in both circumstances, signs of T-cell selection are evident.

In an MCMV bone marrow transplantation (BMT) model of infection, the latent viral load is highest in the lungs, and this also represents the organ where reactivation occurs most frequently [32]. Pulmonary CD8 T cells of many specificities peak during acute infection but then decline during the latent phase of infection. However, CD8 T cells specific for IE1 and m164 become enriched, in both relative and absolute terms, and are predominantly $CD62L^{lo}$, indicative of recent antigen exposure [30].

A similar phenomenon occurs in immunocompetent BALB/c mice, and in both situations IE1- and m164-specific CD8 T cells accounted for only ~20–30% of MCMV-specific CD8 T cells during acute infection but accounted for ~70–80% of MCMV-specific CD8 T cells during chronic infection [30]. Interestingly, IE1-specific CD8 T cells do not simply persist after infection is cleared. Exponential expansion until 10 days post-infection is followed by rapid contraction by day 14, reminiscent of well-characterized CD8 T-cell responses to lymphocytic choriomeningitis (LCMV) and *Listeria monocytogenes*, but then a phase of continuous inflation of the memory CD8 T-cell response occurs for presumably the remainder of the lifetime of the mouse [33].

Interestingly, memory inflation in C57BL/6 mice does not depend on initial immunodominance. The CD8 T-cell response to most epitopes, including the immunodominant M45 epitope, contracts between days 7 and 14 and remains constant throughout the life of the mouse [34, 35] (M. Munks, in preparation). However, CD8 T-cell responses to m139 and M38 epitopes, which rank second and fifth in the acute immunodominance hierarchy, become more prominent during chronic infection, and responses to two epitopes in IE3 that are barely detectable during acute infection increase dramatically during the next several months. Similarly, the CD8 T-cell response to a recombinant LCMV epitope expressed behind the IE1 promoter is just above the threshold of detection at day 7 but undergoes an impressive 10-fold increase in less than 4 months [34]. Together, these results indicate that some feature of these epitopes/antigens and/or the CD8 T cells that recognize them, other than numerical superiority, is responsible for their dominance during chronic infection.

In addition to numerical differences, there are phenotypic differences between CD8 T cells of different specificities. CD8 T cells specific for "inflationary" antigens are CD62L^{lo} [34], reminiscent of "effector memory" T cells [36]. They also lack two receptors for cytokines involved in homeostatic proliferation, CD127 (IL-7Rα) and CD122 (IL-15Rα), and are CD27^{-} and CD28^{-}. In contrast, CD8 T cells specific for "non-inflationary" antigens are CD62L^{hi}, CD127^{+}, CD122^{+}, CD27^{+}, and CD28^{+}.

12.3.2.3 The Impact of Interference With Antigen Presentation by MCMV's VIPRS on the CD8 T-Cell Response

Both HCMV and MCMV encode proteins that interfere with the MHC class I pathway of antigen presentation [3, 28]. It is often presumed that an important function of these viral genes that interfere with antigen presentation (VIPRs) is to prevent or diminish priming of some T-cell specificities. It logically follows then that the T-cell response will focus on those epitopes least affected by the VIPRs. *In vitro* data have mostly supported this assumption; until recently, virtually all memory CD8 T-cell responses described from HCMV were derived from IE1 or from the structural protein UL83/pp65, which is brought into the cell as a part of the virion and can thus be processed and presented before any VIPRs are expressed [37]. In MCMV, the only CD8 T-cell epitope known for more than a decade was derived from IE1/pp89, which is expressed before the VIPRs. Furthermore, the dominance of the m164 epitope, encoded by an E gene, could be explained by its ability to be presented by infected cells despite the effects of the VIPRs [38].

Excellent tools are now available for studying the effect of MCMV VIPRs on CD8 T-cell priming *in vivo*. Surprisingly, this work has not revealed any significant effect of the VIPRs on the size, specificity, or surface phenotype of the responding CD8 T-cell population. Infection with wild-type MCMV or an MCMV mutant lacking *m152* results in equivalent priming of CD8 T cells specific for M45 [31], even though *m152* completely abrogates recognition of infected cells by M45-specific CD8 T cells. This lack of effect of *m152* on CD8 T-cell priming was later confirmed in BALB/c mice. [37]. Even more definitively, the CD8 T-cell response to a mutant virus lacking all three VIPRs (*Δm4+m6+m152)* has now been compared to wild-type MCMV infection. In both BALB/c mice, where seven epitopes were examined [29], and C57BL/6 mice, where 24 epitopes were tested (M. Munks, in preparation), the CD8 T-cell response was virtually identical for these two viruses. The VIPRs failed to influence the order of the immunodominance hierarchy in both the acute and chronic phases of infection.

Is it possible that (1) there are many antigens whose presentation is not affected by these genes, (2) that these genes do not affect antigen presentation *in vivo*, or (3) that they do not function in the cell type responsible for T-cell priming *in vivo*, namely DCs? If none of these statements is true (discussed below), then the simplest explanation is that *in vivo*, CD8 T cells are primed by cross-presentation, not by infected DCs.

1. Of the known BALB/c (H-2^{d}) epitopes, five of seven are not presented by wild-type MCMV-infected cells *in vitro*, with IE1 and m164 representing the exceptions. Additionally, wild-type MCMV prevents presentation of all 15 C57BL/6 (H-2^{b}) epitopes tested to date [39] (A. Pinto, unpublished data). Thus, only a minority of epitopes are able to escape the effects of the VIPRs *in vitro*, and the first explanation can be ruled out.
2. Garamond>In an elegantly designed study using a bone marrow transplantation model of MCMV infection, it was demonstrated that CD8 T cells specific for the immunodominant C57BL/6 M45 epitope did not restrict replication of wild-type MCMV *in vivo* [40]. However, these same CD8 T cells did restrict replication of MCMV lacking the VIPR *m152*. Thus, the VIPRs do indeed affect antigen presentation *in vivo*, and the second explanation can be ruled out as a blanket phenomenon. It has not been excluded that there may be an important antigen-presenting cell *in vivo* in which the VIPRs do not function, but at present there is no positive evidence to support that contention. It should be noted that CD8 T cells of some other specificities do offer some protection against wild-type MCMV *in vivo*, and hence they must be able to recognize infected cells to some degree. However, all available evidence suggests that the VIPRs cause a large quantitative difference in the presentation of most epitopes, especially those from E genes that are expressed later in the replication cycle than the VIPRs.
3. Most studies of VIPR function and biochemistry have been performed in fibroblasts. Although an earlier report suggested that VIPRs might not function in macrophages, this has been disputed [41, 42]. Recent studies from our laboratory also indicate that VIPRs function effectively in bone marrow–derived DCs. Because DCs are the most relevant cell

type for priming the CD8 T-cell response, there is no reason to believe that priming *in vivo* is performed by a cell type immune from the impact of VIPRs.

The cumulative data suggest that VIPRs (1) are effective against most antigens, (2) function *in vivo*, and (3) function in bone marrow–derived DCs. Yet there is no evidence that VIPRs can affect CD8 T-cell priming *in vivo*. The most plausible explanation for this apparent paradox is that *in vivo*, MCMV-infected DCs do not prime CD8 T cells. Rather, infected cells are most likely phagocytosed by DCs, and their antigens are cross-presented to naïve CD8 T cells.

12.3.2.4 The Memory T-Cell Response to HCMV

Primary HCMV infection of immunocompetent individuals is normally asymptomatic. Thus, most of the literature on HCMV-specific T-cell responses describes memory, rather than acute effector, T-cell responses. From 1987 to 1996, the prevailing paradigm was that most CD8 T-cell responses were directed against the tegument protein pp65 [43]. Development of tetramer, ELISPOT, and intracellular cytokine staining assays simplified the analysis and led to more powerful quantitation. Reevaluation revealed that IE-1, previously thought to be very subdominant, was codominant with pp65 in many individuals [43].

A study by Elkington et al. [44] shifted the paradigm yet again. Fourteen potential HCMV antigens, with a variety of cellular functions, were selected for study. From these, more than 200 peptides were synthesized that matched HLA-binding motifs. CD8 T-cell responses were detectable for 9 of 14 antigens, including 31% of all peptides tested. Importantly, nearly half of the HCMV-specific T-cell responses detected were directed against antigens other than pp65 and IE-1. Another study soon confirmed this general finding. Manley et al. [45] derived CD8 T-cell clones from five HCMV-infected subjects. The clones were screened with an HCMV mutant lacking VIPRs, and 385 virus-specific clones were identified. Of these, 46% were pp65 specific or IE1 specific. The remainder were pp150 specific (5%), gB specific (2%), or of unknown specificity (47%). These studies demonstrated that while CD8 T-cell responses to pp65 and IE-1 are numerically important, many other antigens are also recognized that in combination account for a large percentage of the total HCMV response.

This work was extended by Sylwester et al. [46], who definitively examined the immunogenicity of all HCMV proteins across a large number of HLA types. Overlapping 15-mer peptides from each HCMV protein were synthesized and then pooled by protein. Peripheral blood mononuclear cells (PBMC) were isolated from 33 HCMV seropositive subjects, who were selected for their HLA diversity. Each individual peptide pool was tested by cytokine flow cytometry for its ability to stimulate CD4 and CD8 T-cell responses from each subject's PBMC. The most significant finding was that 151 of the 213 HCMV ORFs tested elicited a CD4 and/or CD8 T-cell response in at least one individual, and responses were detected against all functional categories of proteins. The CD8 T-cell antigen most frequently responded to was UL48, which had not been previously described as an antigen, followed, not surprisingly, by pp65 and IE-1. CD4 T cells most frequently responded to gB, pp65, and UL86. A typical subject had an HCMV response that was astonishingly large and complex. The median subject had 4.0% HCMV-specific CD4 T cells recognizing 12 antigens and 4.6% HCMV-specific CD8 T cells recognizing eight antigens. Not surprisingly, all functional classes of ORFs were represented. Interestingly, a significant bias towards highly conserved ORFs was found. Although all kinetic classes were represented, IE antigens were highly over-represented, a finding that has important implications for how memory T-cell "inflation" might be occurring (described below).

12.3.2.5 **Increasing Size and Oligoclonality of the CD8 T-cell Response to HCMV With Age**

There is substantial evidence that the HCMV T-cell response increases in size and becomes more clonal over time. In a study by Khan et al. [47], CD8 T-cell responses were compared between HLA-A2$^+$ and/or HLA-B7$^+$ HCMV-infected adults that were either old (age 60–95) or young (age 20–55). The old group had more than three times as many A2-pp65 tetramer$^+$ CD8 T cells as the young group and more than five times as many B7-IE-1 tetramer$^+$ cells as the young group, on average.

The increased size of T-cell responses to HCMV coincides with increased clonality. Wills et al. [48] initially demonstrated that in some individuals, up to 75% of CD8 T cells specific for pp65 utilized a common TCR Vβ chain. Bitmansour et al. [49] have described a similar phenomenon in CD4 T cells, using a more refined analysis of CDR3 lengths and TCR sequencing. Here, the pp65-specific response in four healthy subjects was limited to 1–3 dominant clonotypes, which accounted for up to 50% of the total CMV-specific CD4 T-cell response. These powerful molecular analyses have recently been reapplied to HCMV-specific CD8 T cells, quite remarkably showing that CD8 T cells frequently share TCR Vβ chains, even when those CD8 T cells are derived from different individuals [43].

Given that the number of HCMV-specific CD8 T cells increases over time, and that both CD4 and CD8 T cells have restricted Vβ usage indicative of massive clonal expansion, it is not surprising that HCMV-specific T cells have a surface phenotype and functional properties characteristic of being highly differentiated [43]. They are usually CD45RA$^+$, CCR7$^-$, lack CD27 and CD28, and are positive for perforin and bcl-2. CD57, a marker of replicative senescence, is also commonly found on HCMV-specific T cells. Taken together, these markers are indicative of an effector memory rather than central memory phenotype.

12.3.2.6 **What Accounts for Memory "Inflation" and TCR Vβ Focusing?**

Although the time scale is very different, we assume that the CD8 T-cell inflation observed in MCMV infection is related to the memory T-cell response described in HCMV-infected humans. But if MCMV inflation is independent of the acute immunodominance hierarchy, a very important question arises: What factors determine which CD8 T-cell specificities will become dominant during chronic MCMV and HCMV infection? Some of the likely explanations will be discussed in the context of the factors discussed in Section 12.1 of this chapter: (1) virus-centric models, (2) APC-centric models, and (3) T cell–centric models.

Virus-centric Explanation 1: Does MHC Class I Immune Evasion Allow Presentation of a Subset of CD8 T-Cell Epitopes?

Despite the instinctive assumption that CMV VIPRs would have a profound impact on immunodominance, the available evidence in both human and mice suggests that this is not the case. As previously stated, Manley et al. [45] isolated 385 HCMV-specific CD8 T-cell clones of varying specificities. Of these, 39% were pp65- or pp150 specific and all of these could recognize wild-type HCMV (AD169) or a mutant HCMV lacking its VIPRs. However, of the 14% that were IE-1- or gB specific, or of the 47% that were of an unknown specificity, none could recognize wild-type HCMV. This demonstrates that the majority of HCMV-specific memory CD8 T cells are specific for antigens that are not presented by HCMV strains with intact VIPRs.

The effect of VIPRs on immune focusing was studied more directly in MCMV infection. No major difference has been observed in the CD8 T-cell response when mice are infected chronically with either wild-type MCMV or a mutant lacking its VIPRs [35] (M. Munks, in preparation). Thus, there is convincing evidence that VIPRs do not have a major effect on either CD8 T-cell priming or the subsequent inflation that occurs during chronic CMV infection.

Virus-centric Explanation 2: Does Abortive Reactivation Lead to Selective Expression of Proteins?

Unlike the gammaherpesviruses, which have well-defined "lytic" and "latent" gene expression programs, there is no evidence in CMV infections of different genetic programs for lytic replication and reactivation from latency. This has been examined in some detail in a BMT model in the lungs of mice latently infected with MCMV [32]. It was found that after a sublethal dose of γ-irradiation, MCMV expressed the transcription factors *IE1* and *IE3* and the virion component *gB* in a sequential manner, with unknown "checkpoints" existing between each of these genes. Yet infectious virus was produced infrequently, indicating that even in an experimental system designed to reactivate MCMV, the vast majority of reactivation events are abortive.

Is abortive reactivation the norm in latent infection? If so, it logically follows that T cells specific for those proteins expressed earliest during reactivation, and therefore most frequently, will have a selective advantage in terms of antigen "hits" and proliferation. If correct, this model predicts an "inflation advantage" for CD8 T cells that recognize IE antigens, then E antigens, with L antigen–specific CD8 T cells trailing behind.

Data from HCMV are largely consistent with this model. When Sylwester et al. [46] examined the expression kinetics of ORFs that elicited the largest and most frequent memory CD4 and CD8 T-cell responses, they indeed found that IE antigens were most favored in both cases, particularly for CD8 T cells. While IE ORFs represent less than 5% of the HCMV coding capacity, they represent more than 10% of both CD4 and CD8 T-cell responses. Additionally, CD8 T-cell responses to IE antigens are more than four times larger than E antigens, on average. However, while memory T-cell responses to E antigens and E-L antigens are least favored, L antigens are slightly favored over E and E-L antigens.

This latter finding is difficult to reconcile with the proposed model. But perhaps this occurs only because some L antigens are strongly favored during initial T-cell priming following primary HCMV infection (e.g., because they are expressed at much higher levels than other genes). We are aware of only one published study of the T-cell response during primary HCMV infection, but it may support this explanation. Using HLA-A2-restricted CD8 T-cell epitopes, previously defined from chronically infected adults, Gibson et al. examined the kinetics and antigen specificity of CD8 T-cell responses in infants [50]. Notably, in infants who responded to both pp65 and IE1, the relative immunodominance of pp65 compared to IE1 evolved as well. Regardless of whether initial CD8 T-cell responses to pp65 were larger or smaller than IE1-specific responses, by one year of age the IE1-specific response predominated over the pp65-specific response in all infants.

In MCMV infection, there is also considerable support for the model that IE antigens undergo preferential inflation. The two BALB/c antigens that predominate in chronic infection (IE1 and m164) are indeed transcribed under IE conditions, although m164 protein has not been demonstrated to appear until E times [38]. In C57BL/6 mice, responses to IE3 are barely detectable during acute infection but then inflate. Finally, recombinant epitopes expressed behind the IE1 promoter also undergo inflation [34]. In contrast, the other epitopes (M38, m139) that inflate in chronically infected C57BL/6 mice are not known to be expressed at IE times. However, by analogy with HCMV, there is some reason to believe that these may be IE genes. Clearly, this needs to be tested.

In conclusion, a circumstantial but strong case can be made that (1) most viral attempts at reactivation are abortive and (2) CD8 T cells specific for antigens expressed most frequently by the virus during abortive reactivation will tend to dominate the T-cell response in chronic infections.

Even if this model is true, it is not the sole explanation for chronic immunodominance. Two MCMV antigens, M38 and m164, contain multiple CD8 T-cell epitopes that are discordant in their ability to undergo inflation. In the case of M38, CD8 T cells specific for $M38_{316-325}$ inflate during chronic infection of C57BL/6 mice, but CD8 T cells specific for $M38_{38-45}$ do not (M. Munks, unpublished data). Similarly, BALB/c CD8 T cells specific for $m164_{257-265}$ undergo inflation, but C57BL/6 CD8 T cells specific for $m164_{267-275}$ and $m164_{283-290}$ do not. Perhaps early expression of antigens during reactivation is required, but is not sufficient, for inducing CD8 T-cell inflation. In this case, one or more of the APC-centric and/or T cell–centric models (described below) may provide an additional mechanistic layer on the process. Clearly, significant work is required to further test this model.

APC-centric Explanation: Does the Predominant APC Cell Type Differ Between Acute and Chronic MCMV Infection?

In the influenza model in C57BL/6 mice, a change in immunodominance in secondary infection has been attributed to differential presentation of two epitopes by different APCs [51]. Responses to an epitope that could only be presented by DCs dominate the primary response, whereas responses to an epitope that could also be presented by macrophages, B cells, and epithelial cells dominate the secondary response. Thus, chronic immunodominance could reflect better presentation by a broader array of APCs. As a variant of this explanation, epitopes that are dependant on immunoproteasome (constitutively expressed by DCs) for presentation may dominate the acute response, whereas those that can be processed by normal proteasomes may be advantaged in the chronic response.

While it may be tempting to speculate that a similar phenomenon occurs after MCMV infection, there is currently no data to support this model. First, most of the epitopes for C57BL/6 mice can be presented from a transgene by fibroblasts without IFN-γ treatment and thus are not dependent on the immunoproteasome. Second, as described, available evidence indicates that because of the function of VIPRs, directly infected cells of all types are impaired in their ability to present most of these epitopes. Hence, the role of directly infected cells in stimulating MCMV-specific CD8 T-cell responses remains to be demonstrated.

T Cell–centric Model: Is There Selection for T Cells With "Optimal" TCRs During Chronic Infection?

In MCMV infection, it is clear that CD8 T-cell "inflation" occurs for only some epitopes. However, even among T cells of the same specificity, there is evidence that some TCRs are "optimal" compared to others. For example, TCR Vβ chain usage by IE1-specific CD8 T cells is more diverse during acute MCMV infection and more focused during chronic infection [33]. In HCMV infection, both CD4 and CD8 T-cell responses can be extremely focused in clonality, as described above, presumably the endpoint of the same phenomenon that has been clearly documented in mice.

So far there are no data to indicate that IE1-, m164-, or m139-specific CD8 T cells from chronically infected mice are significantly more sensitive to low doses of peptide than are CD8 T cells from acutely infected mice (S. Sierro, personal communication, and M. Munks, unpublished observation). While it remains unknown which features of a TCR make it optimal, it is clear that some form of selection is occurring *in vivo* among T cells of a shared specificity, adding an additional layer of complexity to unravel.

12.3.3 HCMV and Immunosenescence

What are the health implications of being HCMV seropositive and having large virus-specific oligoclonal T-cell expansion? It is becoming increasingly clear that

HCMV infection impairs vaccination and antiviral immunity and is associated with higher mortality rates in the elderly (for a review, see Ref. [52]).

For example, it is known that HCMV seropositivity and the high frequency of $CD8^{+}CD28^{-}$ T cells that are associated with HCMV infection are correlated with poor antibody responses following influenza vaccination [52]. Additionally, CMV-specific immune responses appear to suppress the expansion of EBV-specific T cells that occurs in HCMV seronegative individuals, suggesting that HCMV may also impair EBV immunity [53]. In two aging studies of elderly Swedish, an "immune risk phenotype" (IRP) has been defined that accurately predicts mortality. CMV seropositivity is an important indicator of the IRP, and individuals with expanded populations of $CD8^{+}CD45RA^{+}$ $CD27^{-}CD28^{-}CD57^{+}$ T cells are most at risk [52].

It is unknown whether the effect of HCMV on other immune responses is due to active suppression or is simply a result of "overcrowding" of the T-cell compartment. However, evidence is mounting that HCMV is an important player in the immunosenescence process.

12.3.4
T-Cell Cross-reactivity Between Betaherpesviruses

As mentioned in the overview of the betaherpesvirus section, no T-cell antigens have been reported for HHV-6 or HHV-7. While this precludes an analysis of T-cell immunodominance in individuals infected with these viruses, some interesting and relevant data on CD4 T-cell responses do exist.

CD4 T-cell clones, stimulated with antigen from HHV-6, HHV-7, or HCMV, were isolated from seropositive subjects and used to analyze epitope cross-reactivity between the viruses [54]. Of the T-cell clones driven with either HHV-6 or HHV-7 antigen, 28% reacted to both HHV-6 and HHV-7 antigen. In other words, a large percentage of HHV-6- and HHV-7-specific CD4 T-cell clones recognized an antigen shared by the two viruses. In contrast, only 4% of T-cell clones were cross-reactive between HHV-6 and HCMV, and only 2% of HHV-7-specific and HCMV-specific T-cell clones were cross-reactive. Not surprisingly, an even smaller percentage (1%) of all CD4 T-cell clones recognized all three viruses.

In retrospect these results are not completely surprising, given that genomic sequencing efforts later revealed that HHV-6 and HHV-7 are more closely related to each other than either virus is to HCMV. Additionally, the limiting dilution analysis (LDA) data should be considered semi-quantitative compared to contemporary techniques such as peptide–MHC tetramers, ELISPOT, and ICS. Yet these data are instructive in a more conceptual sense. Whereas different HCMV strains could be thought of as siblings, HHV-6 and HHV-7 are analogous to cousins. Within this conceptual framework, it is less surprising that T cells will cross-react between these "different" viruses. Instead, T-cell cross-reactivity between viruses sharing a common ancestor can be interpreted as an indication of the amino acid relatedness of any two viruses.

This may help to explain the observation that the HCMV-specific T-cell response is stronger against conserved genes than against non-conserved genes [46]. Herpesvirus infections are highly prevalent, and co-infection of the same individual is frequent. Thus, it is likely that some HCMV seropositive individuals are infected with multiple HCMV strains, or also with HHV-6 and/or HHV-7. In these individuals, T cells that were initially primed against one strain of HCMV may receive an additional stimulus each time they are super-infected with another HCMV strain or another close viral relative. As the individual becomes infected with more strains, or more viruses, the largest T-cell responses will be towards the genes that are most conserved among all viruses present.

12.3.5 Gammaherpresviruses

The third subfamily of mammalian herpesviruses is the gammaherpesviruses. This subfamily has two genera: gamma-1 and gamma-2. The prototypical member is Epstein-Barr virus (EBV), a gamma-1 herpesvirus [55]. More recently described are Kaposi's sarcoma herpesvirus (KSHV) [56] and the rodent virus MHV-68, both classified as gamma-2 herpesviruses. The orchestration of latency by gammaherpesviruses differs from that of the alpha- and betaherpesviruses in that they encode a separate set of genes that are expressed only during latent infection. They are also distinguished by their ability to cause malignancies. These two features are related: the propagation of viral genome in latently infected cells is achieved in part by driving the cells into proliferative cycle, which can lead to tumors in the host. Latency is established in B cells for each of these viruses. Lytic infection primarily occurs in epithelial cells—in the oropharynx in the case of EBV—although B cells can also undergo lytic infection with production of infectious virus progeny.

From the perspective of immunodominance, the gammaherpesviruses are both interesting and instructive in that the T-cell responses to the two classes of proteins (lytic and latent) follow quite distinct courses. There is a much greater body of literature describing the CD8 T-cell response to EBV than to the gamma-2 herpesviruses; consequently, most of this section will deal with EBV.

12.3.6 Epstein-Barr Virus

EBV infects a majority of people worldwide, although the seroprevalence is lower in industrialized countries [57]. It is spread by saliva. Primary infection is usually asymptomatic, especially when it occurs in the first decade of life, as is the norm in non-industrialized countries. Primary infection later in life, typically during adolescence, can lead to the syndrome of infectious mononucleosis (IM). EBV is the cause of Burkitt's lymphoma, a B-cell malignancy, and of nasopharyngeal carcinoma.

EBV infection of B cells *in vitro* leads to their transformation into lymphoblastoid cell lines (LCLs). LCLs express all six of EBV's latent genes—the leader protein (LP); the Epstein-Barr nuclear antigens (EBNA) 1, 2, 3A, 3B, and 3C—as well as two latent membrane proteins (LMP1 and LMP2). The easy ability to generate autologous LCLs from seropositive individuals means that immunity to these six proteins has been extensively studied for two decades. In contrast, there is no easy model for lytic EBV infection. In contrast to HSV or CMV, EBV does not infect fibroblasts or epithelial cell lines *in vitro*; therefore, data on immune responses to lytic antigens are much more scarce. There are some 60+ lytic cycle genes. Two IE genes are known: BZLF1 and BRLF1, which are transcription factors. There are 30+ early genes: these can be subdivided into transcription factors expressed in the earliest times points of E gene expression, such as BMLF1 and BMRF1, and later E genes. There are also approximately 30 late genes. A proportion of LCLs undergo spontaneous lytic cycle infection, and this is the main model for studying lytic infection.

12.3.6.1 **Interference with the MHC Class I Pathway by EBV**

Both alpha- and betaherpesviruses encode genes expressed during lytic cycle infection that impair presentation of antigens via the MHC class I antigen presentation pathway (VIPRs). Although such genes have not yet been identified for lytic cycle EBV infection, it seems very likely that they exist: as lytic infection progresses, MHC class I is downregulated on the cell surface [58]. Latent cycle genes do not downregulate MHC class I. However, one latent cycle protein, EBNA 1, interferes with its own degradation by the proteasome [59]; this diminishes, but does not abolish, its presentation by MHC class I molecules [60, 61]. For many years, this was thought to completely prevent the generation of EBNA 1–specific responses. However, it has become clear that such responses are in fact generated.

12.3.6.2 **Overview of the Response to Lytic and Latent Proteins**

The CD8 T-cell response to EBV has been studied for more than 20 years by several major groups: this extensive literature has recently been reviewed [62]. Acute infection by EBV is generally identified only if the patient experiences infectious mononucleosis (IM); thus, studies of acute infection are largely limited to IM patients. During this phase of infection, massive CD8 T-cell responses are elicited. These are predominantly directed towards lytic cycle proteins, although responses to latent proteins are also made. As IM resolves, the responses to lytic cycle proteins contract to a small memory population. Responses to latent cycle proteins dominate the repertoire of EBV-specific T cells during latent infection [63].

12.3.6.3 **Immunodominance Hierarchy for the Acute Response to Lytic Cycle Proteins**

The first lytic cycle epitopes identified were from IE genes. Recently, Hislop and colleagues more systematically addressed the question of which classes of EBV lytic proteins are most immunodominant during the primary CD8 T-cell response to EBV [64]. CD8 T-cell clones were derived by limiting dilution from IM patients. To drive the clones, PBMC were exposed to either autologous LCLs, some of which express lytic antigens, or anti-CD3 and IL-2. The latter methodology was an attempt to avoid skewing the repertoire of clones by the nature of the *in vitro* restimulation, and it was reassuring that the antigenic specificities of the derived clones were similar by both methods. The antigens recognized by the clones were then identified using recombinant vaccinia viruses expressing a panel of candidate antigens. Because the clones were screened directly on recombinant vaccinia viruses, the method avoided bias towards clones that would be able to detect lytically infected cells. Unfortunately, the authors did not indicate what percentage of the CD8 T-cell clones they derived were able to recognize one of these antigens. In other words, they did not gain a clear picture of how much of the acute response they had detected. However, this study provides the most systematic description to date of the immunodominance hierarchy to EBV.

The panel of candidate antigens included the two known IE genes. Eleven E genes out of the expected 30+ E genes were chosen. These covered a range of functions: transcriptional activators, which would be expressed early during the E phase; components of viral DNA replication machinery, which are expressed later during the E phase; and several others. Ten L genes, out of the 30+ L genes, were chosen. These included virus capsid and tegument components, glycoproteins, and a secreted IL-10 homologue. This panel of 23 antigens was chosen in an attempt to randomly sample EBV lytic cycle genes. Eleven patients were studied, covering a range of HLA types. As for the CMV study by Picker and colleagues [46], the power of this human study with multiple HLA alleles allows genuine virological determinants of immunodominance to emerge. These authors found that all classes of protein were recognized. However, when the authors asked what percentage of their positive clones recognized different classes of antigens, they found a marked skewing towards IE genes and those E genes that are earliest expressed—namely, those encoding transcriptional activators. This skewing was not based on affinity: when peptides were identified, those expressed from L genes were generally recognized at lower concentrations than those from IE or E genes.

Why are IE and the earliest E genes immunodominant? As has been discussed above, there are two main (virus-centric) explanations for this phenomenon. (1) Antigens from this kinetic class are likely to be least affected by the action of VIPRs. (2) If over time most infected cells are destroyed by host defenses early during the infectious cycle, the earliest expressed antigens will gain an incremental numerical advantage. Using LCLs that spontaneously entered the lytic cycle, these authors determined that IE and the dominant E antigens were generally more efficiently presented by the infected cells. As described above, as the lytic cycle continues, MHC class I expression decreases, presumably because of the action of as yet unidentified EBV VIPRs. This likely explains the much poorer presentation of L genes. These authors therefore concluded that the skewing of the CD8 T-cell repertoire towards IE and E genes is explained by their being better presented by infected cells: explanation 1 above. However, as they acknowledge in their discussion, explanation 2 is also consistent with their data. Given that the same skewing towards IE antigens occurs in CMV infection whether or not the VIPRs are functional, we tend to favor explanation 2.

We note, however, that EBV and CMV are two very different viruses, and extrapolation from CMV to EBV requires a high degree of caution.

12.3.6.4 Kinetics of the Response to Lytic Proteins

The syndrome of IM is accompanied by massive EBV-specific CD8 T-cell expansions. Interestingly, primary asymptomatic EBV infection, which in global terms is probably the norm, does not elicit such massive responses, despite equivalent levels of virus load [62]. The factors that trigger the massive T-cell response in IM are not known. To date, no studies of immunodominance in asymptomatic infection have been reported, and it will be interesting to see whether the same immunodominance hierarchy is reached under those conditions.

With resolution of acute viral loads, the reactive T-cell compartment contracts. Margaret Callan's group has described a fascinating phenomenon that so far appears to be unique to EBV [65]. The specificities that dominate the acute response contract most severely, whereas specificities that are subdominant during IM are less severely culled during the contraction phase. The result is that there is more pronounced focusing of the response to lytic antigens during IM than there is during the chronic phase of infection. These authors suggest that those clones of highest affinity are most expanded during acute IM, see antigen more frequently, and consequently are more activated than subdominant clones. However, as a consequence of their activation, they are more susceptible to activation-induced cell death (AICD) and hence "crash" more profoundly upon antigen withdrawal. Studies of other non-persisting infections in mice have generally found that the hierarchy of the acute response predicts the hierarchy of memory. The reasons for the discrepancy are not yet clear. The situation in IM differs from the model systems in mice in that the acute phase lasts much longer; therefore, the clones that dominate the acute response may have reached the limit of the number of possible divisions. These authors described a mathematical model of T-cell expansion and contraction, which takes cellular senescence into account, and found that this can explain their data. This is thus an entirely T cell–centric explanation for chronic immunodominance, with the interesting twist that over time it is moderate rather than high-affinity responders that become immunodominant. However, the fact that acute EBV that does not elicit the syndrome of IM does not drive these very large T-cell responses, despite equivalent virus loads, suggests that there is something unusual about the nature of the acute CD8 T-cell response in IM. If so, the rules that govern the expansion and contraction of this response may well differ from other model infections.

12.3.6.5 Immunodominance Hierarchy of the Response to Latent Proteins

All six of the latent proteins are known to stimulate responses. However, in most individuals, responses to the EBNA 3 proteins dominate. HLA-A2-positive individuals make a strong response to an epitope in LMP2, an example of the impact of MHC polymorphism on immunodominance. All six latent proteins are expressed in EBV-transformed lymphoblastoid cells lines (LCLs). However, only EBNA 1 is needed to maintain the latent virus episomal DNA, and many latently infected B cells *in vivo* express only EBNA 1. The response to EBNA 1 is usually of very low frequency. EBNA 1 contains a Gly–Ala repeat sequence that impairs the efficiency of its degradation by proteasome to generate CD8 T-cell epitopes. For a long time, it was thought that no EBNA 1–specific responses were generated. However, it has recently been appreciated that such responses are generated. Cross-presentation has been proposed as the means by which these responses are generated [66], although, of course, this has not been proven *in vivo*.

12.3.6.6 Kinetics of the Response to Latent Proteins

The kinetics of the response to latent proteins is quite different from that of the response to lytic proteins [63]. During IM, responses to latent antigens are usually detectable but are much less prominent than responses to lytic antigens. However, responses to latent antigens do not contract markedly in transition to chronic infection. Thus, during chronic infection, responses to latent proteins are more prominent than to lytic antigens. The responses to the two classes of antigens also differ phenotypically. In the memory phase, many lytic antigen–specific CD8 T cells downregulate CD45 RO and re-express CD45 RA; these cells are mostly $CCR7^{-}CD62L^{-}$ "effector memory" cells. In contrast, latency antigen–specific T cells remain CD45 RO^{+} and CD45 RA^{-}, and are $CCR7^{+}CD62L^{+}$ "central memory" in phenotype. Two factors are likely to impact both the size of the memory CD8 T-cell response and its phenotype: antigen abundance and the nature of the antigen-presenting cell. During acute infection and IM, there is abundant expression of lytic cycle proteins in the oropharynx, predominantly in epithelial cells. The immunodominant latent antigens are expressed exclusively in B cells. As is the case for all herpesviruses, the amount of antigen (of both classes) expressed during chronic infection is difficult to ascertain.

The murine gammaherpesvirus MHV-68 is used to model EBV infection, and there are intriguing similarities in the two systems. As with EBV, in MHV-68 infection, lytic cycle epitopes dominate the acute response, and latent cycle–specific responses arise a little later and are more prominent during the later (latent) phase of infection [67]. Also as with EBV, the lytic cycle–specific CD8 T cells are largely effector memory ($CD62L^{-}$), whereas latent antigen specific CD8 T cells are largely central memory ($CD62L^{+}$). However, it was the latent antigen–specific central memory T cells that displayed the greatest *in vivo* cytolytic potential, at odds with the normal relationship between lymph node homing potential and effector capability. Most explanations of memory phenotype relate phenotype to frequency of antigen encounter, more frequent TCR triggering being associated with the effector memory phenotype, and less frequent encounters leading to the central memory phenotype. The picture in EBV and MHV-68 infection is at odds with this: when compared to lytic antigen–specific memory CD8 T cells, latent antigen–specific CD8 T cells are maintained in greater numbers and have greater immediate effector potential, both features that suggest more frequent antigen encounter. However, these latent antigen–specific CD8 T cells are uniformly central memory in phenotype. Because latent antigen–specific CD8 T cells encounter their antigen primarily on B cells, it seems possible that this type of APC influences the memory phenotype of the responding T cells.

Understanding the complex pattern of immunodominance and T-cell phenotype in chronic EBV infection is bedeviled by the same issue that occurs for all the chronic herpesvirus infections that have been discussed in this chapter: despite abundant evidence of T-cell reactivity, viral antigen is rarely directly detectable. We are left making a circular argument, in which a phenomenon that we are trying to understand (T-cell specificity and phenotype) is itself the main evidence for the causative factor we would like to ascribe it to (virus activity).

12.3.6.7 **Immunodominance of the LC13 TCR in Responding to HLA-B8/FLRGRAYGL: An Extreme Example of TCR Immunodominance**

The CD8 T-cell response to latent EBV antigens includes one of the most fascinating immunodominance stories in the literature [68, 69]. Individuals who express the MHC class I allele HLA-B8 make an immunodominant response to an epitope from the EBNA-3A protein FLRGRAYL. Essentially all the T cells responding to this epitope from disparate individuals use an identical, "public" TCR known as LC13. The selective pressure on LC13 is underscored by the fact that different nucleotide sequences are used by different T-cell clones to arrive at the identical LC13 TCR sequence. This is therefore an extreme example of TCR immunodominance. What makes this particular TCR so overwhelmingly preferred? One possibility would be that because of holes in the repertoire induced by negative selection, LC13 is the only TCR sequence capable of recognizing HLA-B8/ FLRGRAYL. However, it is known that this is not the case. Fortuitously, the LC13 TCR recognizes HLA-B44 as an alloantigen. Thus, individuals who are heterozygous for HLA-B8 and HLA-B44 do not express LC13, as this TCR is deleted in the thymus. These individuals make a response to HLA-B8/FLRGRAYL that is polyclonal and utilizes a variety of TCRs. Hence, there appears to be something special about LC13 that makes it the best possible of all TCRs for its cognate ligand.

LC13's advantage does not appear to lie in high affinity: the rough estimate of its K_d is about 50 μm. The LC13 TCR has been crystallized alone and in complex with HLA-B8/FLRGRAYL [69]. The TCR undergoes considerable conformational change upon ligand binding: a proline in the CDR3 loop that is not involved in interaction with the peptide–MHC complex acts as a critical "crumple point" that enables CDR3 to reshape itself to fit the ligand. It is a matter of considerable controversy whether conformational change in the TCR alpha and beta subunits contributes to signal transduction or is merely a means of accommodating the TCR to its diverse ligands. While proponents of the importance of conformational change use the LC13 data to support their model, the argument is not yet resolved. However, the extreme immunodominance of the LC13 receptor should provide important information as to just what the immune system prefers when it is making its immunodominance choices at the TCR level.

12.4
Concluding Remarks

Herpesviruses are large, ancient, complex viruses. It is impossible to study the herpesvirus–host interaction for long without developing a deep respect for the multiple layers of redundancy that underpin this very stable relationship. Given the complexity of immunodominance even to simple model agents, it should not be surprising that we are not yet in a position to explain the basis of immunodominance in most herpesvirus infections.

The seemingly safe prediction—that genes that interfere with antigen presentation would have a major impact on immunodominance—turns out not to be correct, at least for MCMV. This highlights some of the major holes in our knowledge. We do not know, for example, whether direct or cross-presentation accounts for most of the priming of CD8 T cells or for maintaining the response during chronic infection. Furthermore, in the chronic phase of infection, which lasts for the life of the host and is the most studied, we have little knowledge of the nature or frequency of viral gene expression. Nevertheless, mammalian immune systems devote a large amount of their resources to herpesviruses, which have been their companions through 80+ million years of evolution. Understanding the basis of these responses would seem to be an important part of understanding normal immunobiology.

References

1 Roizman B, Pellett PE. The Family *Herpesviridae*: A Brief Introduction. In: Knipe DM, Howley PM, eds. Fields Virology. Vol. 2 (ed 4th). Philadelphia: Lippincott Williams and Wilkins; 2001:2381–2390.

2 McGeoch DJ, Dolan A, Ralph AC. Toward a comprehensive phylogeny for mammalian and avian herpesviruses. Journal of Virology. 2000;74:10401–10406.

3 Yewdell JW, Hill AB. Viral interference with antigen presentation. Nature Immunology. 2002;3:1019–1025.

4 Yewdell JW, Bennink JR. Immunodominance in major histocompatibility complex class I-restricted T lymphocyte responses. Annual Review of Immunology. 1999;17:51–88.

5 Chen W, Anton LC, Bennink JR, Yewdell JW. Dissecting the multifactorial causes of immunodominance in class I-restricted T cell responses to viruses. Immunity. 2000;12:83–93.

6 Zammit DJ, Cauley LS, Pham QM, Lefrancois L. Dendritic cells maximize the memory CD8 T cell response to infection. Immunity. 2005;22:561–570.

7 Posavad CM, Koelle DM, Shaughnessy MF, Corey L. Severe genital herpes infection in HIV-infected individuals with impaired HSV-specific CD8+ cytotoxic T lymphocyte responses. Proceedings of the National Academy of Sciences of the United States of America. 1997;94:10289.

8 Koelle DM, Posavad CM, Barnum GR, Johnson ML, Frank JM, Corey L. Clearance of HSV-2 from recurrent genical lesions correlates with infiltration of HSV-specific cytotoxic T lymphocytes. J Clin Invest. 1998;101:1500.

9 Koelle DM, Chen HB, Gavin MA, Wald A, Kwok WW, Corey L. CD8 CTL from genital herpes simplex lesions:

recognition of viral tegument and immediate early proteins and lysis of infected cutaneous cells. Journal of Immunology. 2001;166:4049–4058.

10 Koelle DM, Liu Z, McClurkan CL, Cevallos RC, Vieira J, Hosken NA, Meseda CA, Snow DC, Wald A, Corey L. Immunodominance among herpes simplex virus-specific CD8 T cells expressing a tissue-specific homing receptor. Proceedings of the National Academy of Sciences of the United States of America. 2003;100:12899–12904.

11 Mikloska ZM, Ruckholdt M, Ghadiminejad I, Dunckley H, Denis M, Cunningham AL. Monophosphoryl lipid A and QS21 increase CD8 T cell lymphocyte cytotoxicity to herpes simplex virus-2 infected cell proteins 4 and 27 through IFN-g and IL-12 production. J Immunol. 2000;164:5167.

12 Fruh K, Ahn K, Djaballah H, Sempe P, vanEndert PM, Tampe R, Peterson P, Yang Y. A viral inhibitor of peptide transporters for antigen presentation. Nature. 1995;375:415–418.

13 Hill AB, Jugovic P, York I, Russ I, Bennink J, Yewdell J, Ploegh HL, Johnson D. Herpes simplex virus turns off the TAP to evade host immunity. Nature. 1995;375:411–415.

14 Roizman B, Knipe DM. Herpes simplex viruses and their replication. In: Knipe DM, Howley PM, eds. Fields Virology. Vol. 2. Philadelphia: Lippincott Williams and Wilkins; 2001:2399–2460.

15 Khanna KM, Lepisto AJ, Decman V, Hendricks RL. Immune control of herpes simplex virus during latency. Current Opinion in Immunology. 2004;16:463–469.

16 Jugovic P, Hill AM, Tomazin R, Ploegh H, Johnson DC. Inhibition of major histocompatibility complex class I antigen presentation in pig and primate cells by herpes simplex virus type 1 and 2 ICP47. Journal of Virology. 1998;72:5076–5084.

17 Rosenthal KL, Smiley JR, South S, Johnson DC. Cells expressing herpes simplex virus glycoprotein gC but not gB, gD, or gE are recognized by murine virus-specific cytotoxic T lymphocytes. Journal of Virology. 1987;61:2438–2447.

18 Martin S, Rouse BT. The mechanisms of antiviral immunity induced by a vaccinia virus recombinant expressing herpes simplex virus type 1 glycoprotein D: clearance of local infection. Journal of Immunology. 1987;138:3431–3437.

19 Martin S, Zhu XX, Silverstein SJ, Courtney RJ, Yao F, Jenkins FJ, Rouse BT. Murine cytotoxic T lymphocytes specific for herpes simplex virus type 1 recognize the immediate early protein ICP4 but not ICP0. Journal of General Virology. 1990;71:2391–2399.

20 Banks TA, Allen EM, Dasgupta S, Sandri-Goldin R, Rouse BT. Herpes simplex virus type 1-specific cytotoxic T lymphocytes recognize immediate-early protein ICP27. Journal of Virology. 1991;65:3185–3191.

21 Salvucci LA, Bonneau RH, Tevethia SS. Polymorphism within the herpes simplex virus (HSV) ribonucleotide reductase large subunit (ICP6) confers type specificity for recognition by HSV type 1-specific cytotoxic T lymphocytes. Journal of Virology. 1995;69:1122–1131.

22 Nugent CT, McNally JM, Chervenak R, Wolcott RM, Jennings SR. Differences in the recognition of CTL epitopes during primary and secondary responses to herpes simplex virus infection *in vivo*. Cellular Immunology. 1995;165:55–64.

23 Wallace ME, Keating R, Heath WR, Carbone FR. The cytotoxic T-cell response to herpes simplex virus type 1 infection of C57BL/6 mice is almost entirely directed against a single immunodominant determinant. Journal of Virology. 1999;73:7619–7626.

24 Messaoudi I, Guevara Patino JA, Dyall R, LeMaoult J, Nikolich-Zugich J. Direct link between mhc polymorphism, T cell avidity, and diversity in immune defense. Science. 2002;298:1797–1800.

25 Coles RM, Jones CM, Brooks AG, Cameron PU, Heath WR, Carbone FR. Virus infection expands a biased subset of T cells that bind tetrameric class I peptide complexes. European Journal of Immunology. 2003;33:1557–1567.

26 Reddehase MJ, Koszinowski UH. Significance of herpesvirus immediate early gene expression in cellular immu-

nity to cytomegalovirus infection. Nature. 1984;312:369–371.

27 Reddehase MJ, Rothbard JB, Koszinowski UH. A pentapeptide as minimal antigenic determinant for MHC class I-restricted T lymphocytes. Nature. 1989;337:651–653.

28 Reddehase MJ. Antigens and immunoevasins: opponents in cytomegalovirus immune surveillance. Nat Rev Immunol. 2002;2:831–844.

29 Holtappels R, Munks MW, Podlech J, Reddehase MJ. CD8 T-cell-based immunotherapy of cytomegalovirus disease in the mouse model of the immunocompromised bone marrow transplant recipient: Horizon Scientific Press; 2005, In Press.

30 Holtappels R, Thomas D, Podlech J, Reddehase MJ. Two antigenic peptides from genes m123 and m164 of murine cytomegalovirus quantitatively dominate CD8 T-cell memory in the H-2d haplotype. J Virol. 2002;76:151–164.

31 Gold MC, Munks MW, Wagner M, Koszinowski UH, Hill AB, Fling SP. The murine cytomegalovirus immunomodulatory gene m152 prevents recognition of infected cells by M45-specific CTL but does not alter the immunodominance of the M45-specific CD8 T cell response *in vivo*. J Immunol. 2002;169:359–365.

32 Reddehase MJ, Podlech J, Grzimek NK. Mouse models of cytomegalovirus latency: overview. J Clin Virol. 2002;25 Suppl 2:S23–36.

33 Karrer U, Sierro S, Wagner M, Oxenius A, Hengel H, Koszinowski UH, Phillips RE, Klenerman P. Memory inflation: continuous accumulation of antiviral CD8+ T cells over time. J Immunol. 2003;170:2022–2029.

34 Sierro S, Rothkopf R, Klenerman P. Evolution of diverse antiviral CD8+ T cell populations after murine cytomegalovirus infection. Eur J Immunol. 2005;35:1113–1123.

35 Gold MC, Munks MW, Wagner M, McMahon CW, Kelly A, Kavanagh DG, Slifka MK, Koszinowski UH, Raulet DH, Hill AB. Murine cytomegalovirus interference with antigen presentation has little effect on the size or the effector memory phenotype of the CD8 T cell response. J Immunol. 2004; 172:6944–6953.

36 Wherry EJ, Teichgraber V, Becker TC, Masopust D, Kaech SM, Antia R, von Andrian UH, Ahmed R. Lineage relationship and protective immunity of memory CD8 T cell subsets. Nat Immunol. 2003;4:225–234.

37 Reddehase MJ, Simon CO, Podlech J, Holtappels R. Stalemating a clever opportunist: lessons from murine cytomegalovirus. Hum Immunol. 2004;65:446–455.

38 Holtappels R, Grzimek NK, Simon CO, Thomas D, Dreis D, Reddehase MJ. Processing and presentation of murine cytomegalovirus pORFm164-derived peptide in fibroblasts in the face of all viral immunosubversive early gene functions. J Virol. 2002;76:6044–6053.

39 Kavanagh DG, Gold MC, Wagner M, Koszinowski UH, Hill AB. The multiple immune-evasion genes of murine cytomegalovirus are not redundant: m4 and m152 inhibit antigen presentation in a complementary and cooperative fashion. J Exp Med. 2001;194:967–978.

40 Holtappels R, Podlech J, Pahl-Seibert MF, Julch M, Thomas D, Simon CO, Wagner M, Reddehase MJ. Cytomegalovirus misleads its host by priming of CD8 T cells specific for an epitope not presented in infected tissues. J Exp Med. 2004;199:131–136.

41 Hengel H, Reusch U, Geginat G, Holtappels R, Ruppert T, Hellebrand E, Koszinowski UH. Macrophages escape inhibition of major histocompatibility complex class I-dependent antigen presentation by cytomegalovirus. J Virol. 2000;74:7861–7868.

42 LoPiccolo DM, Gold MC, Kavanagh DG, Wagner M, Koszinowski UH, Hill AB. Effective inhibition of K(b)- and D(b)-restricted antigen presentation in primary macrophages by murine cytomegalovirus. J Virol. 2003; 77:301–308.

43 Moss P, Khan N. CD8(+) T-cell immunity to cytomegalovirus. Hum Immunol. 2004;65:456–464.

44 Elkington R, Walker S, Crough T, Menzies M, Tellam J, Bharadwaj M,

Khanna R. *Ex vivo* profiling of CD8+-T-cell responses to human cytomegalovirus reveals broad and multispecific reactivities in healthy virus carriers. J Virol. 2003;77:5226–5240.

45 Manley TJ, Luy L, Jones T, Boeckh M, Mutimer H, Riddell SR. Immune evasion proteins of human cytomegalovirus do not prevent a diverse CD8+ cytotoxic T-cell response in natural infection. Blood. 2004;104:1075–1082.

46 Sylwester AW, Mitchell BL, Edgar JB, Taormina C, Pelte C, Ruchti F, Sleath PR, Grabstein KH, Hosken NA, Kern F, Nelson JA, Picker LJ. Immunogenicity of the HCMV proteome: broadly targeted HCMV-specific CD4+ and CD8+ T cells dominate the memory compartments of exposed subjects. J Exp Med. 2005, In press.

47 Khan N, Shariff N, Cobbold M, Bruton R, Ainsworth JA, Sinclair AJ, Nayak L, Moss PA. Cytomegalovirus seropositivity drives the CD8 T cell repertoire toward greater clonality in healthy elderly individuals. J Immunol. 2002;169:1984–1992.

48 Wills MR, Carmichael AJ, Mynard K, Jin X, Weekes MP, Plachter B, Sissons JG. The human cytotoxic T-lymphocyte (CTL) response to cytomegalovirus is dominated by structural protein pp65: frequency, specificity, and T-cell receptor usage of pp65-specific CTL. J Virol. 1996;70:7569–7579.

49 Bitmansour AD, Waldrop SL, Pitcher CJ, Khatamzas E, Kern F, Maino VC, Picker LJ. Clonotypic structure of the human CD4+ memory T cell response to cytomegalovirus. J Immunol. 2001;167:1151–1163.

50 Gibson L, Piccinini G, Lilleri D, Revello MG, Wang Z, Markel S, Diamond DJ, Luzuriaga K. Human cytomegalovirus proteins pp65 and immediate early protein 1 are common targets for CD8+ T cell responses in children with congenital or postnatal human cytomegalovirus infection. J Immunol. 2004;172:2256–2264.

51 Crowe SR, Turner SJ, Miller SC, Roberts AD, Rappolo RA, Doherty PC, Ely KH, Woodland DL. Differential antigen presentation regulates the changing patterns of CD8+ T cell immunodominance in primary and secondary influenza virus infections. J Exp Med. 2003;198:399–410.

52 Pawelec G, Akbar A, Caruso C, Effros R, Grubeck-Loebenstein B, Wikby A. Is immunosenescence infectious? Trends Immunol. 2004;25:406–410.

53 Khan N, Hislop A, Gudgeon N, Cobbold M, Khanna R, Nayak L, Rickinson AB, Moss PA. Herpesvirus-specific CD8 T cell immunity in old age: cytomegalovirus impairs the response to a coresident EBV infection. J Immunol. 2004;173:7481–7489.

54 Yasukawa M, Yakushijin Y, Furukawa M, Fujita S. Specificity analysis of human CD4+ T-cell clones directed against human herpesvirus 6 (HHV-6), HHV-7, and human cytomegalovirus. J Virol. 1993;67:6259–6264.

55 Kieff E, Rickinson A. Epstein-Barr Virus and Its Replication. In: Knipe DM, Howley PM, eds. Fields Virology. Vol. 2 (ed 4th). Philadelphia: Lippincott Williams and Wilkins; 2001:2511–2574.

56 Moore PS, Chang Y. Kaposi's Sarcoma-Associated Herpesvirus. In: Knipe DM, Howley PM, eds. Fields Virology. Vol. 2 (ed 4). Philadelphia: Lippincott Williams and Wilkins; 2001:2803–2834.

57 Rickinson A, Kieff E. Epstein-Barr Virus. In: Knipe DM, Howley PM, eds. Fields Virology. Vol. 2 (ed 4). Philadelphia: Lippincott Williams and Willkins; 2001:2575–2628.

58 Keating S, Prince S, Jones M, Rowe M. The lytic cycle of Epstein-Barr virus is associated with decreased expression of cell surface major histocompatibility complex class I and class II molecules. Journal of Virology. 2002;76:8179–8188.

59 Levitskaya J, Coram M, Levitsky V, Imreh S, Steigerwald-Mullen PM, Klein G, Kurilla MG, Masucci MG. Inhibition of antigen processing by the internal repeat region of the Epstein-Barr virus nuclear antigen-1. Nature. 1995;375:685–688.

60 Lee SP, Brooks JM, Al-Jarrah H, Thomas WA, Haigh TA, Taylor GS, Humme S, Schepers A, Hammerschmidt W, Yates JL, Rickinson AB, Blake NW. CD8 T cell

recognition of endogenously expressed epstein-barr virus nuclear antigen 1. Journal of Experimental Medicine. 2004;199:1409–1420.

61 Tellam J, Connolly G, Green KJ, Miles JJ, Moss DJ, Burrows SR, Khanna R. Endogenous presentation of CD8+ T cell epitopes from Epstein-Barr virus-encoded nuclear antigen 1. Journal of Experimental Medicine. 2004; 199:1421–1431.

62 Callan MF. The evolution of antigen-specific CD8+ T cell responses after natural primary infection of humans with Epstein-Barr virus. Viral Immunology. 2003;16:3–16.

63 Hislop AD, Annels NE, Gudgeon NH, Leese AM, Rickinson AB. Epitope-specific evolution of human CD8(+) T cell responses from primary to persistent phases of Epstein-Barr virus infection. Journal of Experimental Medicine. 2002;195:893–905.

64 Pudney VA, Leese AM, Rickinson AB, Hislop AD. CD8+ immunodominance among Epstein-Barr virus lytic cycle antigens directly reflects the efficiency of antigen presentation in lytically infected cells. Journal of Experimental Medicine. 2005;201:349–360.

65 Davenport MP, Fazou C, McMichael AJ, Callan MF. Clonal selection, clonal senescence, and clonal succession: the evolution of the T cell response to infection with a persistent virus. Journal of Immunology. 2002;168:3309–3317.

66 Blake N, Lee S, Redchenko I, Thomas W, Steven N, Leese A, Steigerwald-Mullen P, Kurilla MG, Frappier L, Rickinson A. Human CD8+ T cell responses to EBV EBNA1: HLA class I presentation of the (Gly-Ala)-containing protein requires exogenous processing. Immunity. 1997;7:791–802.

67 Obar JJ, Crist SG, Gondek DC, Usherwood EJ. Different functional capacities of latent and lytic antigen-specific CD8 T cells in murine gammaherpesvirus infection. J Immunol. 2004; 172:1213–1219.

68 Bankovich AJ, Garcia KC. Not Just Any T Cell Receptor Will Do. Immunity. 2003;18:7–11.

69 Kjer-Nielsen L, Clements CS, Purcell AW, Brooks AG, Whisstock JC, Burrows SR, McCluskey J, Rossjohn J. A Structural Basis for the Selection of Dominant ab T cell receptors for Antiviral Immunity. Immunity. 2003; 18:53–64.

Index

Immunodominance: The Choice of the Immune System. Edited by Jeffrey A. Frelinger

ISBN: 3-527-31274-9

u

v